Physical Perspectives on Computation, Computational Perspectives on Physics

Although computation and the science of physical systems would appear to be unrelated, there are a number of ways in which computational and physical concepts can be brought together in ways that illuminate both. This volume examines fundamental questions which connect scholars from both disciplines: is the universe a computer? Can a universal computing machine simulate every physical process? What is the source of the computational power of quantum computers? Are computational approaches to solving physical problems and paradoxes always fruitful? Contributors from multiple perspectives reflecting the diversity of thought regarding these interconnections address many of the most important developments and debates within this exciting area of research. Both a reference to the state of the art and a valuable and accessible entry to interdisciplinary work, the volume will interest researchers and students working in physics, computer science, and philosophy of science and mathematics.

MICHAEL E. CUFFARO is a postdoctoral research fellow of the Rotman Institute of Philosophy at the University of Western Ontario, and an external member of the Munich Center for Mathematical Philosophy at LMU Munich.

SAMUEL C. FLETCHER is an assistant professor of Philosophy at the University of Minnesota, Twin Cities, a resident fellow of the Minnesota Center for Philosophy of Science, and an external member of the Munich Center for Mathematical Philosophy at LMU Munich.

Physical Perspectives on Computation, Computational Perspectives on Physics

Edited by

Michael E. Cuffaro
University of Western Ontario

and

Samuel C. Fletcher
University of Minnesota, Twin Cities

CAMBRIDGE
UNIVERSITY PRESS

University Printing House, Cambridge CB2 8BS, United Kingdom

One Liberty Plaza, 20th Floor, New York, NY 10006, USA

477 Williamstown Road, Port Melbourne, VIC 3207, Australia

314-321, 3rd Floor, Plot 3, Splendor Forum, Jasola District Centre, New Delhi - 110025, India

79 Anson Road, #06-04/06, Singapore 079906

Cambridge University Press is part of the University of Cambridge.

It furthers the University's mission by disseminating knowledge in the pursuit of education, learning and research at the highest international levels of excellence.

www.cambridge.org
Information on this title: www.cambridge.org/9781316622025
DOI: 10.1017/9781316759745

First published 2018
First paperback edition 2020

A catalogue record for this publication is available from the British Library

Library of Congress Cataloging in Publication data
Names: Cuffaro, Michael E., 1973– editor. | Fletcher, Samuel C., editor.
Title: Physical perspectives on computation, computational perspectives on physics / edited by Michael E. Cuffaro (University of Western Ontario), Samuel C. Fletcher (University of Minnesota, Twin Cities).
Description: Cambridge, United Kingdom; New York, NY: Cambridge University Press, 2018. | Includes bibliographical references and index.
Identifiers: LCCN 2018000029 | ISBN 9781107171190 (hardback) | ISBN 1107171199 (hardback)
Subjects: LCSH: Mathematical physics. | Computational complexity. | Physics–Data processing. | Computer science–Mathematics.
Classification: LCC QC20.P47 2018 | DDC 530.15–dc23
LC record available at https://lccn.loc.gov/2018000029

ISBN 978-1-107-17119-0 Hardback
ISBN 978-1-316-62202-5 Paperback

Contents

Figures

Tables

Contributors

NEAL G. ANDERSON is a professor in the Department of Electrical and Computer Engineering, University of Massachusetts, Amherst.

HAJNAL ANDRÉKA is a researcher and Head of the Algebraic Logic Research Division in the Alfréd Rényi Institute of Mathematics, Hungarian Academy of Sciences.

JACK COPELAND is a distinguished professor of Philosophy at the University of Canterbury, Christchurch, New Zealand.

MICHAEL E. CUFFARO is a postdoctoral research fellow of the Rotman Institute of Philosophy at the University of Western Ontario and an external member of the Munich Center for Mathematical Philosophy at LMU Munich.

ARMOND DUWELL is a professor of Philosophy at the University of Montana.

SAMUEL C. FLETCHER is an assistant professor of Philosophy at the University of Minnesota, Twin Cities, a resident fellow of the Minnesota Center for Philosophy of Science, and an external member of the Munich Center for Mathematical Philosophy at LMU Munich.

DOMINIC HORSMAN is a postdoctoral research associate at the Joint Quantum Centre, Durham-Newcastle.

VIV KENDON is a reader in Computational Quantum Information Theory in the Atmol Section in the Department of Physics, Durham University, and in the Joint Quantum Centre, Durham-Newcastle.

ADAM KOBERINSKI is pursuing his Ph.D. in Philosophy at the University of Western Ontario, and is a member of the Rotman Institute of Philosophy.

JAMES LADYMAN is a professor of Philosophy at the University of Bristol.

ROSSELLA LUPACCHINI is an associate professor in the Department of Philosophy and Communication Studies at the University of Bologna.

JUDIT X. MADARÁSZ is a researcher at the Alfréd Rényi Institute of Mathematics, Hungarian Academy of Sciences.

OWEN J. E. MARONEY is a departmental lecturer in Philosophy of Physics in the Faculty of Philosophy at Oxford University.

ROBERT H. C. MOIR is a postdoctoral researcher in the Department of Computer Science at the University of Western Ontario, affiliated with the Symbolic Computation Laboratory of the Ontario Research Center for Computer Algebra and the Rotman Institute of Philosophy.

MARKUS P. MÜLLER is a group leader at the Institute for Quantum Optics and Quantum Information in Vienna, a Visiting Fellow at the Perimeter Institute for Theoretical Physics, and an affiliate member of the Rotman Institute of Philosophy.

ISTVÁN NÉMETI is an emeritus research professor at the Alfréd Rényi Institute of Mathematics, Hungarian Academy of Sciences.

PÉTER NÉMETI is an independent scholar.

JOHN D. NORTON is a distinguished professor in the Department of History and Philosophy of Science at the University of Pittsburgh.

GUALTIERO PICCININI is a professor in the Department of Philosophy and the Center for Neurodynamics at the University of Missouri, St. Louis.

ORON SHAGRIR is a professor of Philosophy and Cognitive Science at the Hebrew University of Jerusalem.

MARK SPREVAK is a senior lecturer in Philosophy of Mind and Cognition at the School of Philosophy, Psychology and Language Sciences, University of Edinburgh.

SUSAN STEPNEY is a professor of Computer Science at the University of York.

KLAUS SUTNER is a teaching professor of Computer Science at Carnegie Mellon University.

GERGELY SZÉKELY is a researcher at the Alfréd Rényi Institute of Mathematics, Hungarian Academy of Sciences.

CHRISTOPHER G. TIMPSON is a university lecturer and tutorial fellow in Philosophy, Brasenose College, Oxford University.

Preface

Historically, philosophers have devoted a great deal of attention to the interrelations between computational concepts and both mathematical concepts and the concepts of intelligence and mind. There has been much less work, however, on the interrelations between computational concepts and those of the natural sciences. Our goal, in putting this edited volume together, was to begin to fill in this gap with a book dedicated to the connections between computation and physics, and to encourage further work in this area.

To this end, we have gathered together an interdisciplinary group of twenty-three of the world's leading physicists, computer scientists, mathematicians, and philosophers of science and mathematics. Together they have contributed twelve original chapters representing diverse perspectives on various questions related to the general topic. These include questions such as: Is the physical universe computable? Is the universe a computer? Can a universal computing machine simulate every physical process? What is the source of the computational power of quantum computers? Is "hypercomputation" possible? When do physical systems implement a computation? Can general principles for physical computation be given? What can we learn from physics-like models of computation? How do methodologies derived from physical practice illuminate computer science? How does framing physical theories in computational language illuminate their characteristic features? Are computational approaches to solving physical problems and paradoxes always fruitful?

Our volume is addressed to philosophers, physicists, and computer scientists interested in the connections between physics and computation. Among philosophers, this includes several groups. Philosophers of physics will find, in this volume, some of the latest computational perspectives on traditional questions in the interpretation of physical theory. Philosophical logicians and philosophers of computer science, conversely, will gain exposure to the latest perspectives from physics to reimagine the nature and interpretation of computation. Philosophers of science interested in more general questions will see how these two scientific fields have provided and continue to provide cross-fertile ideas. And philosophers of mind will learn about the connections between physics and computation with a view to informing computational theories of mind. Finally, physicists and computer scientists interested in new ideas and approaches to the foundations of their theories will find this volume to be a stimulating font of ideas.

This volume includes many completely new ideas and insights by leading specialists, and we have collected these essays together, in part, with the aim of providing a reference to the state of the art. That said, the book's contributors have all made a substantial effort to keep the technicality of each chapter at a level that is accessible to a reader with no more than an undergraduate training in mathematics, physics, or computer science. This accessibility is aided by the pedagogical introduction and overview that begins the work. We therefore expect that the book will be quite useful as part of a graduate seminar on the topic.

A work of this magnitude could not have been completed without the help of numerous others of our friends and colleagues. Particular thanks are due to Hilary Gaskin, who encouraged and supported this project through all of its various stages; to Stephan Hartmann, whose advice was enormously useful to us as we prepared our final proposal for this volume in the autumn of 2015; to Sophie Taylor, for her helpful advice regarding the finer details of the manuscript's compilation; to Philippos Papayannopoulos, who compiled the index; and to Sona Ghosh, for her advice regarding the same. We are also indebted to Jeffrey Bub, Giuseppe Castagnoli, Martin Davis, Walter Dean, Nicolas Fillion, Leah Henderson, Kevin Kelly, Jan van Leeuwen, Chiara Marletto, Stephan Mertens, Wayne Myrvold, Michael Rescorla, and Jos Uffink for the guidance they provided to us as we prepared this manuscript for publication.

Finally, Michael Cuffaro wishes to acknowledge the generous financial support of the Rotman Institute of Philosophy, the Foundational Questions Institute, the Munich Center for Mathematical Philosophy, and the Alexander von Humboldt Foundation. Samuel Fletcher wishes to acknowledge the support of the Munich Center for Mathematical Philosophy and the European Commission, through a Marie Curie International Incoming Fellowship held from 2014 to 2017.

Introduction

Our modern understanding of computation stems in large part from Alan Turing's formalization of the mathematical activity of following an effective method for computing a function. Thus the roots of computer science (at least as traditionally construed) are in this sense those of an essentially mathematical science. The science of Physics, on the other hand, ultimately aims to describe the characteristics of concrete systems as they exist in the natural world. It is thus a nontrivial question to ask whether, and how, physics can illuminate computer science and vice versa.

Indeed, the possible questions one may ask regarding the connections between computation and physics are many, varied, and multi-faceted. For the purposes of a philosophical investigation into these connections they can be usefully characterized as falling into two main categories. On the one hand, there are those questions related to the connections between computational and physical systems, and on the other hand there are those questions related to the connections between computational and physical theory in general. These two main categories can further be subdivided into two sub-categories each, which together comprise the four major parts of this volume:

- Interrelations between computational and physical systems

  I. The computability of physical systems and physical systems as computers
  II. The implementation of computation in physical systems

- Interrelations between computational and physical theory

  III. Physical perspectives on computer science
  IV. Computational perspectives on physical theory

In the remainder of this introductory chapter, we will summarize each of these parts and the particular contributions of this volume that fall under them. Before we do so, however, it will be useful to review some of the basic concepts which will generally be taken for granted in the rest of the book.

1   Computability Theory and the Church-Turing Thesis

Intuitively, computability theory concerns which tasks can be completed in principle by following a completely explicit set of instructions. These instructions must be definite, in the sense that they allow no procedural interpretation or flexibility, and self-contained, in the sense that they require no input other than what is provided in the description of the task itself. Such a set of instructions, called an *effective procedure*, hence demands no creativity of whoever (or whatever) executes it.

Effective procedures have found their greatest application in the mathematical domain, many of whose problems can be reduced to the computation of a function of natural numbers. A function $f : \mathbb{N}^k \to \mathbb{N}$ is said to be *effectively computable* when there is an effective procedure for calculating its value for any argument. For example, the familiar elementary arithmetic functions of addition and multiplication are clearly effectively computable, as are functions composed from them. Further, the effective computability of a mathematical decision problem, such as "Is n prime?", can be encoded into a function whose range is $\{0, 1\}$, corresponding with the "no" and "yes" answers.

To make this informal concept of effective computation formally tractable, myriad models for computation have been proposed, inspired variously from logic, arithmetic, and mechanics.[1] One might naturally expect that these different proposals, various as they are in their starting points, lead to formalizations of differing computational strength. So it is remarkable that they in fact determine extensionally the same class of functions as being computable.

The most important and influential of these proposals, on which we focus in this introduction, is that of the Turing machine (TM). A TM is a type of abstract state machine, consisting of the following components (an example of which is illustrated in Figure 0.1):

- An arbitrarily long tape, divided into sequential squares that can be blank or contain a mark.
- A read/write head, which sits atop a particular square, can read whether it contains a mark, and can perform the following actions: print a mark on the

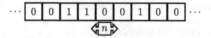

Figure 0.1 A representation of a Turing machine with read/write head in state n and tape entries "0" and "1" representing blank and marked squares, respectively

---

[1] Examples include representability in a formal system, the λ-calculus, recursive function theory, Markov algorithms, register machines, and Turing machines (what we focus on below). See Epstein and Carnielli (2008, ch. 8E) for brief descriptions and references to these various approaches.

square if it doesn't have one, erase the mark on the square if it has one, move one square to the right, and move one square to the left.

- A program, or finite set of instructions, for the read/write head, each of which has the following form:

TM Instruction Form In state n, if the current square is [blank/marked], perform [action] and transition to state m. In abbreviated form: $(n, [0/1], [action], m)$.

The tape is the medium for the input and output to a proposed calculation by the machine, as well as for all the intermediate work required to transition between them. It typically encodes these in binary notation, for example with blank and marked squares representing the numerals "0" and "1," respectively. The read/write head performs the steps leading to the computation through its fixed set of actions. The program encodes an effective procedure for the TM to follow:

1. A TM begins in some pre-specified state at some pre-specified location on the tape.
2. The read/write head reads its current square, then performs the action (if any) specified by the program according to what's read and the current state.
3. It then transitions to another state, according to the program, whereupon step two is repeated.

A program need not have an action and state transition specified for every state and input from the current square. If it does not, then the read/write head halts, indicating the end of the computation, at which point the contents of the tape represent the computation's output.[2]

Suppose now that an encoding of inputs and outputs of natural numbers on the tape and a starting location for the printing head on the tape have been fixed. One then says that a function $f : \mathbb{N}^k \to \mathbb{N}$ is *Turing-computable* if and only if there is a TM that for all $(n_1, \ldots, n_k) \in \mathbb{N}^k$ eventually halts with output encoding $f(n_1, \ldots, n_k)$ when begun with input encoding $(n_1, \ldots, n_k)$.

Because the TM's tape represents the inputs and outputs of the function it computes and the functionality of the read/write head is fixed, the real source of variability amongst TMs comes from the program. TMs may thus be enumerated by their programs, which can be represented by sets of ordered quadruples of the above specified TM Instruction Form. For example, the set $\{(1, 0, 1, 1), (1, 1, \to, 1)\}$ defines a program (i.e., a TM) with only one state, in

---

[2] Despite the physically evocative story involving "components" and so on, a concrete mechanism for implementing or constructing an actual TM is neither provided nor necessary. This is the sense in which the TM is an *abstract* machine providing a *mathematical* definition of computation, rather than a schematic for a physical machine providing an empirical account of computation; cf. Section 4 of this Introduction.

which it writes a mark if the current square is empty and otherwise moves to the right. Since every Turing-computable function must be computable by some TM, it follows immediately that the Turing-computable functions are enumerable. But because there are uncountably many functions of natural numbers, there must be functions that are not Turing-computable – infinitely more, in fact, than those that are.

Two further remarkable facts build upon such an enumeration.[3] First is the existence of *universal* TMs, ones that can simulate the computation of any other TM. In other words, there exist TMs such that, when given input encoding $(m, n_1, \ldots, n_k)$, they eventually halt with output $f_m(n_1, \ldots, n_k)$, where f_m is the Turing-computable function computed by the mth TM. Second is the specification of concrete non-Turing-computable functions. Most famous of these is the halting function $h : \mathbb{N}^2 \to \mathbb{N}$, which is equal to 1 if TM m halts on input n, and is equal to 2 otherwise.

The theory of computability can be developed much further,[4] but to close this section we circle back to the original motivation for TMs: does Turing-computability adequately formalize the concept of effective computability? Clearly every Turing-computable function is effectively computable, for the action of a TM that computes such a function is given by an effective procedure. The statement that the converse is also true is known as either *Turing's Thesis* or the *Church-Turing Thesis*.

Church-Turing Thesis (CTT) Every effectively computable function of natural numbers is Turing computable.

The truth of the CTT would imply that one can identify or replace the extension of the informal concept of effective computability with that of the formal concept of Turing-computability, thereby establishing a completely adequate explication (cf. Carnap 1947, sec. 2; Carnap 1950b, ch. 1) of the former. There is a large literature on the status and interpretation of the CTT,[5] but it is fair to say that it is widely accepted among computer scientists and beyond. That said, the theory of computability and the CTT only make claims about what is possible *in principle* to compute, given the idealizations of arbitrarily large temporal, spatial, and material resources – computing steps, tape squares, and the incorruptible functioning of Turing machinery – that abstract away from their actual abundance. When an accounting of these resources is brought to bear, as in the next section of this Introduction, one can distinguish not just between computable and non-computable functions, but, among the computable ones, those of various degrees of difficulty.

---

[3] For more on TMs, see Barker-Plummer (2016) and references cited therein.
[4] See, for instance, Immerman (2016) and references cited therein.
[5] See, for instance, Copeland (2015) and references cited therein.

2   Computational Complexity Theory

In computational complexity theory, computational problems are classified based on their resource costs, i.e., those in time and space. We will focus on time, which is the more important measure. Arguably the most basic distinction within the theory is that between those decision problems (i.e., yes-or-no questions) that are "easy" (a.k.a. "feasible," "efficiently solvable," "tractable," etc.) and those that are not (i.e., "hard"). According to the Cobham-Edmonds thesis (Dean 2016b), a decision problem is easy if it is solvable in "polynomial time," i.e., if it can be solved in a number of steps bounded by a polynomial function of its input size, n. Problems so solvable on a deterministic Turing machine (DTM) comprise the complexity class P.

Formally, one can conceive of a decision problem as one of determining whether a given string x of length n is in the "language" L. For example, determining whether x is prime amounts to determining whether it is in the language $\{10, 11, 101, 111, 1011, 1101, 10001, 10011, \ldots\}$ (the set of binary representations of prime numbers). Now, call a language L a member of the class DTIME($T(n)$) if and only if there is a DTM for deciding membership in L whose running time, $t(n)$, is "on the order of $T(n)$," or in symbols: $O(T(n))$. Here, $T(n)$ represents an upper bound for the growth rate of $t(n)$ in the sense that, by definition, $t(n)$ is $O(T(n))$ if for every sufficiently large n, $t(n) \leq k \cdot T(n)$ for some constant k.[6] We can now formally characterize P (Arora and Barak 2009, p. 25) as:

$$P = \bigcup_{k \geq 1} \text{DTIME}(n^k). \tag{0.1}$$

A *nondeterministic Turing machine* (NTM) is such that it may "choose," when in a given state, which one of a set of possible successor states to transition to; see Figure 0.2. It is said to accept a string x if and only if there exists a path through its state space that, given x, leads to an accepting state. It rejects x otherwise. NTIME($T(n)$) is now defined, analogously to DTIME($T(n)$), as the set of languages for which an NTM exists to decide, in $O(T(n))$ steps, whether a given string x of length n is in L. The class "NP" is defined as:[7]

$$\text{NP} =_{df} \bigcup_{k \geq 1} \text{NTIME}(n^k). \tag{0.2}$$

---

[6] By "for every sufficiently large n" it is meant that there exists some $n_0 \geq 1$ such that $t(n) \leq k \cdot T(n)$ whenever $n \geq n_0$.

[7] Equivalently (Arora and Barak 2009, p. 42), NP is the set of languages for which one can construct a polynomial-time *deterministic* TM to verify, for any x, that $x \in L$, given a polynomial-length string u (called a "certificate" for x).

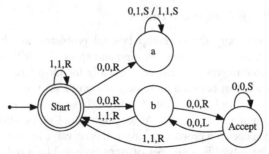

Figure 0.2 This NTM accepts binary strings ending in "00," since for a given such x, there exists a series of transitions which end in "Accept." But this is not guaranteed. The machine is guaranteed, on the other hand, to reject any string not ending in "00." An edge from s_1 to s_2 labeled α, β, P is read as: In state s_1, read α from the tape, overwrite α with β, move the read/write head to P (L = to the left, R = to the right, S = same), and finally transition to state s_2

Exactly how an NTM "chooses" to follow one path rather than another is not defined. In a *probabilistic Turing machine* (PTM), in contrast, we associate a particular probability with each possible transition. We can then define the class BPP (bounded-error probabilistic polynomial time) as the class of languages such that there exists a polynomial-time PTM which, on any given run, will correctly determine whether a string x is in the language L with probability $\geq 2/3$.[8]

It has been conjectured that any language L decidable under a given "reasonable" (i.e., physically realizable) machine model $\mathfrak{M}$ is "efficiently simulable" by a PTM in the sense that a PTM to decide L exists which requires at most a polynomial number of extra time steps compared to a machine of type $\mathfrak{M}$. This is known as the "strong" or "extended" Church-Turing thesis (ECT).[9,10] Over the last three decades, however, evidence has been mounting against the ECT, primarily as a result of the advent of quantum computing (Aaronson 2013, chs. 10, 15), which we will discuss in more detail in the next section.

---

[8] The particular threshold probability 2/3 is inessential. Any probability $p_{min} \geq 1/2 + n^{-k}$, with k a constant, will yield the same class (Arora and Barak 2009, p. 132).

[9] For further discussion of the ECT and related issues, see Dean (2016a,b,c).

[10] ECT is sometimes defined with respect to the TM rather than the PTM model, for there has been mounting evidence that P = BPP (Arora and Barak 2009). For our purposes the choice of TM or PTM is inessential; a TM is a special case of a PTM for which transition probabilities are always either 0 or 1. Moreover, defining ECT with respect to the PTM model is convenient when comparing classical computation in general with quantum computation, which is probabilistic.

3 Quantum Computing

The best-known classical algorithm for factoring arbitrary integers, the number field sieve (Lenstra et al. 1990), requires $O(2^{(\log N)^{1/3}})$ steps to factor a given integer N. Shor's *quantum* factoring algorithm requires only a number of steps that is polynomial in $\log N$ – an exponential speedup over the number field sieve. This and other quantum algorithms provide evidence that the ECT is false, for they seem to show that BQP, the class of languages probabilistically decidable by a quantum computer in polynomial time, is strictly larger than BPP. Note, however, that although the evidence furnished by Shor's algorithm is strong, it is still an open question whether factoring is in BPP.[11]

The state of a classical digital computer, whether deterministic or probabilistic, is describable as a sequence of bits. A bit can be directly instantiated by any two-level classical physical system, such as a circuit that can be open or closed. In a *quantum* computer, the basic unit of representation is not the bit but the qubit. To directly instantiate it, one uses a two-level quantum system such as an electron (specifically: its spin). Like a bit, a qubit can be "on": $|0\rangle$, or "off": $|1\rangle$. In general, however, a qubit's state can be expressed as a normalized linear superposition:

$$|\psi\rangle = \alpha|0\rangle + \beta|1\rangle, \tag{0.3}$$

where the complex "amplitudes" α and β satisfy the normalization condition: $|\alpha|^2 + |\beta|^2 = \alpha\bar{\alpha} + \beta\bar{\beta} = 1$, with $\bar{c}$ the complex conjugate of c. We refer to $|\psi\rangle$ as the "state vector" for the qubit.

Unlike a bit, not all states of a qubit can be observed directly. In particular, one never observes a qubit in a linear superposition with respect to a particular measurement basis. For example, a "computational basis" measurement – "$|0\rangle$ or $|1\rangle$?" – will never reveal a qubit state of the form of Eq. (0.3), aside from the trivial case where one of α or β is 0. In general, given the initial state in Eq. (0.3), such a measurement on a qubit will find it in the state $|0\rangle$ with probability $|\alpha|^2$ and in state $|1\rangle$ with probability $|\beta|^2$.

Quantum computers can efficiently simulate classical probabilistic computers, since it is "easy" in a complexity-theoretic sense to simulate a fair classical coin toss. For example, one can instantiate the transition Q, defined as:

$$Q|0\rangle \to \frac{i}{\sqrt{2}}|0\rangle + \frac{1}{\sqrt{2}}|1\rangle, \qquad Q|1\rangle \to \frac{1}{\sqrt{2}}|0\rangle + \frac{i}{\sqrt{2}}|1\rangle,$$

and then measure in the computational basis.

---

[11] For further discussion, see Cuffaro (in press).

Unlike classical bits, qubits can sometimes exhibit "interference effects." For example, upon applying the "Q-gate" twice to a qubit in the initial state $|0\rangle$, "destructive" and "constructive" interference is exhibited between the complex amplitudes associated with the $|0\rangle$ and $|1\rangle$ components of the state vector, respectively:

$$|0\rangle \xrightarrow{Q} \left(\frac{i}{\sqrt{2}}|0\rangle + \frac{1}{\sqrt{2}}|1\rangle \right) \xrightarrow{Q} \left(-\frac{1}{2}|0\rangle + \frac{i}{2}|1\rangle + \frac{1}{2}|0\rangle + \frac{i}{2}|1\rangle \right) = i|1\rangle.$$

A computational basis measurement on the qubit will now yield $|1\rangle$ with certainty. This ability to exhibit interference effects is held by some to be the key to understanding the source of the power of quantum computers (Fortnow 2003; Aaronson 2013).

The combined state of two or more qubits is said to be separable if it can be expressed as a product state:

$$|\alpha\rangle \otimes |\beta\rangle \otimes |\gamma\rangle \ldots$$

The state

$$\begin{aligned} |\psi\rangle &= |0\rangle \otimes |0\rangle + |0\rangle \otimes |1\rangle + |1\rangle \otimes |0\rangle + |1\rangle \otimes |1\rangle \\ &= (|0\rangle + |1\rangle) \otimes (|0\rangle + |1\rangle) \end{aligned}$$

is an example. Not all states of more than one qubit are separable states. The following is an entangled state; it cannot be expressed as a product state:

$$|\Phi^+\rangle = \frac{|00\rangle + |11\rangle}{\sqrt{2}}.$$

The ability of qubits to form entangled states when combined is another oft-cited source of the power of quantum computers (Steane 2003). Entanglement has been shown to be necessary for achieving quantum speedup when using pure states (Jozsa and Linden 2003). However in the same paper, Jozsa and Linden argue that it may not be a sufficient resource. For an in-depth discussion, see Cuffaro (2017).

Other purported sources of quantum speedup include: massive quantum parallelism achievable through the ability of qubits to realize superposition states (Pitowsky 2002; Duwell, this volume); the same but with an additional ontological posit of many computational worlds (Hewitt-Horsman 2009; Cuffaro 2012); quantum contextuality (Howard et al. 2014); and the structure of quantum logic (Bub 2010). For an overview and further discussion, see Hagar and Cuffaro (2017).

4   Computational Implementation and the Physical Church-Turing Theses

The previous sections of this Introduction concerned what could be computed and how efficiently, largely abstracting from most of the physical, mechanical, and engineering details that would be necessary to describe adequately a concrete computer executing a concrete computation (but noting the possible differences between classical and quantum for complexity theory). This abstraction of computability and computational complexity theory is perfectly unproblematic when the latter are considered as branches of mathematics. But their application to putative concrete computers and computations demands an account of how their abstract objects adequately represent physical objects and processes, or how their descriptions of computation are adequate abstractions from concrete ones. What, in other words, would it take for a physical system to implement a computation?

There is no consensus about the correct account of computational implementation, but it will be helpful for the remainder to keep several different proposals in mind.[12] The *simple mapping* account states, roughly, that a physical system performs a computation when there is a mapping from the sequential states of the system to the computational states of a computational model (say, the state and tape contents of a TM) such that physical state transitions get mapped to computational state transitions. This account is very liberal, in that it designates multitudinous physical processes as implementing multitudinous computations. It is thus often associated with the thesis of (unlimited) *pancomputationalism*, that (nearly) all physical processes implement all (or many non-equivalent) computations, in some sense.[13] Because many take (unlimited) pancomputationalism to be implausible, many other accounts of computational implementation add extra conditions to the mappings of the simple mapping account to make computation less abundant. Causal, counterfactual, and dispositional accounts require that the state transitions support various modal conditions. Semantic and syntactic accounts take seriously the idea of computation as manipulation of meaningful symbols, requiring respectively that the mappings be representational, according to some account of proper representation, or syntactical, according to an account of what it means for states and changes thereof to be syntactically structured. Mechanistic accounts demand that the physical system or process implementing a computation does so in terms of a functional mechanism, an organization of the components of the system suited to the task of manipulating computational vehicles, those components whose states are mapped to computational states.

---

[12] See Piccinini (2017, sec. 2) for a more thorough review of proposals for computational implementation.

[13] See Piccinini (2017, sec. 3) for more on the different types of pancomputationalism.

Regardless of how the issue of computational implementation is settled, it raises the further question of an analog of the CTT for physical computations. Recall from section 1 that the CTT subsumed the extension of an informal concept – effective computability – under a formal one – Turing computability. Effective computability, though vague, concerns in some idealized sense what can be in principle computed by a human agent aided only with simple memory aids such as paper and pencil. Physical computability, by contrast, concerns what can be computed by any physical process made eligible by one of the above accounts. So a physical version of the CTT would subsume a presumably wider set of physical processes under Turing computation, and the converse of a physical version should easily follow from an argument similar to that for the converse of the CTT.

Several versions of a physical CTT have been proposed. Following Piccinini (2011, 2015), it is helpful to distinguish between two classes of physical CTT:

Modest Physical CTT Any function of natural numbers that is physically computable is Turing computable.

Bold Physical CTT Any physical process is Turing computable.

The modest version, like the CTT itself, focuses on the computation of the values of numerical functions. Sometimes Gandy (1980) is interpreted as having advanced such a thesis:

Gandy's Thesis M Any function of natural numbers computable by a discrete deterministic mechanical assembly (DDMA) is Turing computable.

A DDMA is any physical device of which an adequate theoretical description uses discrete dynamics for finitely many parts of bounded complexity that affect each other only locally and deterministically. Gandy (1980) then proves that under a certain formalization of DDMAs, Thesis M follows. This thesis is physical in the sense that DDMAs are intended to be models for arbitrary machines that humans might construct to aid them in computations. However, unless one has an essentially anthropocentric account of computational implementation, it is less plausible that physical computations are exhausted by such machines. Accordingly, Thesis M sits conceptually between the CTT and the modest physical CTT.

The bold version of the physical CTT requires a bit of interpretation: what does it mean for a process to be computable? Typically, this means that an adequate theoretical description of the physical process can be simulated by a TM, in the sense that there is an injective map from the physical states of the system undergoing dynamical evolution to the computational states of a TM. A version of this thesis has been advocated by Deutsch (1985):

Deutsch's Principle Every finitely realizable physical system can be perfectly simulated by a universal model computing machine operating by finite means.

The qualification "finitely realizable" is made in explicit reference to Gandy's formalization of DDMAs: it demands that, at any time, finitely many parts of bounded complexity change their state according to local dynamics. Nevertheless, Deutsch's principle outstrips Thesis M in at least three ways: it does not assume discrete dynamics, nor does it restrict its scope to machines or (more generally) systems adequately described as satisfying the formal requirements for DDMAs; and "universal model computing machine operating by finite means" is intended not to pick out TMs in particular, but a broader class whose operation includes stochastic elements. He argues that this class includes any machines computationally equivalent to universal quantum computers.[14] Unlike the standard CTT, the truth of Deutsch's principle or related versions of the physical CTT could constrain physical theory,[15] ruling out proposals that allowed for the existence of physical systems implementing *hypercomputation*, i.e., the computation of non-Turing computable functions. Various schemes for hypercomputation have been proposed,[16] but it is a matter of controversy whether those proposals are successful at refuting an interesting version of the physical CTT in the context of a plausible account of computational implementation.

5 Landauer's Principle and the Thermodynamics of Computation

Besides providing a possible constraint on physical theory via some version of the physical CTT, computational ideas have also been invoked in explanatory contexts in thermal physics and in investigations of whether a physical computation requires some entropic or energetic cost solely in virtue of its formal properties. Perhaps the main historical origin for these connections is in Maxwell's demon, a thought experiment originally intended to illustrate how an atomic foundation for thermodynamics would entail only statistical validity for its second law – the impossibility of cyclically extracting work solely from the heat energy of a thermal body. In that thought experiment, Maxwell (1871) considered an insulated box of gas separated in two by a partition with a hole just large enough for one gas molecule to pass through. An intelligent demon, equipped with a frictionless shutter for the hole, observes the gas and, whenever a faster-than-average molecule approaches the hole from, say, the left-hand side of the box, opens the shutter to allow it to pass, and similarly for slower-than-average molecules approaching from the right-hand side. Continuing in this way, the gas on the right-hand side of the box becomes hotter and that on the left-hand side cooler, the resulting temperature difference becoming exploitable for extracting work (i.e., by pushing a piston).

---

[14] Deutsch claims that universal quantum computers outstrip the simulating power of TMs at least because of quantum non-locality, but there is no reason why non-local correlations must be simulated by non-local correlations on a classical stochastic system.

[15] For descriptions of some of these, see Piccinini (2017, sec. 4).

[16] See Piccinini (2017, sec. 4.3) for an overview and further references.

Szilard's (1972 [1929]) analysis of this thought experiment focused on what it would take to save the validity of the second law. He considered a simplified version thereof in which the box was not insulated but in contact with a thermal bath and contained only a single molecule of gas. As the molecule bounces around, energetically equilibrating with the walls, an intelligence can frictionlessly place a movable partition in the middle of the box, configured so that when the gas molecule strikes it, it moves, doing work through a series of pulleys. This requires the intelligence to know which side of the box the gas molecule is on and hence measure the gas, so Szilard proposed an energetic cost of $kT \ln 2$, where T is the temperature and k is Boltzmann's constant, to measurement that would balance out the maximum work extractable. Once the energy loss associated with the act of measurement is taken into account, he suggested, the second law is not violated.

Landauer (1961) reversed the operation of Szilard's "engine" to argue that "erasing" a memory unit rather than reading (measuring) it has a minimum energetic cost of $kT \ln 2$. To see how this works, note that one can view the one-molecule gas with the partition as a kind of one-bit memory: if the gas molecule is on the left (resp. right), then the memory reads "0" (resp. "1"). Regardless of the location of the gas molecule, the memory may be reset to "0" by removing the partition, inserting it on the right-hand side of the box, and then doing any work necessary to push it against the gas molecule to the center. What is deemed essential to the "reset" is that the computational process it implements is irreversible, i.e., *not* an injective map on computational states. Thus any adequate physical implementation of the computation cannot be injective either. Landauer argued – for what has become known as his eponymous principle – that the energetic costs to resetting a bit in any physical implementation of a computation must be bounded below by this amount, $kT \ln 2$. Assuming the validity of the second law of thermodynamics (instead of using Landauer's principle to save it), no cyclic process can convert heat energy solely into work, so any thermodynamical process associated with resetting a bit must also generate a minimum amount of heat. Connecting computational processes and concepts of information with thermodynamics, Landauer's principle remains as provocative as it is controversial.[17]

6   Chapter Summaries

Part I: The Computability of Physical Systems and Physical Systems as Computers

Part I (Chapters 1–3) of the present volume addresses the relationship between physical systems and computational systems: In what senses are physical

---

[17] See Maroney (2009a) for a more thorough review of the controversial issues.

systems computable, and which of them are computers or perform computations? The widest positive answer to the latter question – namely, *all* physical systems – is found in the astounding thesis of pancomputationalism (cf. Section 4 of this Introduction). In "Ontic pancomputationalism," Gualtiero Piccinini and Neal G. Anderson clarify and consider arguments for the titular strong form of this thesis, which asserts that all physical systems literally and fundamentally perform a computation. That is to say, according to ontic pancomputationalism, the most basic description and dynamics of a physical system is as a computational system performing a single, specific computation. Piccinini and Anderson point out that ontic pancomputationalism also entails an empirical thesis, that a computational formalism ought to be adequate for describing the physical world. The demands on this formalism depend on whether the computational model is discrete or quantum, both of which Piccinini and Anderson review. They argue that, in spite of the provocative nature of ontic pancomputationalism, its empirical aspects are not well supported by our current evidence, while its metaphysical aspects either collapse into triviality or cannot explain the variety of physical systems actually observed.

Jack Copeland, Oron Shagrir, and Mark Sprevak continue the discussion of the relation between physical and computational systems in their chapter, "Zuse's thesis, Gandy's thesis, and Penrose's thesis." The titular three theses concern various ways in which physical systems are or are not supposed to be computers or computable. Concerning Zuse's thesis, which states that the universe literally is a digital computer – in particular, a cellular automaton – Copeland et al. focus on what they call the implementation problem: What kind of ontology could the hardware implementing the universe's computational process have? Surveying four options to the problem – novel (or even non-physical) entities, instrumentalism, anti-realism, and epistemic humility – they conclude that each is unsatisfactory, either failing to answer the question or doing so only by the exorbitant postulation of new ontology. Gandy's thesis, meanwhile, concerns the computational capabilities of machines (cf. Section 4 of this Introduction). Copeland et al. show that Gandy's axiomatic formalization of machines as DDMAs contains a hidden premise asserting a strong form of determinism, effectively ruling out hypercomputational machines operating in relativistic spacetimes. Thus it is not even successful in characterizing the computability of simple sorts of machines. Finally, they consider the oppositely minded Penrose's thesis, which asserts that the processes of the brain are *not* Turing computable. Penrose's argument for his thesis depends upon a controversial application of Gödel's incompleteness theorems. Copeland et al. point out that if Penrose's argument form is valid, then it in fact can be used to prove a much stronger thesis, that the computational power of the human mind exceeds that of any hypercomputer. They show that this *reductio ad absurdum* holds even when certain plausible modifications are made to the thesis.

But regardless of rationalistic arguments made to support it, they observe, it is ultimately an empirical hypothesis for which we have yet little evidence.

Rossella Lupacchini continues the discussion of the computability of physical systems in "Church's thesis, Turing's limits, and Deutsch's principle" by tracing some of the historical and conceptual threads that lead from Hilbert's program for the foundations of mathematics through the CTT on to Deutsch's principle (for which see Sections 1 and 4 of this Introduction, respectively). She shows how Hilbert's influential quest for objective understanding in mathematics through the formalization of metamathematical notions such as "proof" engaged Gödel, Herbrand, and Church in their early attempts at formalizing the concept of effective calculability. It was only with the work of Post and Turing, which shifted focus from defining which functions are supposed to be computable to which (calculative) processes are supposed to be effective, that the connection between computability and physical processes (however idealized) that could be harnessed by human users was made secure. As Post wrote, "to mask this identification [of effective calculability with recursiveness or λ-definability] under a definition hides the fact that a fundamental discovery in the limitations of the mathematicizing [*sic*] power of *Homo Sapiens* has been made and blinds us to the need of its continual verification" (Post 1936, p. 291). This perspective makes Deutsch's principle – and his suggestion that a quantum computer ought to be the model for a universal finite computer – quite natural once one substitutes a quantum physical substratum for computers for a classical one.

Part II: The Implementation of Computation in Physical Systems

Chapters 4–6 of the volume concern issues related to the way that computations are implemented in physical systems. In his chapter "How to make orthogonal positions parallel: Revisiting the quantum parallelism thesis," Armond Duwell begins this part of the book with a discussion of *quantum computation*, that is, computation as it is implemented in quantum mechanical systems and that takes advantage of the particular physical resources that those systems provide. As was mentioned earlier, there is actually no consensus regarding which physical resources are responsible for quantum speedup. Duwell considers two rival explanations. The first, the *quantum parallelism thesis* (Duwell 2007a), asserts that quantum computers are able to outperform classical computers by computing many values of a function simultaneously. Seemingly opposed to this is the idea that quantum computers can outperform classical computers because the quantum logic associated with the state space instantiated by a quantum system allows it to complete a computational task by performing fewer, not more, computations than a classical system (Bub 2010).

Appealing to Gualtiero Piccinini's (2015) *mechanistic conception* of computation, Duwell argues that the seemingly opposite orientations of these positions stem from conflicting intuitions about how to appropriately describe the quantum systems that perform computational tasks. He argues that these positions do not disagree, however, about the fundamental features of quantum systems that give rise to quantum speedup. Guided by this insight, Duwell argues that the quantum parallelism thesis can be formulated in a way that is both true and does not appeal to controversial computational descriptions.

In "How is there a physics of information? On characterizing physical evolution as information processing," Owen J. E. Maroney and Christopher G. Timpson continue Part II by focusing on the field called "The Physics of Information." One of the core claims of this field is that for every information-processing task there exists a fundamental physical resource cost that is associated with it intrinsically. A significant challenge for the Physics of Information is that the existence of alternative physical models of information processing seems, in fact, to make it impossible to associate such intrinsic physical resource costs. Rather, the existence of alternative models seems to show that the details of physical instantiation cannot be ignored. This threatens to undermine the very basis of an implementation-independent Physics of Information.

To make sense of intrinsic resource cost claims, Maroney and Timpson propose a five-fold criterion for determining when an information-processing task has been physically instantiated: (i) the task's logical states are adequately represented by physical states; (ii) these physical states can be reliably initialized, and in fact have been on any particular occasion in which a computation can be said to have taken place; (iii) the physical states evolve equivalently, in a relevant sense, to the logical states they represent; (iv) the final physical output of the task fixes its logical output and is of a kind that is readable to someone; and (v) the process exhibits a certain amount of error tolerance. Maroney and Timpson take criteria (ii) and (iv) to be especially crucial, and on that basis argue that there is a Physics of Information, not because information itself is physical, but because physically embodied agents are able to carry out information-processing tasks.

Part II ends with "Abstraction/Representation theory and the natural science of computation," by Dominic Horsman, Viv Kendon, and Susan Stepney. As with the previous chapter, Horsman et al. aim to present a number of criteria for determining when a physical process instantiates a computation. However, they also argue that these criteria unite computation with prediction and engineering. Their framework, *Abstraction/Representation (AR) theory*, contains the following components: scientific theory, computational theory, encoding and decoding of a scientific problem in a computational model, and instantiation of a computational problem in a physical system. These components take on different forms depending on the kind of computation one is discussing, so that AR theory is capable of describing vastly different kinds of physical computation. For example, it can describe classical digital computation, including how transistors and other hardware components implement classical computational models such as the von Neumann architecture, and how programming languages, compilers, and other software components can make use of this hardware. And it can also describe the unconventional model of "slime-mold computation," wherein how a slime mold reacts to food sources being placed in its vicinity implements the abstract problem of finding the shortest path through a maze.

According to AR theory, physically carrying out a computation is analogous to doing science, in the sense that both practices involve a representational relation between physical objects and abstract mathematico-logical objects. In each case this presupposes a representational entity. In the natural sciences this can be an experimenter, or perhaps a theorist. In the case of a computational process this can be, for example, a programmer, high-level designer, or end-user. In seeming contrast to the view of Maroney and Timpson, however, the representational entity need not embody a rich conception of agency.

Part III: Physical Perspectives on Computer Science

Chapters 7–9 of this volume concern the question of how physical theory can illuminate the theory of computation. In "Physics-like models of computation," Klaus Sutner argues that insight into the nature of computability theory can be gained by carefully studying what he refers to as "physics-like" computational models. Such models have been constructed with an eye to their possible physical realization and include, for example, the ordered partition automaton and its variants. Sutner's aim is not to argue that computer science should become a part of physics. He rather makes the compelling argument that the study of physical computation is potentially very fruitful for computer science, and that it has not received sufficient attention to date.

Sutner contrasts such studies of physics-like computation with classical recursion theory. He argues that classical recursion theory is isolated not only from physics but also from much of mathematics (including discrete mathematics), in that there has been little exchange of results and techniques between the two disciplines. On the other hand, Sutner argues that there are problems in the theory of computation that stand to profit much from the study of physical computation. He shows, in particular, how a physics-like model of computation (the elementary cellular automaton number 110) produced the first instance of a universal system that was discovered, rather than designed. He also shows how considering the issue of physical realization leads to insight into the old problem of the epistemological status of intermediate recursively enumerable degrees.

In the next chapter, "Feasible computation: Methodological contributions from computational science," Robert H. C. Moir explains how, in contemporary usage, "computational science" has come to refer to a number of different forms of scientific computing. In essence, computational science in this sense is concerned with generating "feasible algorithms" for solving mathematical problems and, unlike traditional computability theory, it can involve numerical methods and hybrid symbolic-numeric methods in addition to symbolic computation. Also, and importantly, unlike traditional computability theory, the feasible algorithms of computational science generally involve the use of approximation extensively.

Moir describes, at length, how the historical roots of approximation methods in computational science can be traced back to the eighteenth- and nineteenth-century techniques developed by physicists to overcome the calculational limitations of their theoretical representations of physical phenomena (for example, in fluid mechanics and astronomy). Moir shows how the use of such approximation methods allowed physicists to extract crucial information from their theoretical models even when exact solutions were not at hand.

Moir reveals how the central strategy used in computational science for developing approximation algorithms itself has an algorithmic structure, and that a version of this strategy underlies symbolic computing. This has consequences for traditional computability theory; Moir argues that it motivates the development of a theory of "higher-order" computation, leads to a different way of thinking about computational complexity, and points the way to a possible expansion of the theory of computation towards encompassing aspects of the methodology and epistemology of scientific inference.

Part III of the book ends with "Relativistic computation," by Hajnal Andréka, Judit X. Madarász, István Németi, Péter Németi, and Gergely Székely. Andréka et al. describe the challenge to the physical CTT that is posed by computational systems which take advantage of the possibilities inherent in the spacetimes allowed by general relativity.

In particular, they describe a thought experiment involving a physical computer composed of a TM, on the one hand, and a spaceship carrying a programmer, on the other, operating in the general relativistic spacetime associated with a huge slowly rotating black hole. Features of the spacetime associated with such a black hole make it so that our programmer can survive entry into its inner event horizon intact. Because of the gravitational time dilation effect of general relativity, which causes clocks in stronger gravitational fields to run more slowly, in a sense, than those in weaker such fields, we can design our computational system so that, in the finite time before the programmer enters the inner event horizon, she is able to receive the results of an infinitely long computation carried out by the distant TM.

Fascinatingly, Andréka et al. argue that such a computer could be realized by a conceivable future civilization. In particular, they argue that such a computer could be made error-tolerant, that limitations with respect to the transfer of information between the TM and the programmer can be overcome, and that the spacetimes which make possible such a computational system are physically realistic.

Part IV: Computational Perspectives on Physical Theory

Part IV (Chapters 10–12) concerns the potential illumination – and distraction – that may result from considering physical theory from a computational perspective. In particular, the first two chapters of this part concern the status of Landauer's Principle (which is adumbrated in Section 5 of this Introduction). James Ladyman brings accounts of computational implementation to bear on the debate over whether Landauer's principle follows from the second law of thermodynamics in "Intension in the physics of computation: Lessons from the debate about Landauer's principle." Ladyman's focus is on the dialectic between the results of Ladyman et al. (2007), which seek to prove a general, sound version of Landauer's principle, and the critique thereof by Norton (2011). In particular, Ladyman argues that Norton's rebuttal to the proof can be blocked once one requires that the implementation of a computation in a physical system be intensional, or modally robust: It must support counterfactuals describing that the same function or logical operation would have been computed, but with a different input, had the initial state of the physical system been different. This also requires a non-dynamical (or "control theory") perspective on thermodynamics that views the theory – and Landauer's Principle – as concerning what human agents can do to systems with their typically limited knowledge of the system's detailed microstates. Thus, an account of computation that concerns human abilities and knowledge reveals how thermodynamics is bound in similar ways.

In contrast with Ladyman, John D. Norton describes how, in the literature on Maxwell's demon (also described in Section 5 of this Introduction), the historical focus on exorcising the demon by considering its computational powers has been a mistake. In particular, he argues that this focus has distracted us, for decades, from a simpler and more satisfying solution which only makes use of concepts from physical theory. His punning title puts it simply: "Maxwell's demon does not compute." Norton begins by reviewing the history of the demon and its naturalization – versions which use an explicit physical mechanism instead of the fanciful demon – arguing that such naturalized demons, lacking any obvious embodiment of intelligent or computational capacity, controvert claims to a general exorcism using computational ideas. Despite this, influential work by Szilard (1972 [1929]) and Landauer (1961) drew the physics community's attention away from more directly physical ideas towards more speculative ones that, Norton argues, have produced more heat than light. Norton then extends previous work (Norton 2013a) exorcising the classical demon using Liouville's theorem to a quantum version, comparing the steps of the arguments in detail. Essential to the argument is an analogy between phase space volume, in the classical case, and Hilbert subspace dimension, in the quantum case. If a putative naturalized quantum demon is to act on an equilibrium state occupying a subspace of dimension almost as large as that of its complete Hilbert space, and the intermediate states into which the demon drives the system occupy instead a small subspace, then the demon will always fail in its task.

In the final chapter, "Quantum theory as a principle theory: Insights from an information-theoretic reconstruction," Adam Koberinski and Markus P. Müller argue that recasting physical theory in terms of principles about information and computation yields additional explanatory power about the nature of the (quantum) physical world. In particular, they explicate a recent derivation of finite-dimensional quantum theory from the following principles: the state and time evolution of every physical system can be reversibly encoded into a number of interacting "universal bits"; the time evolution of these systems is reversible and continuous; states of composite systems are uniquely determined by those of their components and the correlations between them; and one universal bit can carry no more than one binary unit of information. This principle-theory framework (*sensu* Einstein [1954]), they argue, provides a *partial* interpretation of quantum theory in the sense that it describes *in part* what the world would be like if the theory were true: The world is fundamentally restricted with respect to how physical systems can transform information (i.e., perform computations). It is only partial, however, at least because it is agnostic on the interpretation of the quantum state. Accordingly, Koberinski and Müller outline three options for completing the interpretation, but argue that each one poses a challenge for interpreting the quantum state as representing a feature of the world rather than a feature of our knowledge about the world (i.e., for so-called ψ-ontic interpretations rather than ψ-epistemic ones). They conclude by suggesting that the principle of continuous reversible time evolution should be given more attention than it has heretofore received, as it may prove surprisingly central in future efforts to characterize what makes quantum theories different from classical ones.

Part I

The Computability of Physical Systems and Physical Systems as Computers

1 Ontic Pancomputationalism

Gualtiero Piccinini and Neal G. Anderson

1.1 Introduction

Pancomputationalism (PC) is the sensational view that every physical system – including the universe as a whole – is a computing system (Piccinini 2015; Anderson and Piccinini 2017). *Ontic* PC is the increasingly popular version of PC according to which the physical universe is *fundamentally* computational. According to ontic PC, there is a fundamental level of physical reality at which the one and only fundamental computation performed by each physical system can be identified. In addition, that fundamental computation is all there is to the nature of a physical system. Proponents of this view are compelled by the remarkable capacities of computers to simulate physical processes either approximately or exactly, and by the unifying descriptive power of mathematics in both computer science and physics. In this chapter, we analyze arguments for ontic PC and find them far less clear, complete, and plausible than do their proponents.

We will focus on the *computational* formulation of ontic PC, which can be summarized as follows:

Ontic PC Every physical system objectively performs one computation, which exhausts the nature of the physical system.

An alternative, *informational* formulation of ontic PC holds that information is what makes up the universe. Proponents of the computational formulation typically think of computational states as bearers of information, and proponents of the informational formulation typically think of information dynamics as computations. Since typical proponents of ontic PC think of computation as information processing and vice versa, then, the two formulations are roughly interchangeable for present purposes.[1]

Ontic PC includes both an empirical claim and a metaphysical claim.

The empirical claim is that fundamental physical states and state transformations obey specific rules defined within a specific computational formalism,

---

[1] The present authors' perspectives on whether computation is information processing are provided in Anderson (2017) and Piccinini (2015).

which capture the behavior of physical systems exactly and exhaustively. This empirical claim takes different shapes depending on which computational formalism is assumed to describe the universe exactly. In Section 1.2 we will articulate the two most widely held forms of ontic PC: "digital ontic PC," based on the classical formalism of cellular automata, and "quantum ontic PC," rooted in quantum computing. We will point out that, insofar as the empirical claims of ontic PC depart from mainstream physics, there is no evidence to support them.

The metaphysical claim of ontic PC is that computation is what makes up the physical universe. This point is sometimes made by saying that, at the most fundamental physical level, there are brute differences between states – nothing more need or can be said about the nature of the states. This claim requires elucidation. In Section 1.3 we will articulate three versions of this metaphysical claim of ontic PC and argue against each of them.

Before proceeding, three aspects of our study should be noted:

First, in considering ontic PC, we avoid a common formulation of PC to the effect that everything is *a computer*. Our reason is that in both computer science and common parlance the term "computer" is usually reserved for computing systems with special features such as programmability and computational universality (up to their memory limitations) in Turing's (1936) sense. Many computing systems are not programmable or computationally universal, so many computing systems are not *computers* in this sense. Instead, we interpret pancomputationalism as the claim that everything *performs computations* or, equivalently, everything *is a computing system*.

Second, we accept that there is a sense in which a physical system may perform computations even though it has no semantic properties and does not have the *function* to compute. According to semantic accounts of computation, there is no computation without representation; according to mechanistic accounts, computing systems are physical systems whose *function* is to perform computations (Piccinini 2015). Given these accounts, anything that lacks semantic properties or functions, respectively, is ruled out of the class of computing systems. So PC is false according to these accounts. This may be acceptable in other contexts but is too strong for present purposes. Since we wish to entertain ontic PC as a serious possibility and give it due process, we admit accounts of physical computation that do not rule out PC almost from the start.

Third, we will take ontic PC to be a form of *literal* pancomputationalism – the thesis that all physical systems *do* perform computations. Except where otherwise noted, we set aside *metaphorical* pancomputationalist theses to the effect that all physical systems *can be regarded* or *interpreted* as computational. Metaphorical PC may be useful and open perspectives that yield valuable insights in both physics and computation, but it says something fundamentally different and much less sensational than literal PC.

1.2   Forms of Ontic PC and their Empirical Claims

1.2.1   Digital Ontic PC

The earliest version of ontic pancomputationalism is *digital* ontic PC. It says that the entire universe, and everything in it, is a digital computing system, and this is all there is to its nature. Digital ontic PC is due to Konrad Zuse (1967, 1982) and Edward Fredkin (1990, 2003). Fredkin's ontic PC remained unpublished for a long time but influenced a number of American physicists (Feynman 1982; Kantor 1982; Toffoli 1982; Wheeler 1982, 1990; Wolfram 2002). Some of these physicists have hypothesized that the universe is a giant cellular automaton.

A cellular automaton is a lattice of cells; each cell can take one out of finitely many states and updates its state in discrete time steps depending on the states of its neighboring cells. For the universe to be a cellular automaton, it is necessary (but not sufficient) that all fundamental physical magnitudes be discrete, i.e., they take at most finitely many values, and that time and space be fundamentally discrete or emerge from the discrete processing of the cellular automaton.[2] At a fundamental level, continuity is not a real feature of the world – there are no truly real-valued physical quantities. This is, to say the least, at odds with most of mainstream physics – classical and quantum – but it is not obviously false and may even be testable in principle (Fredkin 1990, 2003). The hypothesis is that at a sufficiently small scale, which is currently beyond our observational and experimental reach, (apparent) continuity gives way to discreteness. Thus, all values of all fundamental variables, and all state transitions, can be fully and exactly captured by the states and state transitions of a cellular automaton.

Digital ontic PC eliminates continuity from the universe, primarily on the grounds that eliminating continuity allows classical computational formalisms to describe the universe exactly rather than approximately. Without compelling empirical evidence, digital ontic PC can only be motivated by epistemological concerns. Indeed, many supporters explicitly state that they believe ontic digital PC because they want exact digital computational models of the world. Of course, even someone who shares the desire for exact computational models may well question why we should expect nature to fulfill it.

---

[2] E.g.:

[W]e assume that everything is based on some very simple discrete process, with space, time, and state all being discrete. (Fredkin 2003, p. 195)

In discrete physics (a.k.a. digital philosophy, digital physics and digital cosmology) it is usually supposed that space, time, physical states and quantities and all the microscopic and fundamental physical processes are, ultimately, finite, discrete and deterministic (principally, appearing physically on the Planck scale). (Zahedi 2015, p. 1)

1.2.2   Quantum Ontic PC

Digital ontic PC has inspired bold efforts to model dynamical processes – including quantum mechanical ones – using cellular automata (e.g., 't Hooft 2016). Nevertheless, it is difficult to see how to simulate exactly and exhaustively all the quantum mechanical features of the universe, such as superposition and entanglement, using a classical formalism such as cellular automata (cf. Lloyd 2013). This concern, as originally expressed by Feynman (1982), was one motivation for developing quantum computing formalisms (Deutsch 1985; Lloyd 1996; Nielsen and Chuang 2000).

Instead of relying on digits – most commonly, binary digits or bits – quantum computation relies on qudits – most commonly, binary qudits called qubits. The main difference between a digit and a qudit is that whereas a digit can take only one of finitely many states, such as 0 and 1, a qudit can also take an uncountable number of states that are superpositions of finitely many states in varying degrees, such as superpositions of 0 and 1. Furthermore, unlike a collection of digits, a collection of qudits can exhibit quantum entanglement.

Quantum computing is the basis for the quantum version of ontic PC. According to *quantum* ontic PC, the entire universe, and everything in it, is a *quantum* computational system, and this is all there is to its nature. Several physicists have recently proposed views along these lines, e.g., Lloyd (2006b, 2013), Vedral (2010), and Zizzi (2006).

Quantum ontic PC is perhaps empirically less radical than digital ontic PC, in the sense that it need not eliminate continuity from the universe. Rather, the empirical component of quantum ontic PC requires that any physical process – whose dynamics are presumed to be quantum mechanical – can in principle be exactly simulated by a quantum computer that has a sufficiently large number of degrees of freedom. Simply put, a quantum simulator encodes the quantum state of a target system in its qubits and runs a quantum computation on the qubits that mirrors the dynamics of the target system. Provided that the initial simulator state is isomorphic to that of the target quantum system and provided that the dynamics of the simulator are isomorphic to those of the target system in the right way, the evolved states of the target system will be isomorphic to those of the simulator. Hypothetical measurements performed on the simulator and target system, the statistics of which are state dependent, will thus be unable to distinguish the simulator from the target. Quantum simulations of quantum systems can, in this sense, be exact and exhaustive.

The notion of quantum simulation, and the quantum computers that are posited to run them, gave rise to the research program of showing how to exactly simulate any quantum mechanical process, and more generally any physical process, via quantum computing (Deutsch 1985; Lloyd 1996). Quantum ontic PC takes the additional step of regarding the mother of all target systems – the universe – as a quantum computational system running a quantum simulation of itself.

Below we consider the metaphysical claims of digital and quantum ontic PC to clarify what their acceptance implies.

1.3 The Metaphysical Component of Ontic PC

According to the mainstream view, physical computation is an aspect of a concrete physical system. Physical computation requires a physical medium that implements it. Software presupposes hardware. In other words, software cannot float free without concrete physical hardware with enough degrees of freedom to implement it. Thus, according to the standard view, if the universe were a cellular automaton, the ultimate constituents of the universe would be concrete physical structures that behave like the cells of the cellular automaton. It is legitimate to ask what kind of physical entity such cells are and how they physically interact with their immediate neighbors so as to satisfy their cellular automata rules. For that matter, it is equally legitimate to ask what kind of physical forces constrain such cells to interact only with their neighbors and not with cells that are farther away.

The metaphysical claim of ontic PC reverses all that. Not only can software float free without a further physical medium to implement it – hardware itself cannot "float free" without some sort of software underwriting it. Hardware requires software (Kantor 1982, pp. 526, 534). A physical system is just a system of computational states without any further physical structure to anchor such states. Computation is ontologically prior to ordinary physical processes or is at least on an equal footing. According to this non-standard view, if the universe were a cellular automaton, the cells of the automaton would not be concrete, physical structures that physically interact with one another similarly to subatomic particles. Rather, they would be pure software – purely "computational" entities.

Such a metaphysical claim requires an account of what computation, or software, or physical information, is, such that it gives rise to the physical reality we observe around us. Proponents of ontic PC are not always crystal clear on how they answer this question. Nevertheless, several options can be extracted from their work. We discuss them in turn.

1.3.1 Weak Simulationism

Consider the following principle:

Weak Simulationism: If a physical system is exactly simulated by a computation, then that type of computation is all there is to the nature of the physical system.

If the empirical component of ontic PC is correct – to the effect that every physical system can be exactly simulated by a computation – then the metaphysical component follows from it via weak simulationism.

Weak simulationism is implicit in statements by many supporters of ontic PC. An especially clear case is Seth Lloyd (2013). He begins by asking whether the universe does "nothing more than compute" (2013, p. 8). In order to make that question more precise, he first replaces it with whether "a universal Turing machine [can] efficiently[3] simulate the dynamics of the universe itself" (2013, p. 8), and later whether "a quantum computer [can] efficiently simulate the dynamics of the universe" (2013, p. 12). Regardless of how the question is precisified – in terms of classical or quantum computation, universal or otherwise – Lloyd's underlying assumption is the same: Showing that a physical system can be simulated computationally is enough to conclude that all that the target system does is computing. This is weak simulationism at work.[4]

There are two problems with weak simulationism. First, it does not give a substantive answer to what physical computation is. While it rejects the standard view – to the effect that physical computation is an aspect of an underlying physical hardware – it does not replace it with anything else. It simply insists that computation is all there is to physical systems, without clarifying what that means.

The second problem is that weak simulationism is contradicted by our ordinary experience with computational simulations. Consider an ordinary computational simulation, running on an ordinary electronic computer, of an ordinary physical system – for instance, an atmospheric system producing what we call weather. The simulation is physically constituted by electrical charges arranged in patterns and moving around the circuits of the computer. The physical system is constituted by atmospheric gases characterized by pressure, moisture, and temperature gradients. When the computational states corresponding to the states of the weather take certain values, the simulation predicts sunshine, rain, snow, or hail at certain locations. Given our ordinary experience with both the weather and its simulation, it would be absurd to maintain that all there is to sunshine, rain, snow, or hail is being the output of a computation.

[3] Although Lloyd writes "efficiently," his argument requires "exactness and exhaustiveness" of the simulation. A computation can simulate something efficiently but merely approximately, or it can simulate something exactly and exhaustively but inefficiently. Lloyd's ontic PC requires exact and exhaustive simulation, not necessarily efficient simulation. We will interpret Lloyd charitably on this point.

[4] In fairness to Lloyd, in other works there are passages where he might be endorsing a somewhat more conciliatory view: "The two *descriptions*, computational and physical, are complementary *ways of capturing* the same phenomena" (Lloyd 2006b, p. 28); "The conventional view is that the universe is nothing but elementary particles. That is true, but it is equally true that the universe is nothing but bits – or rather, nothing but qubits" (Lloyd 2006b, p. 153). This more conciliatory view, which raises its own interpretative questions, may not qualify as ontic PC strictly so called; so we leave it aside. We focus on Lloyd (2013) as a representative of weak simulationism.

Computational outputs, such as those yielded by electronic computers, may be taken to *represent* sunshine, rain, snow, or hail, but they make nothing illuminated, wet, or cold. The same holds for other aspects of the weather, or any other physical system being computationally simulated. The simulation may represent their values, but it does not physically reproduce them. In this sense, there is more to the nature of the systems being simulated than mere computations. Their nature is not exhausted by computation – weak simulationism does not hold.[5]

It may be objected that ordinary simulations are merely approximate – they are not exact simulations like those posited by ontic PC. If the outputs of a simulation were to capture the exact values of any measurements that we could possibly perform on the physical system, this objection continues, then we would have captured, within our computational simulation, the entire nature of the simulated system. If we could do this, as ontic PC posits that we can, there would be nothing more to the physical system than a computation.

This objection takes liberties with the ordinary notion of simulation. Degree of exactness of a computational simulation is an epistemic notion; it does not alter the nature of the system being simulated. Even if our weather simulation were so exact as to predict the exact time and location at which individual raindrops hit the ground, it would still fail to reproduce the physical nature of rain. Nothing gets wet and cold because it's hit by the output of a weather simulation – no matter how exact the simulation is. By the same token, nothing burns because it's hit by the output of a fire simulation, and so forth, no matter how exact the simulation is.

To put the point differently, physical systems have *qualitative* features (wetness, hotness, etc.) that are not reproduced by their computational simulations. What computational simulations do is represent the quantitative *values* of those qualitative features. They do not reproduce the qualities themselves.

Here is a third way to put the point: weak simulationism presupposes that isomorphism between physical systems entails exact similarity. As we use the term, two (numerically distinct) systems are exactly similar if and only if they share all their causal properties, intrinsic properties (e.g., shape, size, texture), and compositional properties (what they are made of). We will now argue that the isomorphism that holds between two physical systems, one of which may be a simulation of the other, does not entail that the two systems are exactly similar.

Consider an ordinary physical system, such as a network of neurons within an organism's brain. Some neural networks are isomorphic to spin glasses, in the sense that they obey the same equations (Amit et al. 1984). The evolution

---

[5] A similar point was made long ago by John Searle (1980) with regard to the computational theory of cognition. While Searle's point may or may not apply to cognition – because cognition may well have a computational aspect – the point does apply to ordinary physical systems.

of such systems can also be computationally simulated. In this sense, a neural network, a spin glass, and a computational simulation may be isomorphic to one another. Nevertheless, these three systems have different causal, intrinsic, and compositional properties. The neural network receives and emits neurotransmitters, has a somewhat amorphous and distributed shape, and is made out of neurons. The spin glass responds to changes in the magnetic field and temperature by switching the magnetic poles of its components, which are atoms bonded to one another and arranged in a characteristic shape. Finally, a computational simulation responds to computational inputs and yields computational outputs; a typical computational simulation is made of electrical charges switching between a computer's electronic components and does not have a shape or size in the same sense in which a neural network and a spin glass do (although the circuits on which it runs do). Two numerically distinct neural networks with the same number and type of neurons connected in the same ways and obeying the same equations are exactly similar. Ditto for two spin glasses with the same properties or two computational simulations with the same properties. By contrast, a neural network, a spin glass, and a computational simulation differ in their causal, intrinsic, and compositional properties even though they may well obey the same equations.

Computational simulations are isomorphic to the systems they simulate in the following important sense. For ordinary (approximate) computational simulations, there is a mapping between the computational model's states (including inputs and outputs) and regions of the system's state space (including equivalence classes of initial and final states along state space trajectories), and a corresponding mapping between the computation's state transitions and equivalence classes of the system's state transitions. For exact computational simulations, there is an exact mapping between the computation's states and state transitions and the simulated system's states and state transitions. But notice that the same sort of mapping may occur between a computational simulation and multiple qualitatively different physical systems (e.g., a neural network and a spin glass), as long as the different physical systems follow the same dynamics. Such multiple physical systems may be isomorphic to one another while having different causal, intrinsic, and compositional properties. By the same token, a physical system may be isomorphic to its computational simulation in the present sense while failing to be exactly similar to it. This shows, once again, that the kind of isomorphism that holds between a computational simulation and the physical system it simulates does not entail exact similarity between the two. In other words, once again, weak simulationism does not hold.

Someone might object that in the above examples the two systems are not completely isomorphic but only partly isomorphic. A scale model has the same shape as the system it models but not the same size. A weather forecasting program satisfies the same equations as the weather but it's made out of completely different stuff. Ditto for the spin glass and the neural network. What if the two systems were *completely* isomorphic to one another? How could there be any difference between them?

The answer depends on what we mean by "completely isomorphic," or how many of a system's properties are captured by a given notion of isomorphism. We may all agree that if two numerically distinct systems share all their properties, then they are exactly similar. If a notion of isomorphism captures all properties of two numerically distinct systems, then two isomorphic systems are ipso facto exactly similar. If, by contrast, a notion of isomorphism leaves out some properties of two systems, then two isomorphic systems are not ipso facto exactly similar.

So the relevant question is how many properties are captured by the notion of isomorphism employed by supporters of ontic PC. The notion they employ appears to be the notion of isomorphism that applies to the weather and a weather forecasting system. While the two systems satisfy the same mathematical description, they have different causal, intrinsic, and compositional properties. Thus, they are not exactly similar. By the same token, the universe and a quantum computer that simulates it are isomorphic in that they satisfy the same mathematical description, but in general they are not exactly similar because they have different causal, intrinsic, and compositional properties.

To accept weak simulationism, as it is employed by ontic pancomputationalists, is to accept one of two unpalatable options: either there is nothing more to physical systems than their quantitatively described states, or a physical system and its simulator need only be observationally equivalent in some respects – though not all respects – to be regarded as the same thing. A quantum ontic pancomputationalist might argue that this is a false choice: the quantum computer simulates its target system (i.e., the quantum universe) in every respect since, for her, the simulator is the quantum universe itself. However, with all distinctions between the simulator and its target system erased, this leaves the question of what "simulation" is supposed to mean when a given system can be regarded as both a target and a simulator of itself.

1.3.2   Strong Simulationism

Strong simulationism goes beyond weak simulationism by offering a specific hypothesis about the physical basis of the simulation.

Strong Simulationism: Our universe is a computational simulation run by a physical computer that exists in a separate physical universe.

Assuming for the sake of the argument that ontic PC holds, strong simulationism offers a straightforwardly physical explanation of why it holds:

because the universe we observe is an illusion created by a physical computer running within a real, concrete physical universe. The physical nature of the real, concrete physical universe in which the simulation is run may or may not be similar to the physics of our universe as we conceive it; presumably, we are not in a position to investigate the physical nature of the universe in which the simulation is run beyond the obvious conclusion that computation is possible in that universe. Thus, there is a sense in which strong simulationism is a conservative ontology. That is, strong simulationism does not replace our ordinary physical ontology with a different, non-standard ontology. It simply places genuine physical reality in a different universe from ours, while assigning illusory status to our own universe. Granted, according to strong simulationism we do exist in a Matrix-like universe and – unlike the characters in The Matrix – we don't have concrete physical bodies. But there is still a real, physical universe containing the supercomputer that runs our Matrix-like universe.

Strong simulationism appears to be due to Ed Fredkin (1992, 2003), although others endorse it as well (Schmidhuber 1997, 2000; Bostrom 2003). Here is an especially crisp formulation:

[T]he computation that is physics runs on an engine that exists in some place that we call "Other". There is no reason to suppose that Other suffers the same kinds of restrictive laws present in this universe. Computation is such a general idea that it can exist in worlds drastically different than this one; any number of regular spatial dimensions or almost any kind of spacetime structure with almost any kind of connectivity. There is no reason to think that concepts of matter and energy, of conservation laws and symmetries, and of beginnings and endings are applicable to Other. (Fredkin 2003, p. 193)

There are two problems with strong simulationism. The first problem is that it presupposes something closely analogous to weak simulationism, which is problematic for reasons closely analogous to why weak simulationism is problematic. Specifically, strong simulationism presupposes the following three assumptions: (1) a computational simulation can reproduce all the features of a physical universe, including its causal, intrinsic, and compositional properties, (2) it can generate conscious minds within the simulation, and (3) such conscious minds can perceive the simulated universe to have the features that are being computationally reproduced.

Assumption (1) is close enough to weak simulationism that it faces the same difficulties faced by weak simulationism. Specifically, it presupposes that isomorphism between two physical systems, one of which may be a computational simulation of the other, entails that the two systems are exactly similar (share all their properties). This is not our experience.

Assumptions (2) and (3) presuppose a form of what is known among philosophers of mind as *computational functionalism* (Putnam 1967), namely, that the mind is the software of the brain.[6] The main difference from standard computational functionalism is that, according to strong simulationism, not only our minds but our brains too are mere computational simulations. That is, there is a supercomputer in a different universe, which simulates the existence of human brains together with the phenomenally conscious minds that we normally think are produced by brains. In this scenario, somehow the simulated minds would have phenomenally conscious experiences indistinguishable from those that would be produced by real brains (if there were any).

While some philosophers have defended computational functionalism with a priori arguments, others have argued against it. The main objection is that computational simulation has nothing qualitative within it with which to reproduce the qualitative aspects of our experience. Consider our experience of an itch. We can easily represent the itch by some computational state and simulate its causes and effects. But the simulation does not seem capable of reproducing the itchiness itself. At any rate, there is no empirical evidence supporting computational functionalism, such as actual computer simulations of conscious minds that report being conscious in a credible way – although, to be fair, it is not entirely clear what would count as empirical evidence for or against it.

At this point, a strong simulationist might reply that all of the properties of physical systems that we experience, including their causal powers, intrinsic qualities, and composition, as well as our own experiences, are mere illusions created by the simulation itself. We think of them as features that are not reproduced by an ordinary computational simulation simply because we ourselves are products of the simulation, and we have no way to discern that everything we experience is a manifestation of the giant simulation that is our universe.

The viability of this reply depends on two assumptions: that computational simulations can produce conscious experiences indistinguishable from our own, and that computational simulations can produce experiences of physical systems indistinguishable from those we experience. Neither of these speculative and counterintuitive assumptions is empirically supported at present.

Strong simulationism faces a second problem: Its most basic assumption – that our universe is a computational simulation running in an alternate physical universe – is not empirically testable in any serious way and does not lead to novel predictions. This lack of testability and novel predictions stretches the notion of a scientific hypothesis. For various reasons, philosophers and others sometimes discuss fanciful hypotheses that cannot be tested: whether we are deceived by Cartesian demons, whether we are brains in a vat, whether the universe was created five minutes ago complete with evidence of its past,

---

[6] Computational functionalism is presupposed by weak simulationism too, insofar as weak simulationism is taken to apply to phenomenally conscious physical systems. Other entries into the voluminous literature on computational functionalism include Block (1978), Piccinini (2010), and Chalmers (2011).

etc. These hypotheses may be helpful to sharpen our understanding of certain concepts, but they are not intended or regarded as serious empirical hypotheses. Strong simulationism should be treated as such a hypothesis.

1.3.3   Computational Pythagoreanism

There is one way to avoid the difficulties of weak and strong simulationism without abandoning ontic PC: to replace our ordinary physical ontology with a mathematical one.

Computational Pythagoreanism: Our universe is a mathematical computation.

Pythagoreanism (*simpliciter*, with "small p") is a revisionist ontology, which offers an alternative to our ordinary physical ontology. Instead of ordinary physical objects such as subatomic particles, our universe is made out of mathematical objects, such as numbers or sets. The original version of pythagoreanism, which Aristotle's *Metaphysics* attributes to some ancient followers of Pythagoras, claims that all is number: Every physical entity is constituted by numbers.[7] A more recent version of pythagoreanism claims that all is sets: Every physical system can be reduced to a set (Quine 1976). Computational pythagoreanism adds that the mathematical objects that constitute our universe are computational.

Pythagoreanism faces a dilemma. One option is that the mathematical entities it posits as fundamental are *concrete* entities not unlike chairs, quarks, and galaxies, which exist in spacetime (or whatever concrete substrate replaces spacetime in the ultimate physical theory). When Aristotle attributed the view that physical things are made out of numbers to some ancient Pythagoreans, the numbers themselves were conceived of as *physical* entities (*Metaphysics* 1091a13-18, as cited by Huffman 2015). In this sense, then, the so-called "mathematical" entities posited by Aristotle's Pythagoreans are some sort of concrete entities with concrete properties, which compose ordinary concrete entities and properties in a way analogous to the way subatomic particles and their properties are normally assumed to compose macroscopic objects and their properties.

If this is what computational pythagoreans mean, then the fundamental computations they posit are concrete entities after all.[8] They are something like subatomic particles described computationally. This is not as radical a replacement of our ordinary ontology as it may have sounded at first. Rather, it is a

---

[7] Unfortunately, there is no surviving direct textual evidence – besides what Aristotle somewhat uncharitably says – that ancient followers of Pythagoras held such a view (Huffman 2015).

[8] Perhaps this kind of concrete pythagoreanism is what de Araújo and Baravalle are gesturing towards when they write that "symbols are causally sufficient for the instantiation of any physical (and thus also biological or psychological) process" (2017, p. 13).

redescription of our ordinary physical ontology using computational language. Ontic PC is just ordinary physical theory couched in computational lingo (modulo any revision due to its empirical component). There is nothing overly radical or shocking about it, except perhaps the conception of mathematical entities as physical.

The other option is that the mathematical entities pythagoreans posit are abstract entities, as a platonist about mathematics would have them. According to platonism (with "small p"), mathematical entities are abstract objects, which do not engage in causal or dynamical relationships and exist "outside" spacetime. Abstract entities are nonspatial, nontemporal, and acausal. They don't *do* anything physical and it would be a category mistake to think that they do. They exist in a "Platonic heaven" of abstract entities "outside" spacetime (or whatever concrete substrate replaces spacetime in the ultimate physical theory). According to this platonistic interpretation of computational pythagoreanism, then, our physical universe is constituted by abstract, mathematical computations.[9]

There is one passage in which John Wheeler seems to endorse this platonistic version of computational pythagoreanism: "the building element [of the universe] is the elementary 'yes, no' quantum phenomenon. It is an abstract entity. It is not localized in space and time" (1982, p. 570). Elsewhere, he uses the word "immaterial": "the physical world has at bottom – a very deep bottom, in most instances – an immaterial source and explanation" (1990, p. 5). Regardless of what Wheeler means, Max Tegmark (2008, 2014) is admirably clear and explicit about it. He begins by endorsing a general form of pythagoreanism to the effect that the universe is a mathematical entity ("Our external physical reality is a mathematical structure" [2008, p. 102]), interprets this hypothesis platonistically ("a mathematical structure is an abstract, immutable entity existing outside of space and time" [2008, p. 106]), and later adds the computational component ("The mathematical structure that is our external physical reality is defined by computable functions" [2008, p. 131]).

While we cannot do justice to Tegmark's view in this work, we will briefly assess its prospects within the context of ontic PC. First, here is a brief summary of his argument. Tegmark distinguishes between a purely mathematical description of physical reality (uninterpreted equations) and the terms we normally use to interpret mathematical descriptions and ground them in observation (e.g., "particle" and "observation" [2008, p. 102]). He refers to the latter terms as "baggage." He assumes that, in order to establish that "physical reality is completely independent of us humans" (2008, p. 102), our description of physical reality must be purged of all "baggage." What is left is a purely

---

[9] When Quine (1976) defends his set-theoretic pythagoreanism, he explicitly defends it as the view that everything is an abstract mathematical entity in this sense.

mathematical description, which he then interprets as referring to an abstract mathematical structure. Finally, Tegmark appeals to the kind of reasoning we unearthed behind weak simulationism: If two systems are isomorphic, then they are exactly similar. Therefore, physical reality is just a mathematical structure.[10]

The problem with Tegmark's argument is his assumption that "baggage" makes our descriptions subjective or human-dependent in a problematic way, and therefore must be dispensed with, whereas interpreting our mathematical descriptions in terms of abstract mathematical entities does not. On the contrary, what Tegmark calls "baggage" is all we have to anchor our mathematical descriptions in the physical world. Without "baggage," we are left with either an uninterpreted formalism or an interpretation in terms of abstract mathematical entities. As we will presently see, mathematical entities alone cannot account for physical reality. A better solution is to aim for an interpretation of our mathematical descriptions of physical reality that accounts for its observable properties.

Pythagoreanism understood platonistically is even worse off than weak simulationism and strong simulationism. Both simulationisms lack resources with which to account for the qualitative, causal, intrinsic, and compositional features of the physical universe, and so does platonistic computational pythagoreanism. In addition, the abstract entities posited by platonistic computational pythagoreanism lack resources with which to account for the spatiotemporal and dynamical features of the universe (cf. Martin 1997).

On the first point, recall that, according to both kinds of simulationism, being computationally simulated is sufficient for being wholly computational. As we've seen, however, a computational simulation may be isomorphic to a physical system without reproducing its qualitative features, causal powers, intrinsic properties, and composition. Abstract entities are in the same boat. Abstract entities lack qualitative features, causal powers, intrinsic properties, and composition altogether – or at least they lack the kinds of properties that concrete physical systems have. This is so in virtue of their abstractness. Therefore, abstract objects have no more resources with which to account for such aspects of physical systems than computational simulations do.

In addition, whereas computational simulations at least give rise to some semblance of spacetime and dynamical interactions within the simulation, abstract objects are "outside" spacetime and have no causal or dynamical properties whatsoever. They do not change at all. Thus, it is not clear how they

---

[10] Cf.: "if some mathematical equations completely describe both our external physical reality and a mathematical structure, then our external physical reality and the mathematical structure are one and the same, and then the Mathematical Universe Hypothesis is true: our external physical reality is a mathematical structure" (Tegmark 2014, p. 280).

could possibly account for the fact that the phenomena we experience have spatial locations, are temporally ordered, and change.

To sum up, according to the computational pythagoreanism that we are considering, all there is to a physical system (a cloud, a neural network, a spin glass), at the most fundamental level, is an abstract computational object. But an abstract computational object is supposed to have no spacetime locations, no time flow, no causes, no effects, no intrinsic physical properties, no physical composition, and no qualities. It's just an abstract object existing "outside" spacetime. How is such an abstract object supposed to explain the causal interactions that a physical system goes through as well as its distinctive properties? This is a mystery that supporters of computational pythagoreanism, like supporters of platonistic pythagoreanism more generally, ought to dispel but have not.

1.4   Conclusion

Ontic pancomputationalism – the thesis that the physical universe *is* a computing system – rests on empirical and metaphysical claims. While the nature of these claims differs for the various forms of ontic PC that have been proposed, they are problematic on both counts. On the empirical front, there is no empirical evidence to support any radical departure from standard physics. On the metaphysical front, there is a dilemma: Either ontic PC is just a reformulation of standard physics in computational language, without any change in its underlying ontology, or it is the claim that the ontological basis of ordinary physics is provided by computations in one of the three senses we reviewed (weak simulationism, strong simulationism, and platonistic computational pythagoreanism). But it is not clear how computations alone – which lack the right kinds of causal, intrinsic, compositional, and qualitative properties – can possibly give rise to the multifarious physical systems and phenomena that we commonly experience within spacetime.

Ontic PC requires that formal isomorphism – the kind of isomorphism that obtains between a physical system and its mathematical representation – entails exact similarity, i.e., the kind of similarity that obtains between two numerically distinct systems that share all of their properties. But two systems can be formally isomorphic without being exactly similar, as discussed above. Acceptance of ontic PC requires that we deny as much and accept that two systems that are formally or observationally equivalent in some (though not all) respects are the same. Even more is required for the view that the universe is a computer simulating its own behavior. This erases the distinction between a simulator and its simulated target system, requiring that we accept a system as a simulation of itself.

Finally, although ontic PC may be rich in insights when throttled back to a more metaphorical form of PC, its dramatic pronouncements lose much of their force when the required hypotheses or redefinitions of familiar terms are explicitly acknowledged. A claim that "the universe is a computing system" is not particularly impressive if it simply acknowledges that the freely evolving physical universe qualifies as a computing system under a definition of "computing system" that admits the freely evolving physical universe.

Acknowledgments

The authors thank Ilke Ercan, Nir Fresco, Weibo Gong, John Norton, Jack Mallah, Marcin Miłkowski, Steve Selesnick, and Oron Shagrir for comments on a larger manuscript from which this chapter derives, and the participants in the Nature as Computation Workshop (Tempe, 2015) – especially Paul Davies, Gregory Chaitin, and James Crutchfield – for helpful discussion after a presentation of related work. This material is based upon work supported by the National Science Foundation under grants no. SES-0924527 and SES-1654982 to Gualtiero Piccinini.

2 Zuse's Thesis, Gandy's Thesis, and Penrose's Thesis

B. Jack Copeland, Oron Shagrir, and Mark Sprevak

2.1 Introduction

Computer pioneer Konrad Zuse (1910–1995) built the world's first working program-controlled general-purpose digital computer in Berlin in 1941. After the Second World War he supplied Europe with cheap relay-based computers, and later transistorized computers. Mathematical logician Robin Gandy (1919–1995) proved a number of major results in recursion theory and set theory. He was Alan Turing's only Ph.D. student. Mathematician Roger Penrose (1931–) is famous for his work on spacetime and singularities. What we call Zuse's thesis, Gandy's thesis, and Penrose's thesis are three fundamental theses concerning computation and physics.

Zuse hypothesized that the physical universe is a computer. Gandy offered a profound analysis supporting the thesis that every discrete deterministic physical assembly is computable (assuming that there is an upper bound on the speed of propagation of effects and signals, and a lower bound on the dimensions of an assembly's components). Penrose argued that the physical universe is in part *un*computable. We explore these three theses. Zuse's thesis we believe to be false: The universe might have consisted of nothing but a giant computer, but in fact does not. Gandy viewed his claim as a relatively a priori one, provable on the basis of a set-theoretic argument that makes only very general physical assumptions about decomposability into parts and the nature of causation. We maintain that Gandy's argument does not work and that Gandy's thesis is best viewed, like Penrose's, as an open empirical hypothesis.

2.2 Zuse's Thesis: The Universe is a Computer

Zuse's (1967) book *Rechnender Raum* ("Space Computes") sketched a new framework for fundamental physics. *Zuse's thesis* states that the physical universe is a digital computer – a cellular automaton.

The most famous cellular automaton is the *Game of Life* (GL), invented in 1970 by John Conway (Gardner 1970). GL involves a grid of square cells with four transition rules, such as, "If a cell is on and has less than two neighbors on,

39

it will go off at the next time step," and illustrates an interesting phenomenon: Complex patterns on a large scale may emerge from simple computational rules on a small scale. If one were to look only at individual cells during the GL's computation, all one would see is cells switching on and off according to the four rules. Zoom out, however, and something else appears. Large structures, composed of many cells, grow and disintegrate over time. Some of these structures have recognizable characters: They maintain cohesion, move, reproduce, and interact with each other. They are governed by their own rules. To discover these higher-order rules, one often needs to experiment, isolating the large structures and observing how they behave under various conditions.

The behavior can be dizzyingly complex. Some patterns, consisting of hundreds of thousands of cells, behave like miniature universal Turing machines. Larger cellular patterns can build these universal Turing machines. Yet larger patterns feed instructions to the universal Turing machines to run GL. These in-game simulations of GL may themselves contain virtual creatures that program their own simulations, which program their own simulations, and so on. The nested levels of complexity that can emerge on a large grid are mind-boggling. Nevertheless, everything in GL is, in a pleasing sense, simple. The behavior of every pattern, large and small, evolves exclusively according to the four fundamental transition rules. Nothing happens in GL that is not determined by these rules.

Zuse's thesis is that our universe is a computer governed by a small number of simple transition rules. Zuse suggested that, with the right transition rules, a cellular automaton would propagate patterns, which he called *Digital-Teilchen* (digital particles), that share properties with real particles. More recently, Gerard 't Hooft, Leonard Susskind, Juan Maldacena, and others have suggested that our universe could be a hologram arising from the transformation of digital information on a two-dimensional surface (Bekenstein 2007). 't Hooft says: "I think Conway's Game of Life is the perfect example of a toy universe. I like to think that the universe we are in is something like this" (2002).

GL's four transition rules correspond to the fundamental "physics" of the GL universe. These are not the rules of our universe, but perhaps other transition rules are – or perhaps the universe's rules are those of some non-cellular type of computer: David Deutsch (2003) and Seth Lloyd (2006a) suggest that the universe is a quantum-mechanical computer instead of a classical cellular automaton. If Zuse's thesis is right, then all physical phenomena with which we are familiar are large-scale patterns that emerge from the evolution of some computation operating everywhere at the smallest scales. A description of that computation would be a unifying fundamental physical theory.

Should we believe Zuse's thesis? One can get an idea of how much credence to give it by considering three big problems that a defender of Zuse's thesis needs to overcome. The first is the *reduction problem*: Show that all existing physical phenomena, including those with which we are familiar in physics, could emerge from a single underlying computation. The second is the *evidence problem*: Provide experimental evidence that such an underlying computation actually exists. The third is the *implementation problem*: Explain what possible hardware could implement the universe's computation.

Our focus is on the implementation problem (as we discuss the reduction problem and the evidence problem in Copeland et al. [2017]). What hardware implements the universe's computation? A computation requires some hardware in which to occur. The computations that a laptop carries out are implemented by electrical activity in silicon chips and metal wires. The computations in the human brain (if such there are) are presumably implemented by electrochemical activity in neurons, synapses and their substructures. In Conway's original version of GL, the computation is implemented by plastic counters on a Go board. Notably, the implementing hardware – the medium that every computation requires – must exist in its own right. The medium cannot be something that itself emerges from the computation as a high-level pattern. Conway's plastic counters cannot emerge from GL: They are required in order to play GL in the first place. What then is the medium in the case of the universe?

According to Zuse's thesis, all phenomena with which we are familiar in physics emerge from some underlying computation. The medium that implements this computation cannot be something that we already know of in physics (for example, the movement of electrons in silicon) since, by Zuse's thesis, that is an emergent pattern from the underlying computation. The medium must be something outside the realm of current physics. But what could that be? In what follows we present four options. None leave us in a happy place.

The first option is *weird implementers*. This option asserts that something outside the current catalog of physical entities, hence "weird," implements the universe's computation. In principle, a weird implementer could be anything: ectoplasm, angelic goop, or the mind of God. A weird implementer could also emerge from another computation that has its own weird implementers, which in turn emerge from another computation, and so on. Different versions of the weird implementers response posit different specific entities to implement the universe's computation. Weird implementers are objectionable not because we can already rule them out based on current evidence but because they offend principles of parsimony and the usual standards on evidence. Positing a specific new type of entity should be motivated. If it can be shown that positing some specific type of entity does essential work for us – for example, explanatory work that cannot be done as well any other way – that would be an argument for its existence. But positing a specific weird implementer merely to solve the implementation problem seems ad hoc and unmotivated.

An alternative version of the weird implementers response is to repurpose some *non-physical* entity, which we already know to exist (so avoiding the charge of adding to our ontology), as hardware for the physical universe. What would remain is to show that this entity does indeed stand in the implementation relation to the universe's computation. Max Tegmark (2014) has a proposal along these lines. Tegmark's "Mathematical Universe Hypothesis" claims that the implementing hardware of the physical universe consists in abstract mathematical objects. The existence of abstract mathematical objects is, of course, controversial. But granted that one accepts that those objects exist, Tegmark's idea is that those objects can be repurposed to run the universe's computation. Among the mathematical objects are abstract universal Turing machines. Tegmark proposes that the physical universe is the output of an abstract universal Turing machine run on random input. A similar suggestion is made in Schmidhuber (2013).

Many objections could be raised to this proposal. The most relevant for us is that abstract mathematical entities are not the right kind of entity to implement a computation. Time and change are essential to implementing a computation: Computation is a process that unfolds through time, during which the hardware undergoes a series of changes (flip-flops flip, neurons fire and go quiet, plastic counters appear and disappear on a Go board). Abstract mathematical objects exist timelessly and unchangingly. What plays the role of time and change for this hardware? How could these Platonic objects change over time to implement distinct computational steps? And how could one step "give rise" to the next if there is no time or change? Even granted abstract mathematical objects exist, they do not seem the right sort of things to implement a computation.

The second solution is *instrumentalism* about the underlying computational theory. This replays Mach's treatment of nineteenth-century atomic theories in physics. Mach argued that atomic theories, while predictively successful, do not aim at truth: The atom "exists *only* in our understanding, and has for us only the value of a *memoria technica* or formula" (Mach 1911, p. 49). A scientific theory need not aim at giving a true description of the world. Its value may lie in the instrumental goods it delivers: making accurate predictions, unifying diverse results, aiding calculation, grouping phenomena together in perspicuous ways, and prompting useful future inquiries.

If we are instrumentalists about the computational theory that underlies our universe, then we avoid the implementation problem. An instrumentalist does not care about the computational theory being true, only about its instrumental utility. An instrumentalist sees no problem in positing things that do not exist (the Coriolis force, mirror charges, positively-charged holes, etc.) to achieve her ends. The implementers of the universe's computation could therefore, for an instrumentalist, be anything real or imagined. The implementers could even be notional: assumed for the nonce to generate predictions. An instrumentalist would lose no sleep over the existence or non-existence of implementers since she has no investment in the theory being true.

Instrumentalism may be a reasonable attitude to adopt towards some scientific theories (for example, geocentric planetary theories used for navigation but known to be false). However, it takes a strong stomach to be an instrumentalist about a fundamental physical theory. Zuse's thesis is usually couched as a claim about the *true nature* of the universe: The universe really is a giant computer. Our question was why we should believe this. The instrumentalist responds by changing topic: not by showing that Zuse's thesis is true, but by arguing that it is useful (and even that much has not yet been shown).

The third solution is *anti-realism* about the fundamental physical theory. Anti-realism is the idea that some features of the universe that may appear to be objective features are, in fact, mind-dependent. Zuse's thesis claims that a computation takes place. This claim is presumably made true by the implementers of the computation behaving one way rather than another – by them satisfying a specific pattern described by that computation. On a Go board with plastic counters, whether GL is taking place or not is made true by the implementers behaving in one way rather than another: If a plastic counter is on a specific square at a given moment, the cell is "on"; if it is not, the cell is "off." But what if there were no implementers and the decision about whether an implementer is behaving *this* way rather than *that* way lay inside the head of an agent? GL does not need to involve a Go board and plastic counters. It could for example take place by the agent keeping track of appropriate sequences of "yes" or "no" decisions that settle the question of whether a specific counter is on a specific square. Like Dr B in Stefan Zweig's *Schachnovelle*, an agent might generate a sequence of decisions that implement GL in her head. This may not be easy or convenient, but there is no reason it could not be done. In this case, the hardware that implements the computation is *mind-dependent*.

There is nothing problematic about this considered as a proposition about GL. The anti-realist tries to play the same trick for the computation postulated by Zuse's thesis. John Wheeler's "It from bit" doctrine can be viewed as a move in this direction:

[T]hat which we call reality arises in the last analysis from the posing of yes-no questions and the registering of equipment-evoked responses; ... all things physical are information-theoretic in origin and ... this is a *participatory universe*. (Wheeler 1990, p. 5)

We are participators in bringing into being not only the near and here but the far away and long ago. (Wheeler 2006)

The idea is that the fundamental informational "yes"/"no" states that underlie the physical universe are somehow generated by observers. It is not clear how broad the category of "observer" is: whether it includes simple devices like photographic plates as well as conscious humans. But no matter how broad or narrow the class, this anti-realist solution to the implementation problem should produce a sense of disquiet. As was mentioned above, the hardware that implements a computation cannot emerge from that computation. But this is precisely what is required here. An anti-realist says that the implementation of the universe's computation lies in the registering of a sequence of bits by agents or other observers. But the anti-realist solution also requires that those agents and other observers be physical parts of the universe – they need to be in order to interact causally with the rest of the universe. Therefore, agents and other observers play a dual role: implementing the universe's computation *and* being among the high-level products that emerge from that computation. This contradicts our principle that the hardware that implements a computation cannot emerge as a high-level product from that computation. We have no model of how implementation could work in this case. Anti-realism about computations that take place *inside* the universe (such as GL) is unproblematic. Anti-realism about the computation that generates the *entire* physical universe (including all agents and other observers) seems mysterious and incoherent. At best, it would require significant reworking of existing ideas of implementation.

The fourth solution to the implementation problem is *epistemic humility* about the implementers. This is the suggestion that we trim our ambitions regarding knowledge of the implementers. We know that *something* must implement the universe's supposed computation, but according to this response we say that we know nothing – and can know nothing – about that shadowy substratum. Our proper aim should be to describe the universe's computation; we should remain *silent* about the nature of the implementing medium. Unlike the weird implementers option, epistemic humility makes no positive claim about the specific nature of the implementers other than that some implementer must exist. Unlike instrumentalism, epistemic humility says that Zuse's thesis aims at delivering truth and not just instrumental benefits. Unlike anti-realism, epistemic humility makes no claim that minds or observers are part of the implementing medium.

There are precedents for this kind of humility. Henri Poincaré argued that science can tell us only about the "true relations" between "real objects which Nature will hide forever from our eyes" (1952 [1902], p. 161). Bertrand Russell argued that science can tell us only about the structure of matter, not about its "intrinsic character" (1927, p. 227). These expressions of epistemic humility share the idea that the world contains some sort of shadowy substratum (although neither author says that the substratum implements a computation). Following this line of thought, an advocate of Zuse's thesis might argue that we should not be troubled about committing to the view that a substratum exists – even if knowledge of the nature of that substratum is forever beyond us.

The problem with epistemic humility is that it does not so much answer the implementation problem as admit that we cannot answer it. If one was motivated by the implementation problem at all, one is unlikely to find this a satisfying solution. If the universe is a computer, one might feel that we should be able to say something positive about the implementing medium. Epistemic humility requires that we surrender all ambitions on this score.

Epistemic humility deals with the implementation problem by saying that we can never solve it, instrumentalism changes the topic from truth to usefulness, anti-realism is of dubious coherence, and proponents of weird implementers either shoehorn unsuitable entities into the role of implementers or else indulge in unjustified speculation. These options are not meant to be exhaustive and the considerations raised are not intended to refute Zuse's thesis. But we have at least put some hard questions on the table (and we say more in Copeland et al. [2017]).

One potential route forward for advocates of Zuse's thesis is to combine instrumentalism, anti-realism, and epistemic humility in a way described by Dennett (1991) and Wallace (2003).[1] On such a view, whether something counts as real or not depends on how useful it is to admit it into our ontology. If a computational theory in fundamental physics were to prove sufficiently useful, then, on this view, we should regard the computation described by the theory as real and adopt an attitude of epistemic humility towards the implementing medium. It remains to be seen, of course, how useful Zuse's thesis will prove in fundamental physics.

Even if the universe is not a computer it may nevertheless be comput*able*. We turn next to Gandy's thesis.

2.3 Gandy's Thesis: Turing Computability is an Upper Bound on the Computations Performed by Discrete Deterministic Mechanical Assemblies

In his 1980 article "Church's thesis and principles for mechanisms," Gandy advanced and defended a proposition that he termed "Thesis M": "What can be calculated by a machine is computable" (1980, p. 124).

Gandy said that by *computable* he means "computable by a Turing machine," and he takes the objects of computation to be functions over the integers (or other denumerable domains). It is less clear what he meant by *calculation* and *computation* (we ourselves will use these terms interchangeably) and by *machine*. He said that he was using "the fairly nebulous term 'machine' " for the sake of "vividness," and he made it evident that discrete

---

[1] Thanks to Michael Cuffaro for this suggestion.

deterministic *mechanical* assemblies are his real target, where the "only physical presuppositions" made about a *mechanical* system are that there is "a lower bound on the linear dimensions of every atomic part" and "an upper bound (the velocity of light) on the speed of propagation of changes" (1980, p. 126). We will refer to discrete deterministic mechanical assemblies as DDMAs. Gandy emphasized that the arguments in his paper apply only to DDMAs and not to "*essentially* analogue" systems, nor systems "obeying Newtonian mechanics" (1980, pp. 126, 145). His thesis – which we call Gandy's thesis – is that the functions able to be computed by DDMAs are Turing computable.

Like his teacher Turing, Gandy took an axiomatic approach to characterizing computation. But whereas Turing's classic 1936 paper gave an analysis of *human* computation (Turing 1936; Copeland 2004a, 2015), Gandy's aim was to provide a wider analysis. He pointed out that Turing's analysis does not apply to machines in general: Turing assumes, for instance, that the computer (a human being) "can only write one symbol at a time," an assumption that clearly does not apply to parallel machines, since these can change "an arbitrary number of symbols simultaneously" (1980, pp. 124–125). Gandy formulated the general concept of a DDMA in terms of precise axioms, which he called Principles I–IV. These four axioms define a set of mechanisms – "Gandy machines" – and Gandy proved that the computational power of these mechanisms is limited to Turing computability (with a simplified version of the proof provided by Sieg and Byrnes [1999]).

Principle I, which Gandy referred to as giving the "form of description," sets out a format for describing DDMAs. A DDMA is described by an ordered pair $\langle S, F \rangle$, where S is a potentially infinite set of states and F is a state-transition operation from S_i to S_{i+1} (for each member S_i of S). Gandy chose to define the states in terms of subclasses of the hereditarily finite sets (HF) over a potentially infinite set of atoms (closed under isomorphic structures). These subclasses are termed "structural classes," and the state-transition operation is defined in terms of structural operations over such classes. Putting aside the technicalities of Gandy's presentation, Principle I can be approximated as:

Principle I: Any DDMA M can be described by an expression $\langle S, F \rangle$, where S is a structural class, and F is a transformation from S_i to S_j. Thus, if S_0 is M's initial state, then $F(S_0), F(F(S_0)), \ldots$ are its subsequent states.

Each (non-atomic) state S_i of S is assembled from parts, and these can be assemblies of other parts, etc. Principles II and III place boundedness restrictions on the structure of the states. They can be expressed informally as:

Principle II: For each machine, there is a finite bound on the complexity of the structure of its states. (In Gandy's terminology, this comes down to the requirement that the states of a machine are members of a fixed initial segment of HF.)

In GL, for example, the grid can be arbitrarily large but the complexity of the structure of each state is very simple and can be described as a list of pairs of cells – or, more generally, as a list of lists of cells, since each listed pair of cells is itself a list of cells. In general we can picture a Gandy machine as storing information in a hierarchical way, such as lists of lists (Gandy 1980, p. 131), but Principle II lays down that for each machine there is always a finite bound on the structure of this hierarchy.

Principle III: There is a bound on the number of types of basic parts (atoms) from which the states of the machine are uniquely assembled.

For example, the grid of GL can be assembled from pairs of consecutive cells and their symbols (e.g., ["on," "off"], ["on," "on"], etc.). We need only a limited number of pairs like these to construct any configuration of the grid.

Principle IV puts restrictions on the structural operations that can be involved in state transitions: Each state transition must be determined by the *local* environments of the parts of the assembly that change in the transition. Gandy called this the "principle of local causation" and described it as "the most important of our principles" (1980, p. 135). He explained that the axiom's justification lies in the two "physical presuppositions" governing mechanical assemblies (mentioned above). If the propagation of information is bounded, then in bounded time an atom can transmit and receive information in a bounded neighborhood, and if there is a lower bound on the size of atoms, then the number of atoms in this neighborhood is bounded. Taking these together, we can informally express the principle as follows:

Principle IV: The parts from which $F(S_i)$ is assembled are causally affected only by their bounded "causal neighborhoods": The state of each part is determined solely by its local neighborhood.

For example, in GL the grid is assembled from parts – cells – each of which is either "on" or "off" at any given moment. A cell's state – "on" or "off" – is determined only by the bounded causal neighborhood consisting of its eight adjacent cells.

Gandy's proof that any assembly satisfying Principles I–IV is Turing computable goes far beyond the (relatively trivial) textbook reduction of the actions of some number of Turing machines working in parallel to the action of a single Turing machine. There are Gandy machines with arbitrarily many processing parts that work on the same regions (e.g., printing on the same region of tape), and also Gandy machines whose state transitions involve simultaneous changes in an unbounded number of parts. In GL, for example, there is no upper bound on the number of cells that are simultaneously updated.

To what extent does Gandy's analysis capture machine computation? Wilfried Sieg contends that Gandy provided "a characterization of computations by machines that is as general and convincing as that of computations by human computors given by Turing" (2002, p. 247). We challenge Sieg's contention. It is doubtful that Gandy's analysis even encompasses all cases of *physical computation*, not to mention computation carried out by other, notional, machines. Moreover, even Gandy himself thought that not all physical computing machines lie within the scope of his characterization, and for this reason he explicitly distinguished between "mechanical devices" and "physical devices," saying that he was considering only the former (Gandy 1980, p. 126). As we explained above, Gandy said that his analysis aims only at machines conforming to the principles of Relativity, and he expressly excluded some machines that obey Newtonian mechanics – e.g., machines involving "rigid rods of arbitrary lengths and messengers travelling with arbitrary large velocities, so that the distance they can travel in a single step is unbounded" (1980, p. 145).

More importantly still, we argue that Gandy's characterization does not even cover all cases of computation that are in accord with the principle of local causation and his two overarching physical presuppositions (an upper bound on the speed of propagation of effects and signals, and a lower bound on the dimensions of the assembly's components). We consider discrete mechanical systems that infringe Thesis M in the next section, but we begin with some general considerations about physical computation.

2.4   Is the Physical World Computable?

The issue of whether every aspect of the physical world is Turing computable was raised by several authors in the 1960s and 1970s, and the topic rose to prominence in the mid-1980s. In 1985, Wolfram formulated a thesis that he described as "a physical form of the Church-Turing hypothesis": This says that the universal Turing machine can simulate any physical system (1985, pp. 735, 738). In the same year David Deutsch (who laid the foundations of quantum computation) formulated a principle that he also called "the physical version of the Church-Turing principle" (1985, p. 99). Other formulations were advanced by Earman (1986), Pour-El and Richards (1989), Pitowsky (1990), and Blum et al. (1998).

In the 1990s Copeland coined the term "hypercomputer" for any system – notional or real, natural or artifactual – that computes functions, or numbers, that the universal Turing machine cannot compute (Copeland and Proudfoot 1999; Copeland 2002b). A processing system – either a computing system or a system of some other sort – is said to be "hypercomputational" if the information-processing that it performs cannot be done by the universal Turing machine (Copeland 1996, 2000). Scott Aaronson has suggested (in correspondence) that the physical Church-Turing thesis be called simply the *anti-hypercomputation thesis.* The term "physical Church-Turing thesis" is far from ideal, since the Church-Turing thesis as Turing and Church put it forward concerned only the scope and limits of human computation (Copeland 2015); however, we will continue to use the term here (since many do use it).

We use the term "physical" to refer to systems whose operations are in accord with the actual laws of nature. These include not only actually existing systems but also idealized physical systems (systems that operate in some idealized conditions) and physically possible systems that do not actually exist, but that could exist, or did exist (e.g., in the universe's first moments), or will exist. Of course, there is no consensus about exactly what counts as an idealized or possible physical system, but this is not our concern here.

Gualtiero Piccinini (2011, 2015) distinguishes between what he calls "bold" and "modest" versions of the physical Church-Turing thesis. (The distinction applies equally to versions of the anti-hypercomputation thesis.) Bold versions concern physical systems and processes in general, while modest versions are about systems that themselves compute and processes that themselves qualify as computation. Wolfram's thesis is an example of a bold version:

Wolfram's bold physical Church-Turing thesis: "[U]niversal computers are as powerful in their computational capacities as any physically realizable system can be, so that they can simulate any physical system" (Wolfram 1985, p. 738).

The formulations of Deutsch and others are also bold: Their formulations concern physical systems in general and not just computing systems. (Piccinini emphasizes, though, that the bold versions proposed by different writers are often "logically independent of one another," and exhibit "lack of confluence" [2011, pp. 747–748].) Modest versions of the physical Church-Turing thesis, on the other hand, concern physical systems that themselves compute, and assert that the computational power of any physical computer is bounded by Turing computability. Gandy's thesis is an example. His Thesis M is about calculating machines and his talk about functions that are calculated (or computed) by machines – DDMAs – implies that the mediating processes are calculations (computations).

Are these physical versions of the Church-Turing thesis true? We will discuss modest versions first. There have been several attempts to cook up constructions of highly idealized physical machines that compute functions that no Turing machine is able to compute. Perhaps the most interesting ones have been "supertask" machines – machines that complete infinitely many computational steps in a *finite* span of time. Among such machines we find accelerating machines (Copeland 1998a, 2002b; Copeland and Shagrir 2011), shrinking machines (Davies 2001), and relativistic machines (Pitowsky 1990; Hogarth 1994; Andréka et al. this volume).

Relativistic machines operate in spacetime structures having the property that the entire endless lifetime of one machine is included in the finite chronological past of another machine (called "the observer"): Thus the first machine could carry out an infinite computation, such as calculating every digit of π, in what is from the observer's point of view a finite timespan, say one hour. (Such structures, sometimes called Malament-Hogarth spacetimes, are in accord with Einstein's General Theory of Relativity.)

A relativistic machine RM consists of a pair of communicating Turing machines T_A and T_B: T_A, the observer, is in motion relative to T_B, a universal machine. RM is able to "compute" the halting function. When the input (m, n) – asking whether the mth Turing machine (in some enumeration of the Turing machines) halts or not when started on input n – enters T_A, T_A first prints 0 (meaning "never halts") in its designated output cell and then transmits (m, n) to T_B. T_B simulates the computation performed by the mth Turing machine when started on input n and sends a signal back to T_A if and only if the simulation terminates. If T_A receives a signal from T_B, it deletes the 0 it previously wrote in its output cell and writes 1 there instead (meaning "halts"). After one hour, T_A's output cell shows 1 if the mth Turing machine halts on input n and shows 0 if the mth machine does not halt on n.

RM is of interest since arguably it complies with Gandy's principles. RM is discrete, since it consists of two standard digital computers in communication, and (as a relativistic machine) the speed of signal propagation in RM is bounded by the speed of light. Nonetheless, RM cannot be a Gandy machine if it computes a function that no Gandy machine is able to compute. So what is going on? Our answer is that RM violates an *implicit assumption* that underlies Gandy's Principle I (Copeland and Shagrir 2007). Principle I requires that the process can be described as a sequence $S_0, F(S_0), F(F(S_0)), \ldots$ (where S_0 is the initial state and F is the state-transition function). But it is also assumed that the configuration of each stage $\alpha + 1$, described by S_{i+1}, is to be *uniquely determined* by the configuration of the previous stage, α, described by S_i (i.e., that $S_{i+1} = F(S_i)$). We will call this the assumption of *Gandy determinism*. However, this assumption is not necessarily satisfied by RM. Consider the end stage of T_A: if T_A receives a signal from T_B, then its subsequent behavior is Gandy-deterministic, but if it receives no signal from T_B, its behavior is no longer Gandy-deterministic. To count as Gandy-deterministic, the end stage of T_A-halting-on-0 should be determined, in part, by the no-signal message of the last stage of T_B. However, T_B, a non-halting Turing machine, does not have a last stage: there is no stage of T_B that is the one coming just before the end-stage of T_A-halting-on-0 (since after each stage of T_B, there are infinitely many others at which no signal is sent to T_A). Thus the stage of T_A-halting-on-0 is not Gandy-deterministic.

This implicit assumption is the weak point in Gandy's argument, since not every deterministic assembly need be Gandy-deterministic. Moreover there is an extremely reasonable account of determinism according to which RM is deterministic. It is deterministic in that the end stage of T_A-halting-on-0 is uniquely determined by the initial stage of the machine. This is because the end stage of T_A-halting-on-0 is a *limit* of previous stages of T_B (and T_A), of which the relevant feature is their not sending a signal to T_A. This sense of determinism is in good accord with physical usage where a system or machine is said to be deterministic if it obeys laws that invoke no random or stochastic elements. T_A's halting on 0 is completely determined by the fact that it initially wrote 0 in its designated output cell and the fact that at no stage of the computation was a signal sent by T_B.

RM is not a Gandy machine but it is a DDMA (although not a Gandy-deterministic DDMA). Is it a counter-example to the modest thesis? This depends on whether the machine is *physical* and on whether it really *computes* the halting function.

Is RM physical? Németi and his colleagues provide the most physically realistic construction, locating machines like RM in setups that include huge slow rotating Kerr black holes (Andréka et al. this volume) and emphasizing that the computation is physical in the sense that "the principles of quantum mechanics are not violated" and RM is "not in conflict with presently accepted scientific principles" (Andréka et al. 2009, p. 501). They suggest that humans might "even build" their relativistic computer "sometime in the future" (2009, p. 501). Naturally all this is controversial. Earman and Norton (1993), Aaronson (2005), Piccinini (2011), and others argue that this relativistic physical setup faces serious problems: However, Németi and his colleagues reply resourcefully to these objections (Etesi and Németi 2002; Németi and Dávid 2006; Andréka et al. 2009; Andréka et al. this volume); see also Shagrir and Pitowsky (2003).

Does RM *compute* the halting function? The answer depends on what is included under the heading *physical computation*. We cannot possibly cover here the array of differing accounts of physical computation found in the current literature. But we can say that RM computes in the senses of "compute" staked out by several of these accounts: the semantic account (Shagrir 2006; Sprevak 2010), the mechanistic account (Miłkowski 2013; Fresco 2014; Piccinini 2015), the causal account (Chalmers 2011), and the BCC (broad conception of computation) account (Copeland 1997, p. 695). According to all these accounts, RM counterexamples the modest thesis *if* RM is physical. However, RM does *not* compute if computation is construed as the execution of an algorithm in the classical sense. The classical notion of an algorithm does not accommodate the limit stages found in relativistic computation (although it does accommodate all sorts of nondeterministic processes, e.g., probabilistic processes).

We conclude that Gandy's principles do not provide a general and comprehensive analysis of machine computation. We do not wish to downplay the contribution that his analysis has made to the current understanding of machine computation, but it is important to realize that his analysis is limited in its scope. In fact, its scope is more limited than is suggested by Gandy's own exclusion of analog machines and some types of discrete Newtonian machines: His analysis does not even cover all instances of non-hypercomputational discrete physical computation. For instance, his Principle I does not directly apply to probabilistic algorithms and asynchronous algorithms (Copeland and Shagrir 2007; Gurevich 2012).

We turn now to the bold thesis, which says in effect that the behavior of every physical system can be *simulated* (to any required degree of precision) by a Turing machine. Speculation that there may be physical processes whose behavior cannot be calculated by the universal Turing machine stretches back over several decades; for a review, see Copeland (2002b). Early papers by Scarpellini (1963), Komar (1964), and Kreisel (1965, 1967) made this point. Georg Kreisel stated that "There is no evidence that even present day quantum theory is a mechanistic, i.e. recursive theory in the sense that a recursively described system has recursive behaviour" (1967, p. 270). More concretely, Marian Pour-El and Ian Richards (1981) showed that the familiar three-dimensional wave equation produces non-Turing-computable output sequences for some Turing-computable input sequences. But their result is at the mathematical level: It is an open question whether the requisite input sequences can obtain physically. *RM* (if physical) provides another counterexample to the bold thesis (since the bold thesis implies the modest).

To summarize the discussion so far: The bold thesis is clearly an empirical hypothesis, and at the present stage of physical inquiry it is unknown whether this hypothesis is true. However, it can at least be said that to date there is no empirical evidence against the hypothesis (so far as we know). The modest thesis also seems to be an empirical hypothesis, although here matters are more complex, since a conceptual issue also bears on the truth or falsity of the thesis – the issue of what counts as physical computation. As with the bold thesis, it is currently unknown whether the modest thesis is true or false.

Next we introduce a new, stronger form of the physical Church-Turing thesis and examine some recent work on undecidability in physics. We call this new form the "super-bold" physical Church-Turing thesis. Unlike the bold thesis, it concerns not only the ability of the universal Turing machine to simulate the behavior of physical systems (to any required degree of precision) but also further physical questions about this behavior. Examples are decidability questions such as: "Is the solar system stable?" and "Is the motion of a given system, in a known initial state, periodic?" (Pitowsky 1996).

The super-bold physical Church-Turing thesis: Every physical aspect of the behavior of any physical system is Turing computable (to any desired degree of accuracy).

In 1986 Robert Geroch and James Hartle argued that *un*decidable physical theories "should be no more unsettling to physics than has the existence of well-posed problems unsolvable by any algorithm been to mathematics," and they suggested that such theories may be "forced upon us" in the quantum domain (1986, pp. 534, 549). Arthur Komar raised "the issue of the macroscopic distinguishability of quantum states" in 1964, asserting that there is no effective procedure "for determining whether two arbitrarily given physical states can be superposed to show interference effects" (1964, pp. 543–544). More recently Jens Eisert, Markus Müller, and Christian Gogolin showed that "the very natural physical problem of determining whether certain outcome sequences cannot occur in repeated quantum measurements is undecidable, even though the same problem for classical measurements is readily decidable" (2012, p. 260501-1). This is an example of a problem that refers unboundedly to the future but not to any specific time (as in Itamar Pitowsky's examples mentioned earlier). Eisert, Müller, and Gogolin suggest that "a plethora of problems" in quantum many-body physics and quantum computing may be undecidable (2012, pp. 260501-1–260501-4).

Dramatically, a 2015 *Nature* article by Toby Cubitt, David Perez-Garcia, and Michael Wolf outlined their proof that "the spectral gap problem is algorithmically undecidable: there cannot exist any algorithm that, given a description of the local interactions, determines whether the resultant model is gapped or gapless" (Cubitt et al. 2015, p. 207). Cubitt describes this as the "first undecidability result for a major physics problem that people would really try to solve" (Castelvecchi 2015).

The spectral gap, an important determinant of a material's properties, refers to the energy spectrum immediately above the ground energy level of a quantum many-body system (assuming that a well-defined least energy level of the system exists): The system is said to be gapless if this spectrum is continuous and gapped if there is a well-defined next least energy level. The spectral gap problem for a quantum many-body system is the problem of determining whether the system is gapped or gapless, given the finite matrices describing the local interactions of the system.

In their proof Cubitt et al. encode the halting problem in the spectral gap problem, so showing that the latter is at least as hard as the former. The proof involves an infinite family of two-dimensional lattices of atoms, but they point out that their result also applies to finite systems whose size increases: "Not only can the lattice size at which the system switches from gapless to gapped be arbitrarily large, the threshold at which this transition occurs is uncomputable" (Cubitt et al. 2015, pp. 210–211). Their proof offers an interesting countermodel to the super-bold thesis, involving a physically relevant example of a finite system of increasing size such that there exists no Turing-computable procedure for extrapolating the system's future behavior from (complete descriptions of) its current and past states. (For discussion of such systems, see Geroch and Hartle [1986] and Copeland [2002b, 2004a].)

It is debatable whether any of these quantum models corresponds to real-world quantum systems. The Komar model involves a system with an infinite number of degrees of freedom, and Cubitt et al. admit that the model invoked in their proof is highly artificial, saying "Whether the results can be extended to more natural models is yet to be determined" (Cubitt et al. 2015, p. 211). There is also the question of whether the spectral gap problem becomes computable when only local Hilbert spaces of realistically low dimensionality are considered. Nevertheless, these results are certainly suggestive. The super-bold thesis cannot be taken for granted – even in a finite quantum universe.

We turn next to Penrose's speculations concerning physical uncomputability.

2.5   Penrose's Thesis: Uncomputability and the Brain

Penrose's thesis is the claim that the action of the brain is hypercomputational (Penrose 2013, p. xxxiii). Penrose holds that the brain's uncomputability is key to explaining the phenomenon of consciousness (Penrose 1989, 1990, 1994; Hameroff and Penrose 2014). According to Penrose, the brain's hypercomputational action, and the role this plays in generating conscious experience, will not be fully understood until the advent of what he calls the New Theory in physics: He says that "hypercomputational actions" in the brain are the "non-computable effects of [the] New Theory" (Penrose 2013, p. xxxiii). This "presumed New Theory," he says, goes "beyond current quantum mechanics": It is "presently unknown in detail" and involves "hitherto undiscovered laws" (2013, pp. xxxii, xxxiii).

Penrose's argument for his thesis is based on Gödel's incompleteness theorems, which he "regard[s] as providing a strong case for human understanding being something essentially non-computable" (2013, p. xxviii) – understanding being "one manifestation of human consciousness" (2011, p. 347). This general line of argument, made famous in an article by the philosopher John Lucas (1961), is often called the "Gödel Argument," although in fact it was anticipated by Emil Post as early as 1921 (Post 1965, p. 417). Penrose calls it the "Gödel-Turing Argument" (e.g., in his [2011]), and Turing himself dubbed it the "Mathematical Objection" (1950, p. 450), giving the following elegant summary of it:

Recently the theorem of Gödel and related results ... have shown that if one tries to use machines for such purposes as determining the truth or falsity of mathematical theorems and one is not willing to tolerate an occasional wrong result, then any given machine will in some cases be unable to give an answer at all. On the other hand the human intelligence seems to be able to find methods of ever-increasing power for dealing with such problems "transcending" the methods available to machines. (Turing 1948, pp. 410–411)

Turing by no means endorsed the "Gödel-Turing Argument." His subtle objection to it, involving what we call his "multi-machine theory" of mentality, is described in Copeland and Shagrir (2013) – and is very different from the objection that Penrose imputes to Turing, in our view mistakenly (e.g., in his [1997, p. 112]). We shall return briefly to Turing's views below.

Gödel's view, as he expressed it in his 1951 Gibbs lecture, was that the incompleteness results establish a *disjunction*: *either* "there exist absolutely unsolvable diophantine problems" (where, Gödel explained, "the epithet 'absolutely' means that they would be undecidable, not just within some particular axiomatic system, but by *any* mathematical proof the human mind can conceive"), *or else* "the human mind ... infinitely surpasses the powers of any finite machine" (1951, p. 310). (For a fuller study of Gödel's views, see Copeland and Shagrir [2013].)

Later, at the beginning of the 1970s, Gödel in effect recast this disjunction into an implication:

If my result [incompleteness] is taken together with the rationalistic attitude which Hilbert had and which was not refuted by my results, then [we can infer] the sharp result that mind is not mechanical. This is so, because, if the mind were a machine, there would, contrary to this rationalistic attitude, exist number-theoretic questions undecidable for the human mind. (Gödel in Wang 1996, pp. 186–187)

What Gödel called Hilbert's "rationalistic attitude" was summed up in the latter's celebrated remark that "in mathematics there is no *ignorabimus*" – there is no mathematical question that in principle the mind is incapable of settling (Hilbert 1900, p. 445).

Gödel's position, then, was that his incompleteness results do *not* entail that the mind is not mechanical, but if these are coupled with the rationalistic attitude that there are *no* absolutely undecidable problems – an attitude that, he emphasized, "remains entirely untouched" by his negative results (Gödel 193?, p. 164) – then it does indeed follow that the mind is not mechanical. In a note[2] written in 1963 (Figure 2.1), Gödel explained where he sat in this debate (at any rate at that time). Referring to his 1951 disjunction he said: "I believe, on ph[ilosophical] grounds, that the sec[ond] alternative is more probable."[3]

Clearly the success of the Gödel Argument turns on whether this "rationalistic attitude" could ever be established to be correct – i.e., whether it could

---

[2] Thanks to Juliette Kennedy for assistance in locating this document in the Firestone Library.
[3] See van Atten and Kennedy (2003) for a discussion of Gödel's 1963 note.

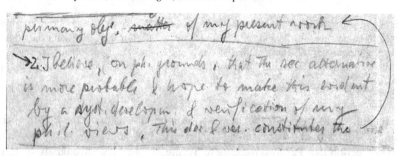

Figure 2.1 Extract from notes Gödel made for a letter that he sent to TIME Inc in the summer of 1963. The extract reads: "I believe, on ph[ilosophical] grounds, that the sec[ond] alternative is more probable & hope to make this evident by a syst[ematic] developm[ent] & verification of my phil[osophical] views. This dev[elopment] & ver[ification] constitutes the primary obj[ect] of my present work." Credit: Unpublished works of Kurt Gödel (1934–1978) are Copyright Princeton Institute for Advanced Study and are used with permission. All rights reserved by the Princeton Institute for Advanced Study. Thanks also to the Firestone Library, Rare Books and Special Collections, Princeton University

ever be established that there are no absolutely undecidable problems. It is, to be sure, difficult to see how this could ever be done. But, in any case, Gödel's "sharp result" is undercut by Turing's rebuttal of the Mathematical Objection (Copeland and Shagrir 2013). Moreover, numerous other objections have been raised to the Gödel Argument, and to the detailed formulation of it endorsed by Penrose (for example, in Penrose [1990] and the commentaries that follow). Rather than attempting to survey these many objections here, we will focus on what seems to us to be the absolutely central difficulty with Penrose's argument, namely that the argument appears to reduce to absurdity (Copeland 1998b; Copeland and Proudfoot 2007).

The *reductio ad absurdum* is this. Let us suppose Penrose's argument does successfully establish that (as he puts it) human mathematicians do not use a knowably sound Turing-machine algorithm in order to ascertain mathematical truth. If so, then his argument shows with equal success that human mathematicians, in ascertaining mathematical truth, do not use any knowably sound procedure that is capable of being executed by an *oracle machine*. Turing's oracle machines (or *o*-machines) are the result of equipping a universal Turing machine with at least one additional basic operation that no Turing machine proper can simulate (Turing 1939). Turing called these new basic operations "oracles," saying that oracles work by "some unspecified means" (1939, p. 156).

As Turing explained, oracle machines form a hierarchy that extends ever upwards. Let the *first-order o*-machines be those whose oracle produces the values of the Turing-machine halting function $H(x,y)$. The *second-order o*-machines are those with an oracle that can say whether or not any given first-order o-machine eventually halts if set in motion with such-and-such a number inscribed on its tape, and so on for third-order and in general α-order o-machines. Penrose's argument was originally marketed as showing that human understanding does not consist in any process that a Turing machine can execute (see, e.g., Penrose 1994, ch. 2), but his argument is so powerful that it equally supports the conclusion that human understanding does not consist in any process that the richly hypercomputational oracle machines can execute. (This applies even to the "cautious oracles" that Penrose introduces in his [2016].) Penrose's argument moves relentlessly up through the orders, stopping nowhere.

Penrose noted this difficulty in his 1994 book *Shadows of the Mind* (p. 380). He also suggested a way out:

[I]t need not be the case that human mathematical understanding is in principle as powerful as *any* oracle machine at all. ... Thus, we need not necessarily conclude that the physical laws that we seek reach, in principle, beyond every computable level of oracle machine (or even reach the first order). We need only seek something that is not equivalent to *any* specific oracle machine. (1994, p. 381)

What does Penrose mean here? It is customary in recursion theory to say that problems of equal "hardness" are of the same *degree*: Problems that are solvable by Turing machines are said to be of degree 0. Penrose seems to be suggesting that physical laws occupy a position *in between* degree 0 and degree 1 (the degree of problems that are solvable by a first-order oracle machine but not by a Turing machine). It is indeed known that there are degrees between 0 and 1 (Friedberg 1957; Sacks 1964) and this seems to make sense of what Penrose is suggesting: For some degree between 0 and 1, the "physics of mind" is exactly that hard. This is certainly a coherent position – and for all that anyone presently knows, it may in fact be true.

However, this suggestion does not prevent the *reductio ad absurdum* that we are discussing (Copeland 1998b). Let i (for "intermediate") be a degree between 0 and 1 and let I be the class of o-machines that are able to solve problems of degree i (and no harder problems). Do mathematicians use, in ascertaining mathematical truth, a knowably sound procedure able to be executed by a machine in I? Not if Penrose's argument is sound, since it applies equally to the o-machines in I. To borrow a phrase of Penrose's (from his [2013, p. xxxiv]), the Gödel Argument involves a "never-ending capability of being able to 'stand back' and contemplate whatever structure had been considered previously": Whatever structure – whatever physical system – is contemplated, the argument deems it not to be the mind.

In his more recent work Penrose does not repeat the suggestion just discussed. But nor does he offer any way of avoiding the *reductio ad absurdum* that he noted in his 1994 book. Commenting on the fact that, no matter what device D is specified, the Gödel Argument entails that the mind is more powerful than D, Penrose says only that the mind is "something very mysterious" and that its theory must involve "something very subtle" (1996, sec. 13.2; 2013, p. xxxiv). John Lucas was happy to conclude from the Gödel Argument that "no scientific enquiry can ever exhaust the ... human mind" (1961, p. 127), and Gödel thought that the brain must be "a computing machine connected with a spirit" (in Wang 1996, p. 193). Unlike Gödel and Lucas, Penrose seems to think that there must be a fully physical account of consciousness, but he has failed to make it clear what physical conception of consciousness can possibly remain for one who endorses the Gödel Argument.

It is a pity that Penrose chose to support his thesis by means of the Gödel Argument, since the argument is ultimately a distraction and moreover tends to mask the fact that Penrose's thesis is – like the various forms of the physical Church-Turing thesis considered above – a thoroughly empirical thesis. It is a serious hypothesis that, far from requiring a radical New Theory, might even be consistent with current quantum mechanics, as the undecidability of the spectral gap problem perhaps tends to indicate. There is, so far as we are aware, not a shred of empirical evidence for Penrose's thesis, but this situation might change in the future. One can only keep an open mind.

We conclude with a comment on the relationship between Penrose's view of the brain and Turing's. Penrose says: "It seems likely that he [Turing] viewed physical action in general – which would include the action of a human brain – to be always reducible to some kind of Turing-machine action" (1994, p. 21). Penrose even named this claim *Turing's thesis*. Yet Turing never endorsed this thesis and was aware that it might be false. Turing was in fact an important forerunner of the modern debate concerning the possibility of uncomputability in physics and uncomputability in the action of the human brain (as was first pointed out in Copeland [1999] and Copeland and Proudfoot [1999]). In a 1951 lecture on BBC radio Turing suggested that it may not be possible for a computer to simulate the human brain because of the brain's quantum-mechanical nature (Turing 1951; Copeland 1999, pp. 448, 451–452). Far from subscribing to what Penrose called Turing's thesis, Turing in this lecture contemplated the possibility that the physics of the brain might be uncomputable. (Even Andrew Hodges, who used to maintain that Turing claimed "that the action of the brain must be computable" [2004, p. 51], now seems to have accepted that his previous view of Turing was wrong [Hodges 2012].)

2.6   Summary

We have discussed a number of theses concerning the relationship between physics and computation. We began with the thesis that the physical universe is a computer (Zuse's thesis) and moved on to the thesis that the behavior of all discrete deterministic mechanical assemblies is Turing computable (Gandy's thesis) and then the more general physical Church-Turing thesis (which is also known as the "anti-hypercomputation thesis"). We distinguished three versions of the physical Church-Turing thesis: the modest, the bold, and the super-bold versions. We ended with the thesis that some actions of a specific physical system – the human brain – are not Turing computable (Penrose's thesis).

Acknowledgments

Shagrir's research was supported by the Israel Science Foundation grant 1509/11.

3   Church's Thesis, Turing's Limits, and Deutsch's Principle

Rossella Lupacchini

A common opinion holds that David Hilbert was a "formalist" who wished to purge intuition from mathematics in favor of a mere game with symbols. As a result, even physical concepts expressed mathematically would be divested of all meaning. Quantum mechanics in particular, whose mathematical methods originated in Göttingen, is taken as an emblematic example of a puzzle despite its extraordinary predictive power. "Actually, Hilbert was a great believer in intuition" (Stillwell 2010, p. 91n). For the benefit of mathematics itself, he strongly encouraged the flow of new material from the world outside mathematics and relations with physics and epistemology. His conviction was that mathematics bridges the gap between logic and experience, between abstraction and intuitive understanding, and the *axiomatic method* provides the indispensable tool for grounding the foundations of mathematics. Proceeding axiomatically from geometry to logic *via* mathematical physics, Hilbert clearly did not pursue a mere game with symbols. Rather he wished to reach through the *forms* of mathematical thought and grasp its mechanism.

The procedure of the axiomatic method, as he explained, "amounts to *a deepening of the foundations* of the individual domains of knowledge – a deepening that is necessary for every edifice that one wishes to expand and to build higher while preserving its stability" (Hilbert 1918, p. 1109). It was a deepening beyond the individual domains of knowledge that created the demand for full clarity about "the essence of mathematical proof itself." It was a sharpening of the methodological reflection on the essence of mathematical proof that focused on the concept of *computability*. Established as a distinctive concept of modern logic, it would also reveal a distinctive physical character and, therefore, its genuine mathematical fabric.

3.1   Hilbert's Mathematical Forms

In the late summer of 1930, Königsberg offered a synoptic view on Hilbert's formalism. While Kurt Gödel's announcement of his first incompleteness theorem pointed to the sunset of Hilbert's program, Hilbert's address on *Logic and the Knowledge of Nature* revived the Kantian matrix of his philosophical attitude.

Gödel outlined the proof of his first incompleteness theorem at the Second Conference for the Epistemology of the Exact Sciences. John von Neumann, who participated in the Conference to talk about "the formalist foundations of mathematics," promptly realized the significance of Gödel's argument. Nevertheless, there was disagreement between Gödel and von Neumann as to the impact of the incompleteness theorems on Hilbert's finitary consistency program. Unlike von Neumann, who took for granted that Hilbert's program could not be accomplished, Gödel remained confident of a happy ending as long as he believed that incompleteness would be provable only for "*Principia Mathematica* and related systems." He was cautious about a general definition of *formal system.*

The problem of mathematics' consistency emerged in the nineteenth century, when the discovery of non-Euclidean geometry and the increasing *abstractness* of algebraic structures seriously stretched the *intuitive understanding* of the mathematical notions. At that point the distinctive character of the mathematical objects and procedures became an issue to be clarified. It was the tension between Dedekind's structuralist abstraction, on one side, and Kronecker's insistence on the effective construction of mathematical objects from the natural numbers, on the other side, that oriented Hilbert toward his "formalism" (Sieg 2009, p. 537). But unlike Dedekind and Kronecker, who wished to purify the science of numbers from foreign geometric ideas, Hilbert wished to safeguard "the use of geometrical signs as a means of strict proof" (1900, p. 1100) through a rigorous axiomatic investigation of their conceptual content.

The ideal constructions of geometry as well as the imaginary elements of number theory help Hilbert illustrate the way in which his "proof theory" (*Beweistheorie*) could surpass the domain of finite logic, and obtain *provable* formulae that are the images of the transfinite theorems of ordinary mathematics.

In my proof theory, the transfinite axioms and formulae are adjoined to the finite axioms, just as in the theory of complex variables the imaginary elements are adjoined to the real, and just as in geometry the ideal constructions are adjoined to the actual. The motivation and the success of the procedure is the same in my proof theory as it is there: that is, the adjoining of the transfinite axioms results in the simplification and completion of the theory. (Hilbert 1923, p. 1144)

Hilbert's proof theory provides a perspective picture of mathematics: on the one hand, mathematics in the strict sense is projected onto the *formal* plane and becomes a "stock of formulae," which are the *ideal objects* of the theory; on the other hand, a new *metamathematics* takes shape on the projective plane, *distinct* from the formalized mathematics. It is this metamathematics that allows mathematical proofs to become *objective* forms for mathematical understanding. The very notion of proof ought to be built from scratch:

Kant already taught – and indeed it is part and parcel of his doctrine – that mathematics has at its disposal a content secured independently of all logic and hence can never be provided with a foundation by means of logic alone ... Rather, as a condition for the use of logical inferences and the performance of logical operations, something must already be given to our faculty of representation [in der Vorstellung], certain extralogical concrete objects that are intuitively [anschaulich] present as immediate experience prior to all thought. (Hilbert 1967 [1926], p. 376)

Thus a proof is conceived of "as a figure that must intuitively [anschaulich] appear to us as such" (Hilbert 1923, p. 1137), and to detect possible contradictions in which mathematical thought may incur, one can dispense with a longish chain of uncertain steps, since they will result in an erroneous "constellation of signs." In this sense, the fundamental idea of Hilbert's proof theory can also be traced to Leibniz's design of a *calculus ratiocinator*, primarily conceived to discredit Descartes's skepticism about the reliability of the deductive method. Both Leibniz and Hilbert clearly recognized that our knowledge is molded from symbols, for only the use of symbols allows the construction of a *representation space* where a mathematical proof becomes "something concrete and displayable."

In Leibniz's thought, however, no symbolic process could help achieve the full clarity of pure intuition. By contrast, Hilbert's dream was nothing less than to distill the *virtus visiva* of intuition into finitary methods. In pursuing this goal, his strategy was to gain evidence of the reliability of mathematical inference from the constant interplay between two ways of "proving":

between the formalized ("mathematical") proving within a formal system and the contentual ("metamathematical") proving about the system. While the former is an arbitrarily defined logical game (that must be, however, to a large extent analogous with classical mathematics), the latter is a chaining of immediately evident contentual insights. (von Neumann 1961 [1927], pp. 256–257)

Hilbert's road from the foundations of geometry to metamathematics amounts to a sharpening of the axiomatic method. According to Wilfried Sieg, "Hilbert uses the axiomatic method to shift the Kroneckerian effectiveness requirements from the mathematical to the 'systematic' metamathematical level" (2009, pp. 537–538). To let the "formalist" Hilbert come to full light, however, it might be also worth considering his devotion to the rigor of abstract axiomatics in dialectic contrast with his concerns for intuitive understanding. Hilbert's demand for finite means aimed at providing not solely an answer to Kronecker's pressure on effective constructions, but also a revision of the Kantian boundaries on the domain of the a priori.

Deconstructing the Kantian forms of a priori intuition, Hilbert's extra-logical concrete objects – intuitively given before all thought – correspond to Kant's conditions for "the possibility of conceptual knowledge." Indeed, for Hilbert "the axiomatic method is basic to his anti-metaphysical account of the nature of mathematics" (Hallett 1990, p. 242). The "pre-established harmony" that he saw in the striking correspondence between physical facts and mathematical thoughts has to be taken "in a sense different from that in Leibniz":

We can understand this agreement between nature and thought, between experience and theory, only if we take into account both *the formal element and the mechanism that is connected with it*; and we must do this both for nature and for our understanding. (Hilbert 1930, p. 1160, emph. added)

3.2   From a Logical Point of View

Hilbert concluded his address in Königsberg (1930) with the commanding words: "We must know, we shall know." His belief that for the mathematician "there are absolutely no unsolvable problems" was seriously shaken when Gödel showed that there are relatively simple problems in the theory of integers that cannot be decided through a formal system like *Principia Mathematica* or Zermelo-Frankel set theory and, also, that the formal system cannot prove its own consistency. Despite his incompleteness theorems, Gödel was not inclined to dispense with the importance of a finitary consistency proof, not expressible within a formal system like *Principia Mathematica*. It was probably a letter from Jacques Herbrand (April 7, 1931) that encouraged him to see the significance of his incompleteness theorems for Hilbert's finitary program with a more critical eye and to focus on the link between decidability of logical relations and *calculability* of number theoretic functions to clarify the issue (Sieg 2009, pp. 557–559).

In December 1933, in his lecture to the Mathematical Association of America in Cambridge (MA), Gödel distinguished two parts in the problem of the foundations of mathematics: first, reducing the totality of the methods of proof actually used by mathematicians to a minimum number of axioms and primitive rules of inference; and second, giving a justification for these axioms and rules of inference.

The first part of the problem has been solved in a perfectly satisfactory way, the solution consisting in the so-called "formalization" of mathematics, which means that a perfectly precise language has been invented, by which it is possible to express any mathematical proposition by a formula. Some of these formulas are taken as axioms, and then certain rules of inference are laid down which allow one to pass from the axioms to new formulas and thus to deduce more and more propositions, the outstanding feature of the rules of inference being that they are purely formal, i.e., refer only to the outward structure of the formulas, not to their meaning, so that they could be applied by someone who knew nothing about mathematics, or by a machine. (Gödel 1933, p. 45)

Instead, with regard to the second part of the problem, the situation was still extremely unsatisfactory: "Our formalism works perfectly well and is perfectly unobjectionable as long as we consider it as a mere game with symbols, but as soon as we come to attach a meaning to our symbols serious difficulties arise." A kind of Platonism unquestionably disputable, Gödel admitted, necessarily influenced all the attempts to interpret axioms. However, he conjectured, "we might at least be able to prove their freedom from contradiction by unobjectionable methods." As for the methods of proof, he noticed that they become less convincing as we ascend from the strictest form of constructive mathematics, which is based exclusively on the method of complete induction in its definitions as well as in its proofs, towards ordinary non-constructive mathematics. Hilbert and his students tried to use only the strictest constructive methods. "But unfortunately, the hope of succeeding along these lines has vanished entirely in view of some recently discovered facts." Gödel's hope was then to work out the details of those "unobjectionable methods" on the edge of Hilbert's finitism.

It was most likely in Princeton in the spring of 1934 that the notion of "effective calculability" clearly appeared in the guise of *recursiveness*. In a series of lectures at the Institute for Advanced Study, Gödel noted that *primitive recursive functions* "have the important property that, for each given set of values of the arguments, the value of the function can be computed by a finite procedure." As to the converse, namely, that a *finite computation* equals a recursive procedure of some sort, he claimed: "This cannot be proved, since the notion of finite computation is not defined, but it serves as a heuristic principle."

In Princeton, at that time, Alonzo Church and his students were working on the concept of λ-definability. In the fall of 1933, Church proposed that the λ-definable functions are all the effectively calculable functions (Davis 1982, p. 8). But Gödel's reaction was very critical and he deemed Church's proposal as "thoroughly unsatisfactory."[1] Thus, in his talk at a meeting of the American Mathematical Society held in New York City (April 1935), Church did not define the notion of *effective calculability* in terms of λ-definability, but rather in terms of "Herbrand-Gödel general recursiveness."

Following a suggestion of Herbrand, but modifying it in an important respect, Gödel has proposed ... a definition of the term *recursive function*, in a very general sense. In this paper a definition of *recursive function of positive integers* which is essentially Gödel's is adopted. And it is maintained that the notion of an effectively calculable

---

[1] Cf. Church's letter to Kleene dated November 29, 1935 (quoted in Davis 1982, p. 9).

function of positive integers should be identified with that of a recursive function, since other plausible definitions of effective calculability turn out to yield notions which are either equivalent to or weaker than recursiveness. (Quoted in Davis 1982, p. 10)

In the published version of his paper Church wrote: "This definition is thought to be justified by the considerations which follow, so far as positive justification can ever be obtained for the selection of a formal definition to correspond to an intuitive notion" (1936, p. 100). For Gödel, however, all the considerations hitherto adduced were not sufficient to show that the "generally accepted properties" of the notion of effective calculability necessarily lead to *define* some particular class of functions, either recursive or λ-definable. He saw the essential point in defining what a *finite procedure* is. It was only Turing's work that persuaded him that a mathematically satisfactory definition of the notion of "mechanical procedure" and (therefore) of effectively calculable function of integers had been achieved. As he wrote:

Turing's work gives an analysis of the concept of "mechanical procedure" (alias "algorithm" or "computation procedure" or "finite combinatorial procedure"). This concept is shown to be equivalent with that of a "Turing machine." (Gödel 1934, pp. 369–370)

How could a "Turing machine" clear Gödel's doubts as to Church's thesis?

3.3   Struggling to Make Sense of Church's Thesis

The two basic ingredients of a *Turing machine* are a potentially infinite one-dimensional *tape* divided into *squares* and a "reading-writing" mechanism. Each square can carry one *symbol* from a finite alphabet $\{S_0, S_1, S_2, \ldots, S_n\}$; taking the symbol S_0 as the blank square, the minimal binary alphabet $\{S_0, S_1\}$ suffices. As for the mechanism, at a given instant, it can be in one of a finite number of *states* $\{q_1, q_2, \ldots, q_m\}$ and scan one square. Depending on the symbol it reads on the square and its state, it can change the symbol on the square and/or move (left or right) to an adjacent square. In doing so, it is bound to follow an instruction described by a *quintuple* of the form "$q_i\ S_j\ S_k\ d\ q_l$," where the first two characters describe the initial condition at the beginning of a computational step and the remaining three characters describe the effect of the instruction to be executed in that condition (q_i is the current state, S_j is the symbol currently scanned, q_l is the state to enter next, S_k is the symbol to replace S_j, and d indicates motion of one square to the right, or one square to the left, or staying fixed, relative to the tape). A *Turing machine* is defined by a finite list of quintuples.

What makes it the "most satisfactory way" of defining computability to Gödel's eye?

3.3.1   Human Limits and Natural Law

Church (1936) used the word "definition" for what became known as *Church's thesis*.[2] As he himself remarked, the difficulty of selecting a formal definition corresponding to an intuitive notion lies in its "positive justification." Emil Post, who proposed another version of computability "virtually identical" to Turing's (cf. Davis and Sieg 2015), saw his own formulations and Gödel-Church's as a "working hypothesis" and stressed that the success of the program at issue would "change this hypothesis not so much to a definition or to an axiom but to a *natural law*" (1936, p. 291). With regard to "Church's identification of effective calculability with recursiveness," he added in a note:

Actually the work already done by Church and others carries this identification considerably beyond the working hypothesis stage. But to mask this identification under a definition hides the fact that a fundamental discovery in the limitations of the mathematicizing [*sic*] power of *Homo Sapiens* has been made and blinds us to the need of its continual verification. (Post 1936, p. 291)

The notion of computability introduced independently by Post and Turing in 1936 has a character remarkably different from Gödel-Church's. Neither Post nor Turing try to identify some particular *class of functions* as effectively calculable, but rather what makes calculability *effective*. They both focus on "finite processes," though they are possibly unlike in their respective philosophical attitudes. Not unlike Gödel, Post seems to be concerned with the boundary between what in mathematics can be achieved by purely formal means and what requires understanding and meaning. In tune with Hilbert, Turing seems rather to be tracking "some extra-logical concrete objects" as conditions for the possibility of effective calculability. Further research into Post-Turing's concept would dig out an implicit *structural* link at the root of pure mathematics and theoretical physics.

Turing's analysis of effective calculability proceeds, by elimination of irrelevant details, to distill the bare essentials of the notion. For Turing, a computation is a finite sequence of reading-writing processes carried out by a "computer." Considering a human being as a computer, he tries to figure out which human capabilities are critical to carry out those processes. On the reasonable assumption that "human memory is necessarily limited," the answer turns out to be nothing but *distinguishing* and *replacing* symbols. For this reason, some boundedness conditions are to be imposed on the configurations of symbols the computer has to manipulate. All such configurations must be "immediately recognizable" by the computer and, therefore, cannot incorporate infinitely many symbols: "if we were to allow an infinity of symbols, then

---

[2] The word "thesis" was later introduced by Kleene (1943, p. 274).

there would be symbols differing to an arbitrarily small extent" (Turing 1936, p. 75). Connecting the effectiveness of calculability to the *limitations of sight and motion* of the computer involved, Turing was able to design "a machine to do the work of [the] computer." The mathematizing power of *Homo Sapiens* appears to be projected onto a one-dimensional space and captured by a finite state system.

It should come as no big surprise that later Gödel (1972) argued for "a philosophical error in Turing's work." His fascination with the mathematical ability to reach more and more abstract concepts[3] led him to assume that Turing (1936) had argued that "mental procedures cannot go beyond mechanical procedures" and, therefore, to worry whether it did not imply some limitations on the human mind.

What Turing disregards completely is the fact that *mind, in its use, is not static, but constantly developing,* i.e. that we understand abstract terms more and more precisely as we go on using them, and that more and more abstract terms enter the sphere of our understanding ... although at each stage the number and precision of the abstract terms at our disposal may be *finite,* both (and, therefore, also Turing's number of *distinguishable states of mind*) may *converge towards infinity* in the course of the application of the procedure. (Gödel 1972, p. 306)

What is of interest in the present context is not so much whether or not Gödel misunderstood Turing's argument, but rather to contrast Gödel's vision with Turing's and Post's. Indeed, Gödel's passage highlighted an important point, namely, how to conjugate the abstractness and (increasing) complexity of mathematical terms with mechanical procedures. But Turing did not disregard such issues.

3.3.2   Computability and Unsolvability

"Mathematical reasoning," Turing (1939) observed, "may be regarded rather schematically as the exercise of a combination of ... *intuition* and *ingenuity.*" Although in pre-Gödel times one might have thought that all the intuitive judgments of mathematics could be replaced by a finite number of formal rules, "we have gone to the opposite extreme and eliminated not intuition but ingenuity." From the rules of a formal logic it is always possible to obtain "a method of enumerating the propositions proved by its means. We then imagine that all proofs take the form of a search through this enumeration for the theorem for which a proof is desired. In this way ingenuity is replaced by patience." But no formal logic, Turing observed, could wholly eliminate intuition. This suggests

---

[3] In "the ability to form objects of *higher type* than the natural numbers," Gödel also saw the "true reason" for incompleteness. See Stillwell (2010, p. 74).

seeking a "non-constructive" logic, which allows discerning when a step in a proof makes use of intuition, and when it is purely formal. What makes it possible to define the character of steps in a mathematical proof? What distinctive processes are relevant?

In his 1954 paper, Turing regards a mathematical proof as a *puzzle*. The task of "proving a given mathematical theorem within an axiomatic system," he explains, "is a very good example of puzzle." Indeed, Hilbert's *Entscheidungsproblem*, which he himself had proved to have a negative answer by means of his computing machine, could be viewed as the mother of all puzzles. Now his goal is to dissect the very machinery of that proof so as to seal off the conceptual analysis of finite procedures from the machine model. Thus, Turing defines a *systematic* procedure as "a puzzle *in which there is never more than one possible move in any of the positions which arise and in which some significance is attached to the final result.*" If the (position) state of a puzzle is described by sequences of symbols in a row, in order to fix the appropriate state transformations the question to be answered is:

"What sort of rules should one be allowed to have for rearranging the symbols or counters?" In order to answer this one needs to think about what kinds of processes ever do occur in such rules, and, in order to reduce their number, to break them up into simpler processes. Typical of such processes are counting, copying, comparing, substituting. (Turing 1954, p. 588)

The answer turns out to be just *substitutions*. Therefore, a *normal form* for puzzles can be stated as a "substitution-type puzzle." This means that: *"Given any puzzle we can find a corresponding* substitution puzzle *which is equivalent to it in the sense that given a solution of the one we can easily use it to find a solution of the other."* Then, by *reductio ad absurdum*, it is shown that there is no systematic procedure for deciding whether substitution puzzles are solvable. In more general terms, this means: "no systematic method of proving mathematical theorems is sufficiently complete to settle every mathematical question, yes or no."

Turing's argument for *puzzle solving* traces Post's line of thought from his *Formulation I* (1936) to his work on *normal systems* in the 1920s. Consequently, the outstanding similarity between the Turing and Post models of computation[4] appears from a dual perspective. While for Turing that model

---

[4] In Post's *Formulation I*, a *symbol space* – consisting of a two-way infinite sequence of spaces or boxes – corresponds to the Turing machine's tape, and *a set of directions* corresponds to the machine's instruction list of quintuples. A "problem solver" or "worker" plays the part of Turing's *computer*: "The problem solver or worker is to move and work in this symbol space, being capable of being in, and operating in but one box at a time. And apart from the presence of the worker, a box is to admit of two possible conditions, i.e., being empty or unmarked, and having a single mark in it, say a vertical stroke" (Post 1936, p. 289).

was the way to prove the undecidability of the *Entscheidungsproblem*, for Post it was a way of making explicit the character of unsolvability.

As a matter of fact, Post had devised a general concept of formal systems in the early 1920s. He proved that the theorems of any formal system can be produced by a simplified system, which he called a *normal system*. Then he realized that, as normal systems, all systems for generating theorems are enumerable. Therefore, by applying the diagonal argument, he was able to conclude that for certain formal systems there is no finite process determining whether a given proposition is a theorem. It is of philosophical interest that while Post, in the early 1920s, did not finalize his unsolvability results because computability was still unclear as a logical concept, in the mid-1930s the demand for a formal definition of the concept emerged from the attempts to obtain negative results. This also unveils computability as a structural matrix of the diagonal argument.

Unlike Gödel, Post saw incompleteness as a consequence of unsolvability, which stems from the diagonal argument (Stillwell 2010, p. 73). Seemingly, Post envisaged Gödel's incompleteness results in "full generality" almost a decade before Gödel, but he also thought that "for full generality a complete analysis would have to be made of all possible ways in which the human mind could set up finite processes for generating sequences" (Post 1965, p. 408). In his opinion, that was "not a matter for mathematical proof, but of psychological analysis of the mental processes involved in combinatory mathematical processes." As regards mental procedures, Post's sensibility appears closer to Gödel's than to Turing's.

3.3.3 Stretching Human Limits

Another question worth exploring is whether these limits of computability still hold when one purges Turing's computer of its *human* pedigree. In his analysis of computability, Robin Gandy (1980) replaced Turing's computer, which he called "computor," with a discrete mechanical device. Fixing principles for mechanisms as general as possible, he conceived a logical model of computability in terms of a *Gandy machine*. Unlike a Turing machine, whose actions are set by human limitations, a Gandy machine is made to depend on a *physical* limitation, namely, a *principle of local causality*, whose "justification lies in the finite velocity of propagation of effects and signals," according to relativity theory (Gandy 1980, p. 135). Not unlike Turing's computability, the states of computation must be "immediately recognizable." A Gandy machine is a deterministic discrete mechanical device, which operates *in parallel* on an arbitrary number of bounded parts. Because of the parallelism involved, the machine has to recognize "patterns" in a given state and act on them *locally*. Again, the restrictions imposed on computability rest upon the computer's capacities.

Dilating the visual field and the velocity of the computer does not change the scope of computability. Thus, Gandy's work does not modify the Church-Turing thesis as to the extension of "what is effectively calculable," but makes it crystal clear that understanding computability in "full generality" is not a matter for *psychological analysis* of the mental processes involved. Going deeper into Turing's conceptual argument, the link between *computability* and *observability* betrays its indifference to the capacities of the agent involved.

3.4   From a Physical Point of View

In contrast to Gödel's concern on possible limitations of the mathematical mind, Turing rather aimed at improving his argument that *any* computational task that can be carried out by a human being can as well be carried out by a Turing machine (cf. Davis 2000, p. 163). Since the Turing machine was to show that no systematic procedure can solve the general decision problem, Turing pursued a more general notion and designed a *universal* machine U as a single (computing) machine capable of performing any computation that can be done by any Turing machine M and, therefore, of executing all possible computations. The decisive idea was to encode the list of instructions (the "program") of a machine M on the tape of the machine U. This makes U *computationally equivalent* to any M: When U is set going it will perform the same task as M. "The importance of the universal machine is clear. We do not need to have an infinity of different machines doing different jobs. A single one will suffice" (Turing 1948, p. 414).

Meditating further on the universal machine as an "automaton" operating by finite means capable of imitating any one-purpose machine, Turing was led to explore to what extent the various kinds of search involved in the idea of "intellectual activity" might also be expressed in mechanical terms. He devoted much attention to the nature of mental processes, but – unlike Gödel – he was inclined to see them as consequences of physical processes. In some sense, by connecting the effectiveness of calculability to the limitations of the computer involved, Turing's computing model took Hilbert's requirement for finite means to its limits. Those limits shed light upon an underlying symbolic structure connecting mechanical procedures and physical processes. Classical physics, however, is regarded as continuous whereas the universal Turing machine is discrete; as a consequence, the logical model is not fit to represent classical physical systems.

It may be surprising that a serious challenge to Hilbert's finitism and Turing's limits emerged from disquiet over the *discontinuity* of quantum measurement. Pondering over the problem of quantum measurement, Hermann Weyl emphasizes that "measuring means application of a sieve or grating"

(1949, p. 259). No grating can define more than n distinct characters.[5] Classical physics, however, assumes that measurement could be made infinitely accurate, at least in principle, by enhancing the resolving power of the instrument (observer), whereas quantum theory recognizes that every measurement or observation rests upon "distinguishability" limits. Moreover, the classical presupposition that for each physical quantity one can assign a well-defined value to a system at a given time is defied by *incompatible* quantum observables. According to Heisenberg's *uncertainty principle*, their values are "correlated" but cannot be simultaneously precise.

The evolution of a quantum system over a period of time is captured by the time-dependent Schrödinger wave function, or quantum state. This function is not only *continuous* in time, but also *complex* vector-valued. The probability that a measurement will find the system to be in some (vector-valued) state at some time is given by the square modulus of a complex number, which expresses the *probability amplitude* of the event. A quantum state is always a *coherent superposition* of some finite number of distinguishable states given by a unit vector[6] in a complex space. In the simple case of a two-state system, like a photon transmitted or reflected by a semi-transparent mirror (beam splitter), the two distinguishable states can be denoted by $|\rightarrow\rangle$ and $|\uparrow\rangle$. Their respective complex probability amplitudes, α and β, coalesce into the unit vector $|\psi\rangle$,

$$|\psi\rangle = \alpha |\rightarrow\rangle + \beta |\uparrow\rangle,$$

which determines the state of the photon. When a measurement concerning transmission or reflection is performed, the state $|\psi\rangle$ is projected either onto the state $|\rightarrow\rangle$, with probability $|\alpha|^2$, or onto the state $|\uparrow\rangle$, with probability $|\beta|^2$.

Viewing quantum mechanics as a theory "from the outside," which does not apply to the observer, there should be no contradiction between the description of non-observed quantum systems and the "projection postulate," which applies to quantum measurements. The ideal determinism of the classical world conjugates the continuous evolution of a physical system with the complete description of its state at any given time, so classical probability arises as a consequence of "incompleteness." By contrast, quantum probability arises as a consequence of Heisenberg's *uncertainty relations*, i.e., it is inherent in the conceptual structure of the theory (as described, for example, by Pitowsky [2006]).

As is well known, Hilbert remained hesitant about the new theory. Von Neumann's formulation of quantum mechanics as an operator calculus in what he called the "Hilbert space" has been interpreted as an attempt to make quantum

---

[5] A grating can also be viewed as a splitting of the vector space associated with a physical system into *mutually orthogonal* subspaces, each of which corresponds to a character (*quantum state*).

[6] As a vector of probability amplitudes, the norm of $|\psi\rangle$ must equal unity.

physics palatable to Hilbert (Macrae 1992, p. 129). Studying von Neumann's *Mathematical Foundations of Quantum Mechanics* (1932), even Turing was intrigued by the process of quantum measurement (Hodges 2004, p. 54). He was not so much interested in the continuous evolution of non-observed quantum systems as in the continuity breaks by projection operators. His interest was likely motivated by his dream of "constructing a brain" out of the appropriate transformations of the universal machine. An intelligent machine, in his opinion, ought to be a device capable of interacting with the outside and evolving accordingly. In this perspective, Turing's dream can be traced back to Leibniz's visionary project for a *universal characteristic* and forward to the *universal quantum computer*.

3.5   The Universal Computer

Speculating on the possibility of turning reasoning into calculation, Leibniz was prompted to pursue a more general and ambitious goal, namely, to establish an *algebra of concepts*: neither a mere communication technique, nor a means to support natural memory, but rather a method of proof (*ars dijudicandi*) and of discovery (*ars inveniendi*). The crucial problem, in Leibniz's plan, was not so much to express a *calculus ratiocinator* in mechanical terms as to distill the essential constituents of reasoning, namely, a set of *universal characters* to be used as an *alphabet of thought*. Because the elements of such an alphabet should make it possible to grasp the "substance" of ideas, one had to go beyond binary digits and seek special symbols representing concepts. For Leibniz, universality pertains not so much to the (syntactical) structure as to the symbols of his *characteristica*. It might spring reasonably from the nature of *monads*.

His *Monadology* (1714) encourages a suggestive reversal of the conceptual analysis of computability. Rather than making the infinite accessible to resolution by finite means, the issue becomes drawing meaningful "perceptions" from the infinite unity of each monad. A monad is a *simple substance* and a *unity of perceptions*. As a simple substance, its constant change comes from an *internal principle*. As a unity of perceptions, each monad contains the whole universe. Perception plays a rather subtle role: "the passing state which involves and represents a multitude in the unity or in the simple substance is nothing other than what one calls *perception*" (Leibniz 1714, sec. 14). Thus, perception performs the inner constant change, and also, as a function of correlation, enables monads to "express" each other. As Leibniz explains:

§56. This *interconnection* or accommodation of all created things to each other, and each to all the others, brings it about that each simple substance has relations that express all the others, and consequently, that each simple substance is a perpetual *living mirror* of the universe.

§57. Just as the same city viewed from different directions appears entirely different and, as it were, multiplied *perspectively*, in just the same way it happens that, because of the infinite multitude of simple substances, there are, as it were, just as many different universes, which are, nevertheless, only perspectives on a single one, corresponding to the different points of view of each monad.

It would be not far wrong to say that Leibniz's monads were the first appearance of *qubits* (quantum bits) on the stage of "natural philosophy" (though couched in metaphysical terms, as a mathematical idealism required at the dawn of the eighteenth century). In contrast to the Cartesian separation between *res cogitans* and *res extensa*, which implies the separation between mathematical thought and physical world, "logical subject" and "physical object" overlap in Leibniz's concept of *being*. And his system of nature originates from an inner *principle of continuity*.

3.5.1 Quantum Computability

Like a monad, a qubit, which is the basic unit of quantum computation, involves an infinite multitude. As a two-state quantum system, it can be prepared in a coherent superposition $|\psi\rangle$ of two distinguishable states $|0\rangle$ and $|1\rangle$,

$$|\psi\rangle = \alpha \, |0\rangle + \beta \, |1\rangle,$$

for any pair of complex numbers α and β such that $|\alpha|^2 + |\beta|^2 = 1$. There is an infinite number of choices of basis with respect to which the state of a qubit can be decomposed. Taking qubits as elements of its alphabet, quantum computation proceeds through *unitary transformations* on qubits.

At first glance this looked like a turning point: "instead of being limited to shuffling a finite collection of [classical] states through permutation, one can act on qubits with a continuous collection of unitary transformations" (Mermin 2003, p. 27). Although in principle this seems to question Turing's limits, in fact there is no way to extract information from qubits other than by *measuring* them with yes-no questions (classical bits). Indeed, according to David Mermin, the quantum theory of computation "provides dramatic proof that the abstract analysis of computation cannot be divorced from the physical means available for its execution." But then the point at issue is the nature of that marriage. Does it make sense to connect a logical notion with *physical means*?

Gandy's analysis of computability had shown that Turing's limits cannot be surpassed by increased ability on the part of a classical computer. In 1985, David Deutsch took a further decisive step. He replaces the *abstract* model, which is embodied in either the human computer of Turing or the discrete mechanical device of Gandy, with a *physical* model, namely, a *quantum computer*. This allows him to give a precise *physical* description of the class of processes that can be carried out by *finite means*, clarifying the relation between physical and computational operations.

Investigating how the computing power of a physical system could be related to what is or is not computable in mathematical terms, Deutsch was led to establish that a "perfect simulation" of a physical system S on a computing machine M is possible if there exists a program for M that renders M *computationally equivalent* to S. Hence, he stated what he called "the physical version of the Church-Turing principle" (*Deutsch's principle*) as follows: *Every finitely realizable physical system can be perfectly simulated by a universal model computing machine operating by finite means.* Here "finitely realizable physical systems" must include any physical object upon which experimentation is possible, and "finite means" can be defined axiomatically (cf. Gandy 1980), without restrictive assumptions about the form of physical laws (Deutsch 1985, p. 99).

Deutsch's physical approach to computability contrasts with the background of the logical works considered above. While the concept of computability emerged in mathematics from the concept of *proof*, the concept of quantum computability emerged in physics from the concept of *predictability*. One main reason was that the efficiency of the Turing model was challenged by the surprising predictive power of quantum physics. The very notion of "probability" pointed out a deepening of foundations.

3.5.2   Computability and Predictability

The first insight into *quantum* computability is due to Richard Feynman (1982), who posed the question of how to simulate efficiently a quantum evolution. The task was reckoned beyond the capacities of any Turing machine, even *probabilistic*, for the amount of information involved in describing the evolution of quantum states in classical terms grows exponentially with time. However, considering the huge computation required to predict the result of a multiparticle interference experiment, Feynman realized that the very act of performing the experiment could be tantamount to a complex computation. This led him to conjecture that it might be possible to simulate efficiently a quantum evolution, provided the simulator itself is a quantum system. Deutsch's principle was instrumental in proving Feynman's conjecture. Deutsch also stated that a *universal quantum computer* (universal simulator) could perform any computation that any other quantum computer could perform.

The impossibility of efficiently simulating quantum measurements compared with a probabilistic Turing machine brings about a conflict between *quantum interference* effects and Kolmogorov's additivity axiom for probability theory. According to quantum theory, when an event can occur in several mutually exclusive ways the total probability for the event is the square modulus of the sum of the *probability amplitudes* for each way considered separately. Therefore, if there are only two mutually exclusive ways, like in the case of a photon transmitted or reflected by a beam splitter, then the classical theory provides the event with probability $p = p_1 + p_2$, while the *quantum probability* for the event is:[7]

$$p = |\alpha|^2 = |\alpha_1 + \alpha_2|^2$$
$$= |\alpha_1|^2 + |\alpha_2|^2 + |\alpha_1| |\alpha_2| \left(e^{i(\theta_1 - \theta_2)} + e^{-i(\theta_1 - \theta_2)} \right)$$
$$= p_1 + p_2 + 2\sqrt{p_1 p_2} \cos(\theta_1 - \theta_2).$$

Here the classical probability, $p_1 + p_2$, is refined by an "interference term," which marks the difference from the classical view. Depending on the *relative phase* $(\theta_1 - \theta_2)$, the interference term can be either negative (*destructive* interference) or positive (*constructive* interference), and the total probability p decreases or increases accordingly.

This makes no difference when a photon encounters a (single) beam splitter. The quantum prediction agrees with the classical one that the photon will be transmitted or reflected with the same probability, 1/2. But it does make a difference when a photon, initially traveling horizontally, passes through a sequence of two beam splitters, and reaches either the detector D_H or the detector D_V (in the *horizontal* or *vertical* directions).[8]

Two equivalent paths of the same length between the two beam splitters determine four alternative ways in which the event can occur. The photon

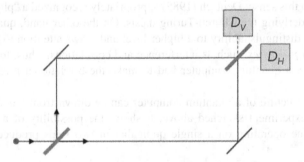

Figure 3.1 A representation of a Mach-Zehnder interferometer

---

[7] Recall that a complex number α can also be written in the polar form $\alpha = |\alpha| e^{i\theta}$, where the *phase* θ expresses the angular distance from the real axis in the complex plane.

[8] This experimental set-up, known as a "Mach-Zehnder interferometer," can be realized by placing two ordinary mirrors behind the first beam-splitter in the directions of the transmitted and reflected beams, as is shown schematically in Figure 3.1.

can be transmitted or reflected twice, transmitted by the first beam-splitter and reflected by the second, and vice versa: *TT, RR, TR,* and *RT.* In this case, the classical prediction is that the photon will arrive either at the detector D_H or at the detector D_V with the same *uncertainty* (probability 1/2), whereas quantum theory predicts with *certainty* (probability 1) that the photon will arrive at the detector D_H. How can it be obtained?

In the quantum world, the interaction with the first beam splitter transforms the initial state of the photon into a superposition state, which remains unchanged until the interaction with the second beam splitter. The quantum system evolves *continuously* until it is actually measured. Hence, a sequence of two independent evolutions (the interactions with the beam splitters) transforms the initial state according to the product of two "unitary matrices" (a generalization of complex numbers with unit modulus) representing the actions performed by the two beam splitters. Due to constructive interference, the result will be the identity transformation, which returns the initial state.[9] Leaving invariant the angular separation between distinguishable states, unitary transformations preserve information about probability amplitudes and let two uncertainties produce a certainty.

The challenge was to let it break through computation. If Turing's limits spotted an implicit "grating" underlying the concept of "effective calculability," quantum computability cannot dispense with that grating. In fact, quantum measurement provides conclusive evidence that "boundedness conditions" are also crucial in characterizing the *observability* of a physical property. However, while Turing's conceptual argument brought a logical notion down to earth by showing its link with the distinguishability of symbols, it is the distinguishability of quantum observables that imposes restrictions on quantum computing. In this sense, Deutsch (1985) appropriately recognized a "physical principle" underlying the Church-Turing thesis. On the other hand, quantum physics takes distinguishability to a higher level and draws attention to peculiar *relational processes*, such as interference and correlations. Thus, to meet that challenge, a quantum computer had to make these relational processes *effective.*

The basic structure of a quantum computer can be drawn from the simple interference experiment sketched above. It shows the possibility of a computing machine operating on a single qubit that in two steps produces the

---

[9] The square root of the matrix I representing the total evolution provides the matrix U_H representing the action of the beam-splitter:

$$I = \begin{pmatrix} 1 & 0 \\ 0 & 1 \end{pmatrix} = \frac{1}{\sqrt{2}} \begin{pmatrix} 1 & 1 \\ 1 & -1 \end{pmatrix} \frac{1}{\sqrt{2}} \begin{pmatrix} 1 & 1 \\ 1 & -1 \end{pmatrix}.$$

logical identity.[10] The two beam splitters, acting independently as *logical gates*, implement the two computational steps. Each of them corresponds to a unitary operation U_H that evolves the qubit basis states $|0\rangle$ and $|1\rangle$ as

$$|0\rangle \overset{U_H}{\Longrightarrow} \frac{1}{\sqrt{2}}(|0\rangle + |1\rangle), \quad |1\rangle \overset{U_H}{\Longrightarrow} \frac{1}{\sqrt{2}}(|0\rangle - |1\rangle).$$

Thus, applying two consecutive operations U_H to the initial state $|0\rangle$ gives the final state:

$$|0\rangle \overset{U_H}{\Longrightarrow} \frac{1}{\sqrt{2}}(|0\rangle + |1\rangle) \overset{U_H}{\Longrightarrow} \frac{1}{\sqrt{2}}\left(\frac{1}{\sqrt{2}}(|0\rangle + |1\rangle) + \frac{1}{\sqrt{2}}(|0\rangle - |1\rangle)\right) = |0\rangle.$$

It should be noted that the logical operation U_H has no classical analog. The extraordinary efficiency of *quantum algorithms* arises from the possibility of exploiting quantum interference to amplify the probability of correct results.[11] Although in certain cases quantum computers perform more efficiently than Turing machines, they do not increase the class of "computable" functions. Speeding up the process of computation, however, quantum algorithms also zoom in on the "general accepted properties" of the concept of computability. Quantum speedup relies upon interference among alternative ways, which may be regarded as computational paths.[12] Should one add *quantum interference* to the general accepted properties of computability? Can a physical property filter through the abstractness of a mathematical concept?

3.5.3 Mathematical Forms and Physical Meaning

When classical physics urged mathematics to open its timeless realm of numbers and geometric figures to motion, calculus provided the instrument to describe the evolution of a physical system. Bringing about the essential relational character of physical quantities, quantum physics discloses the twofold root, logical and geometric, of complex numbers. Thus, quantum physics prompts computability to "animate" its mechanisms over a *complex representation space.*

[10] Note that, with a "phase shift," it is also possible to produce the logical negation.

[11] A collection of n qubits constitutes a quantum register of size n. Since one qubit can store both 0 and 1 in a superposition, a quantum register of two qubits can store all four numbers $|00\rangle$, $|01\rangle$, $|10\rangle$, and $|11\rangle$ in a superposition. Adding more qubits to the register, one can increase its capacity for storing numbers exponentially. Given a quantum register of size k, a quantum computer can handle 2^k numbers in one computational step and result in a superposition of all the corresponding outputs. (To accomplish the same task, a Turing machine should repeat the computation 2^k times, or use 2^k different processors working in parallel.)

[12] As for the rationale behind the quantum speedup, interpretations differ. For extensive discussions, see Steane (2003), Saunders et al. (2010), and Cuffaro (2012).

The *logical* definition of a quantum computer, a *quantum Turing machine*, revises a Turing machine in the light of the mathematical formalism of quantum theory:

- bits (the minimal two-symbol alphabet) are replaced by qubits;
- quintuples of program instructions are replaced by unitary operators;
- a potentially infinite one-dimensional tape is replaced by a (separable) Hilbert space.

A unit sphere, the so-called "Bloch sphere," provides a convenient representation of a qubit. Mapping the state $|0\rangle$ to the "North pole" $N = (0,0)_{\varphi,\theta}$ and the state $|1\rangle$ to the "South pole" $S = (\pi, 0)_{\varphi,\theta}$, the unit vector $|\psi\rangle$ describing a qubit state is:

$$|\psi\rangle = \cos\frac{\varphi}{2}\,|0\rangle + e^{i\theta}\sin\frac{\varphi}{2}\,|1\rangle\,.$$

Unitary operators can be visualized as rotations on the unit sphere, but they require complex matrices. Indeed, the unitary requirement follows from the interpretation of the matrix elements as probability amplitudes. Therefore, the rotation angles must be halved to turn into phases. This reveals a distinct character of the complex representation space:[13] any unitary transformation in the complex space has its *dual*, and both correspond to the same rotation in the real space. Does any physical meaning attach to the structure of such a space?

Leibniz's *Monadology* (1714) may help find an answer insofar as it suggests considering a physical system as a "point of view," rather than as a "material point." In contrast to the more familiar description of a physical system from the outside, it conveys a picture "from the inside" through a monad's eye. If a monad is a unity of perceptions in constant change according to (what we would now call) a unitary transformation, a perception can be thought of as a ray connecting two monads. To chart its universe, a monad should select one of its perceptions and set its "visual field" accordingly. Since getting the opposite (sense of) direction would require a rotation of $\pm\pi$, a perception's "angle of view" equals 2π. Note that the orthogonal direction requires a rotation of $\pm\pi/2$ and, therefore, sizes an angle of π. Then, rotating its visual plane about an axis, the monad could map all perceptions into rays of a *two-dimensional complex* space. Doubling the angles of a monad's internal view brings about the Bloch sphere on the three-dimensional real space. The universal character that Leibniz wanted for the symbols of his alphabet of thought has been captured by complex numbers. Insofar as monads find embodiment in qubits, Leibniz's "universal characteristic" may find an echo in the universal quantum computer.

---

[13] Recall that the appropriate representation space for a two-state quantum system is a two-dimensional complex space.

3.5.4   The Computability of Nature and the Nature of Computability

Against the background of Leibniz's metaphysical perspective, Deutsch's *constructor theory* stands out as its specular image. Leibniz's system of knowledge is guided by the law of continuity and proceeds, through geometry, from logic to the theory of nature. The rationale of nature is drawn from the variety of monads' perspectives. The passage from the infinite multitude of possible universes to the real one conforms to a "criterion of perfection" which requires *maximizing existence*: a sort of "constructive interference" between *ideal elements*. It follows that knowing means projecting the ideal vision into the real one. By contrast, Deutsch's philosophical attitude may be parallel to Kant-Hilbert's. His constructor theory entirely relies on the laws of physics and "seeks to express all fundamental scientific theories in terms of a dichotomy between *possible* and *impossible* physical transformations – those that can be caused to happen and those that cannot" (Deutsch 2013, p. 4331). Consequently, the *principle of computability of nature* "must be that a computer capable of simulating any possible physical system is *physically possible*."[14] Beginning from experience, the construction of knowledge proceeds from the real world to the possible (multiverse). Both Leibniz's, and Deutsch's visions grow out of "complex interference" and neither of them could avoid Turing's limits.

If the cogency of Turing's computability model stems from the move from arithmetically meaningful steps to *general symbolic processes* underlying them (Sieg 2009, p. 604), quantum computability harmonizes those processes with complex interference. Adjoining a *geometrical* component to the logic of computability, it takes Hilbert's advice on "the use of geometrical signs as a means of strict proof." Remaining within Turing's limits, it adjoins a physical meaning to Hilbert's demand for finite means. As a consequence, one can say that the *quantum computability of nature* fulfills Post's program of changing "Church's hypothesis" to a "natural law." Does it also support Post's consequent claim as regards the limitations of human mathematical power?

Turing would probably agree in viewing the relation between mental and mechanical procedures as a relation between physical and computational processes. And probably Gödel would not disagree in viewing the continuity of quantum systems as a "shadow of the mind." If Gödel (1946) saw the precise mathematical definition of mechanical procedure as a "kind of miracle," quantum measurement helps unveil an underlying physical reason for that miracle. On the one hand, mathematical thought constantly develops as a coherent superposition of an infinite multitude of perspectives; on the other hand, it

---

[14] "This is a slightly improved version of the Turing principle (Deutsch 1985)" (Deutsch 2013, p. 4342).

must rely on distinguishable symbols to be projected into objective forms for mathematical understanding. Thus, it can be viewed either "from the outside" as a scrutiny of the infinite by finite means, or "from the inside" as a dissection of continuity into distinguishable states. Computability gives shape to the structural matrix of this dual vision.

Acknowledgments

I am grateful to David Deutsch, John Stillwell, and the referees for helpful comments and corrections.

Part II

The Implementation of Computation in Physical Systems

4   How to Make Orthogonal Positions Parallel: Revisiting the Quantum Parallelism Thesis

Armond Duwell

4.1   Introduction

The discovery and development of quantum computers have raised a number of interesting philosophical issues. Perhaps the most important issue is how to explain why quantum computers appear to be faster than classical computers for some computational tasks.[1] In the literature, one can find, arguably, two polar opposite positions espoused. One position is that the source of quantum speedup, at least in part, is due to the fact that quantum computers can compute many values of a function in a single step (Deutsch 1997; Jozsa 2000; Duwell 2004, 2007a). Following Duwell (2007a), call this the quantum parallelism thesis (QPT). The alternative position is that quantum computers don't do this at all, and in fact perform the requisite computational task by doing *fewer* computations than classical computers performing the same task (Steane 2003; Bub 2010). In this paper, I want to utilize the mechanistic account of computation, most fully developed in Piccinini (2015), to shed light on the debate.

I will argue that the mechanistic account helps us understand that the polar opposite positions described above can be seen as consequences of conflicting intuitions about the appropriate computational *description* of quantum systems that perform computational tasks, and not a disagreement about the fundamental *features* of quantum systems that allow for quantum speedup. The problem arises in large part because of a confusion about what a computation is. The mechanistic account provides us with a principled means of understanding the usefulness and limits of computational descriptions of phenomena, and also helps us understand the essential differences between classical and quantum computers. Taking these things into consideration, I will argue for the truth of a thesis related to the QPT that doesn't appeal to problematic computational descriptions:

QPT′: A quantum computer can generate a system in a state correlated to multiple values of a function in a single computational step.

---

[1] There is no proof that the class of computational problems efficiently solvable on a quantum computer exceeds those efficiently solvable on classical digital computers. That said, in what follows I will not make that qualification repeatedly and will write as though quantum speedup is a fact about computation and not one about our inability to find efficient classical algorithms.

I will also examine a potent objection that Steane (2003) leveled at the QPT which can be applied to the QPT'. Steane argues that a certain kind of quantum computer, a measurement-based quantum computer (MBQC), just doesn't satisfy the QPT. So, even if the QPT is true of certain algorithms run on quantum circuit computers, it is not true in cases where MBQCs simulate those algorithms, and cannot be used to explain quantum speedup in those cases. While it doesn't attack the truth of the QPT, it does call into question its usefulness in explaining quantum speedup, which is an issue for both the QPT and the QPT'.

In Section 4.2, I will describe the mechanistic account of computation. In Section 4.3, I will describe evidence for the QPT and a major objection to it. In Section 4.4, I will argue that the two opposing views of quantum speedup described above are actually not in conflict when one eliminates confusion regarding the appropriate computational description of the quantum parallelism process, which underwrites the QPT. I will also argue for the truth of the QPT'. In Section 4.5, I will introduce MBQCs in order to deal with Steane's objection to the QPT and the corresponding objection to the QPT' regarding MBQCs. In Section 4.6, I will argue that when appropriately understood, MBQCs simulating the quantum parallelism process are an instance of the QPT'. In Section 4.7, I argue that the QPT' can play an important role in explaining certain cases of quantum speedup.

4.2   The Mechanistic Account

In this section I will describe the mechanistic account of computation. I will emphasize how computational cost is measured, how computational tasks are associated with functions over strings of letters from an alphabet, how this relates to the computations a system performs, and finally what a computational explanation consists in. I will use a simple quantum circuit computer to illustrate the view.

The Mechanistic Account of Computation: A physical computing system is a mechanism whose teleological function is performing a physical computation. A physical computation is the manipulation (by a functional mechanism) of a medium-independent vehicle according to a rule. A medium-independent vehicle is a physical variable defined solely in terms of its degrees of freedom (e.g., whether its value is 1 or 0 during a given time interval), as opposed to its specific physical composition (e.g., whether it's a voltage and what voltage values correspond to 1 or 0 during a given time interval). A rule is a mapping from inputs and/or internal states to internal states and/or outputs.[2] (Piccinini 2015, p. 10)

---

[2] "Mapping" here must be understood in a colloquial way rather than the more specific mathematical sense of "function." Some computing components operate probabilistically, and the action of these components cannot be described by a mathematical function from inputs and/or internal states to internal states and/or outputs.

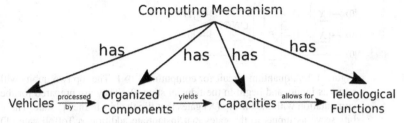

Figure 4.1 A representation of computing mechanisms

A mechanism is a system consisting of components that have a specific organization, and the components that compose the mechanism have certain capacities and properties. The capacities of a mechanism are due to and explained by the organization and capacities of its components. A computing mechanism will have a teleological function, and that teleological function is to perform a computation or computations.[3] A computation is a manipulation of a computational vehicle. Many of the components of a computational system will have, individually, the teleological function of manipulating a vehicle in a particular way. See Figure 4.1.

One way to familiarize oneself with the mechanistic account in this context is to apply it to a quantum circuit computer. The vehicles in this model of computation are qubits, which always begin in some fiducial state, $|0\rangle$. These vehicles can be realized by the spin of an electron, or the polarization of a photon. Such systems have other degrees of freedom, but these degrees of freedom are not relevant to the computation that a system performs.[4] Components of a quantum circuit computer consist of quantum gates and wires. Gates have the function of manipulating sequences of qubits, sometimes individually, and sometimes jointly. The wires have the function of transmitting qubits from gate to gate unchanged. The spatiotemporal arrangement of the vehicles and the components determine the capacities of the computing device, and the particular computation that they perform.

Consider an example. The quantum circuit in Figure 4.2 can be used to perform binary addition modulo 2. The computation proceeds from left to right. The top two qubits are the input bits while the third plays the role of an ancilla.

---

[3] Exactly what a teleological function is and how a system can come to have a teleological function is a controversial matter. See Piccinini (2015, ch. 6) for an overview. It is outside the scope of this paper to delve into the relevant details.

[4] Many other degrees of freedom can be relevant for the proper operation of the computer, such as the charge or lack thereof of the system that realizes the computational vehicle, but these won't be degrees of freedom that the components of a system are meant to manipulate in order to effect the particular computation the system performs. Hence they are not computationally relevant.

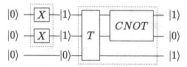

Figure 4.2 A quantum circuit for computing $1 \oplus 1$. The top two qubits will serve as inputs and begin in the fiducial state $|0\rangle$. They are prepared for the computation with the bit-flip (X) gates that flip the bits to the $|1\rangle$ state. These bits serve as inputs to the gates that instantiate addition, a Toffoli gate (T) and controlled-not gate (CNOT). The output of the gate is encoded in the middle qubit. A third qubit is required for reversibility. The dotted box that encloses the gates on the left side of the figure corresponds to a preparation of arguments to be evaluated, and the dotted box that encloses the gates on the right side of the figure corresponds to the evaluation of the arguments

All qubits begin the computation in a fiducial state, $|0\rangle$. The computation can be seen as proceeding in two conceptually different steps. The first is to prepare the computational vehicles that are considered as representing arguments to the addition function. It is natural to take a qubit in state $|0\rangle$ to represent 0, and in state $|1\rangle$ to represent 1. In this case, the action of the bit-flip (X) gates is to perform a rotation in the computational basis that takes $|0\rangle$ to $|1\rangle$ and vice versa, preparing the arguments 1 and 1 to be added together. The second step is to proceed with the evaluation of the arguments prepared. The Toffoli (T) and controlled-not (CNOT) gates can be interpreted as evaluating binary addition modulo 2. They manipulate the qubits in such a way that the final state of the middle qubit can be interpreted as encoding the evaluation of $1 \oplus 1$.

The mechanistic account of computation affords a robust view of computational explanation. That is to say, the capacities of a computational system can be *explained* in virtue of the computations that they perform. A computation is just a manipulation of a computational vehicle. In a computing device there will be a description of a system that involves only primitive computing components. Primitive computing components are those components whose behavior can be described as the processing of a computational vehicle, but such components cannot themselves be mechanistically described as having components that have the teleological function of processing vehicles. The gates associated with the quantum circuit described above can be considered primitive computing components. The vehicle operations that these components execute constitute the most basic computational steps that a computational system can execute. *By definition*, the action of a primitive computing component on a computational vehicle is a computation. Providing the computational description of a system in terms of its primitive computing components provides the most fundamental explanation of the capacities of a system possible in terms of computations. That said, one need not restrict computational explanation to only include primitive computing components. One can also cluster primitive computing components together to effect higher-level computational explanations. For example, for the circuit in Figure 4.2, one could group the T and CNOT gates, as they are in the dashed box, and consider them a complex computing component which one might label "ADD" for conceptual convenience. The behavior of the ADD component on computational vehicles allows it to be interpreted as instantiating binary addition modulo 2.

A careful reader might have noticed that I have been rather deliberate about how I have used the word "function." Components of a computational system can have a *teleological* function associated with them; they may be designed to manipulate vehicles in certain ways. These manipulations might be described using mathematics by associating certain states with letters from an alphabet and a function that maps letters or sequences thereof to other letters or sequences thereof. The action of components *need not* be describable by mathematical functions, especially when the action of a component has an indeterministic effect on computational vehicles. Furthermore, even when components are describable by mathematical functions, they will generally not be *uniquely* describable by a particular function.

For example, suppose that instead of interpreting the $|0\rangle$ as representing 0, and the state $|1\rangle$ as representing 1, one could take state $|0\rangle$ to represent 1, and state $|1\rangle$ to represent 0. In this case, the function that would describe the action of the rightmost set of gates in Figure 4.2 is no longer binary addition modulo 2. This illustrates the problem of computational underdetermination.

The problem of computational underdetermination arises because computational *tasks* or *problems* are often defined in terms of functions from strings to strings over some alphabet. Concrete computing devices are not alphabets, so one has to take certain states of a concrete computing device to *represent* letters or strings of letters from alphabets. Different choices for what the input and output states represent – call these interpretations – can lead to different functions being associated with a computing system.

A widely agreed-upon constraint on interpretations is that, for a computing device to compute a particular function f, the computing device has to associate any input with its appropriate output, regardless of which input is actual (Maudlin 1989; Copeland 1996; Scheutz 1999; Duwell 2004; Rescorla 2014). That said, that restriction won't pick out a *unique* function that is computed by a machine. I concur with Piccinini's assessment of the situation when he writes, "It is certainly possible to label the digits processed by a digital computing mechanism using different letters. But these do not constitute alternative computational descriptions of the mechanism; they are merely notational variants, all of which attribute the same computation to the mechanism" (2015, p. 12). The primitive computing components and their arrangement determine what computations a computing device performs. The function one interprets the device to compute does not. Similar remarks can be made for the functions that primitive computing components can be considered to compute as well. Looking forward, what matters to computational explanation will be the action of primitive computing components on vehicles, not what functions they might be interpreted as computing.

The mechanistic account provides us with several important facts about computation. First, primitive computing components perform one computation each. Second, the computation performed by a computing device is not individuated by the functions that it can be interpreted as computing, but by its computational description. There is some freedom for users of a computing device to interpret states as is convenient for them. Finally, proper computational explanations of a computing system's capacities are provided by the computational description of a system. These facts will be helpful in deciding the truth of the QPT.

4.3   The QPT

The evidence for the QPT consists in certain kinds of processes associated with several efficient quantum algorithms of the following sort. Consider a system of $n + m$ qubits. Let n qubits be the input register for the quantum computer under consideration and m qubits be the output register. Each qubit can represent a single binary number, 0 or 1. The n-qubit input register can represent any integer from 0 to $2^n - 1$. The input register can be also put into a superposition of all quantum states representing the integers from 0 to $2^n - 1$. For convenience, let us adopt the following notation: $|x\rangle$ will stand for the n-qubit state representing the number x in binary. For example, suppose that $n = 2$. The number three would then be represented by the state $|3\rangle = |1\rangle|1\rangle$. With this notation, an n-qubit system in state $|\psi\rangle$ that is a superposition of all numbers from 0 to $2^n - 1$ is given by $|\psi\rangle = \frac{1}{\sqrt{2^n}} \sum_{x=0}^{2^n-1} |x\rangle$.

Suppose we have a unitary gate, U_f, that performs the following evolution:

$$|x\rangle|y\rangle \xrightarrow{U_f} |x\rangle|f(x) \oplus y\rangle, \tag{4.1}$$

where $f : \{0, \ldots, 2^n - 1\} \to \{0, \ldots, 2^m - 1\}$ and $\oplus$ is addition modulo 2. Using the linearity of unitary evolution and the fact that quantum systems can be put into superpositions, it appears that all values of a function can be computed in one pass of a unitary gate if we put the input register into a superposition of all points in the domain of the function:

$$\frac{1}{\sqrt{2^n}} \sum_{x=0}^{2^n-1} |x\rangle|0\rangle \xrightarrow{U_f} \frac{1}{\sqrt{2^n}} \sum_{x=0}^{2^n-1} |x\rangle|f(x)\rangle. \tag{4.2}$$

Refer to this process as the quantum parallelism process. The action of the gate ensures that the right counterfactual behavior is present to consider the gate as computing all values of f. A minor qualification is that quantum states need to be interpreted ontologically, so that any distinct states correspond to distinct ontological differences in the systems they are assigned to (Duwell 2007a). This requirement is met in almost all interpretations save for the quantum Bayesianism of Caves et al. (2007), the pragmatic view of Healey (2012), and perhaps other epistemic views of the quantum state.[5] I have argued (2007a) that there is no essential connection between the QPT and the many-worlds interpretations.

As indicated in the introduction, there is dispute over whether a process like this can be considered as a computation of multiple values of f in a single computational step. The primary objection to this claim is that the values of the function purportedly computed are *inaccessible* on the basis of a single measurement on the qubits that have undergone the quantum parallelism process. Where might this intuition that accessibility is required come from?

The science of classical digital computing can be described as the construction and consideration of models that manipulate strings of letters of an alphabet, e.g., Turing machines, finite state automata, and classical circuits. Importantly, these manipulations are such that they always take one string in an alphabet to another string in the alphabet. The computational process can be described as instantiating a function from finite strings of an alphabet to finite strings of an alphabet. For circuit computers, these are just strings of ones and zeros.

In order for a concrete computing device to instantiate one of these models of computing, it needs to instantiate the components of the model and their arrangement. So, the states of the concrete computational vehicles need to be interpreted as encoding individual letters of an alphabet or finite strings. Moreover, the components of a concrete computing device need to modify the concrete computational vehicles in ways that are interpretable as corresponding to the action of the components in the computational model. Insofar as classical models of computation are instantiated in classical concrete systems, their computational vehicles will be *distinguishable* from one another.

Not so in the quantum case. In the quantum case, we individuate computational tasks or problems just like we do in the classical case, as computing a particular function from strings of an alphabet to strings of an alphabet. That said, what differs dramatically in the quantum case is that quantum vehicles are not always distinguishable from one another. For example, take a problem associated with a binary alphabet. We can transparently label the

---

[5] Depending on how one interprets the ontological significance of the wavefunction in Bohm's theory, it may or may not be met.

computational basis states of qubits as $|0\rangle$ and $|1\rangle$, but primitive computing components can take these states to superpositions of states which are not distinguishable from the basis states in a single measurement. The situation is more dramatic when we use multiple qubits which can be in entangled states. The analogy between classical and quantum computing breaks down.

It's not clear whether in these kinds of cases one should demand that the results of a computation be accessible or not. It seems to me that there are reasonable conflicting intuitions at play. On the one hand, the physical state of the computer is correlated with all values of a function being computed, which supports a claim that these values have been computed. On the other hand, we cannot access those values by a single measurement on that state, which raises doubts regarding claims of computation. I don't see how philosophical analysis will decide the matter. A more fruitful approach is to inquire whether there is anything in common between those who have conflicting intuitions regarding accessibility as a necessary requirement to consider the values of a function computed.

4.4   Common Ground

In this section, I will forge common ground between those on alternative sides of the debate regarding the QPT. Common ground is established by recognizing that the difference between these sides is largely terminological rather than substantive.

Consider some things that have been written by opponents of the QPT. Bub (2010, p. 244) writes:

Rather than "computing all values of a function at once," a quantum algorithm achieves an exponential speed-up over a classical algorithm precisely by avoiding the computation of *any* values of the function at all.

Steane (2003, p. 476) writes:

A quantum computer can be more efficient than a classical one at generating some specific computational results because quantum entanglement offers a way to generate and manipulate a physical representation of the correlations between logical entities without the need to completely represent the logical entities themselves.

What Steane has in mind by "representation of correlations between logical entities without the need to represent the entities themselves" is an entangled state correlated to all values of a function being computed. The values of the function evaluated are not completely represented insofar as they are not encoded in distinguishable systems, hence are not accessible.

For example, suppose one wanted to evaluate the function $f(x) = x$ on $\{0, 1\}$ using a quantum gate whose action is defined by the evolution in Eq. (4.1).

In this case, one can utilize the quantum parallelism process to effect the transition:

$$\frac{1}{\sqrt{2}} \sum_{x=0}^{1} |x\rangle |0\rangle \xrightarrow{U_f} \frac{1}{\sqrt{2}} \sum_{x=0}^{1} |x\rangle |f(x)\rangle. \tag{4.3}$$

The final state in Eq. (4.3) will be $(|00\rangle + |11\rangle)/\sqrt{2}$. The points of the function evaluated are not completely represented insofar as they are not encoded in distinguishable systems; one cannot determine them both using measurements in the computational basis. That said, when measurements of the qubits are made in the computational basis, the outcomes will always be correlated to one input-output pair for $f(x) = x$.

The statements made by Bub and Steane were made in the context of comparing classical and quantum computers. So, one can focus on what features are present in the classical context when one computes many values simultaneously.

Confine ourselves to a classical circuit model. If a classical circuit computer were to compute many values of a function, f, simultaneously, it would require representing all of the values to be evaluated, and providing all of the requisite components to evaluate those values. If we assume that the function to be evaluated is $f : \{0, 1\}^n \rightarrow \{0, 1\}^n$, then each input to the function requires n two-dimensional systems, which we may refer to as *cbits*, distinguishing them from bits, which are binary numbers. In order to represent all of the possible elements in the domain of the function, one needs 2^n n-cbit systems. If the evaluation of a single element of the domain of f requires k gates, then one requires $2^n k$ gates to alter the state of the vehicles representing all of the elements of the domain of the function. So, the computational resources required to effect the desired transition in cbits is high, and this is ignoring the resources required to prepare the cbits in the desired state. But what the classical circuit offers are *distinguishable* systems that encode each value of the function. In order for a quantum computer to offer *distinguishable* systems that encode each value of the function, it would require the *same* amount of resources as in the classical case. So, it seems that what Bub and Steane have in mind is that in order to claim that one has computed any value of a function, it is *necessary* that that value is encoded in an accessible way. As was indicated above, whether accessible results are necessary to claim that a function has been computed is at issue. That said, it appears that the point that Bub and Steane are making can be made without appealing to a controversial view about what it takes to compute a function:

A quantum algorithm achieves an exponential speed-up over a classical algorithm precisely by avoiding encoding the values of a function in distinguishable systems.

Deutsch (1997), Jozsa (2000), and Duwell (2004, 2007a) obviously reject that it is necessary to have a value of a function accessible in order to consider it computed, but they would have no truck with the above claim.

This game can be played in the other direction of the debate as well. Advocates of the quantum parallelism thesis know well that the function evaluations in question are not accessible. So, one can try to take out what is objectionable about the quantum parallelism thesis by trying to state it without an appeal to the concept of computation:

QPT': A quantum computer can generate a system in a state correlated to multiple values of a function in a single computational step.

It seems to me that this does not conflict at all with the concerns that Bub and Steane have raised.

Even if my analysis is correct regarding the truth of the QPT', Steane (2003) has raised an important challenge to the claim that this is a principle that can be appealed to in order to satisfactorily explain quantum speedup generally. He suggests that it isn't true of MBQCs. If correct, this would demonstrate a substantial limitation of the view. To understand the challenge and my response, the next section introduces the reader to MBQCs.

4.5   MBQCs

MBQCs utilize quantum systems to compute significantly differently than quantum circuit computers. MBQCs are composed of an entangled cluster of qubits, a set of measurement devices that make measurements on individual qubits, and a classical computer that keeps track of the results of individual measurements and sets measurement directions as the computation proceeds.

The basic idea of MBQCs is to exploit entanglement to steer systems into particular mixtures via measurements on individual qubits. Suppose Alice and Bob each possess a member of an entangled pair in the state $(|00\rangle_{AB} + |11\rangle_{AB})/\sqrt{2}$, and Alice wants to alter Bob's qubit to the state $|0\rangle$. By measuring her qubit in the computational basis, $\{|0\rangle, |1\rangle\}$, Alice can force Bob's qubit into a proper mixture of the computational basis states with equal probability. Given the results of the measurement, Alice knows which, if any, local transformations Bob would have to perform on his qubit to transform the state of his qubit to $|0\rangle$: either the identity operation, I, or the bit-flip operation, X. So, Alice can steer Bob's system into whatever state she pleases. If Alice is interested in doing computational work with Bob's system, it is enough that it is a trivial rotation away from the desired state. In that case, Alice can alter her computational procedure to systematically compensate for the difference between the desired state and the state of Bob's system.

Raussendorf and Briegel (2001) demonstrated that measurements on suitably entangled quantum systems can simulate any quantum circuit. For example, one can generate a system in a state that is related to an input state by any rotation using a sequence of five qubits. Qubit 1 is in an arbitrary state, $|\psi\rangle$. The other four qubits begin in a fiducial state $|+\rangle$, where $|+\rangle = (|0\rangle + |1\rangle)/\sqrt{2}$. All qubits are subsequently entangled using the CPHASE gate, defined by $|x\rangle|y\rangle \xrightarrow{CPHASE} (-1)^{xy}|x\rangle|y\rangle$, where $x, y \in \{0, 1\}$, applied to nearest neighboring qubits, which in this case has the structure of a one dimensional lattice. The first four qubits are sequentially measured.

Let the Euler representation of the intended rotation be $U_R(\xi, \eta, \zeta) = U_z(\zeta)U_x(\eta)U_z(\xi)$, where $U_x(\alpha) = e^{(-i\alpha\sigma_x/2)}$, and $U_z(\alpha) = e^{(-i\alpha\sigma_z/2)}$. Qubits $j \in \{1, 2, 3, 4\}$ will be subjected to measurements with bases $\mathcal{B}_j(\phi_j) = \{(|0\rangle_j + e^{i\phi_j}|1\rangle_j)/\sqrt{2}, (|0\rangle_j + e^{-i\phi_j}|1\rangle_j)/\sqrt{2}\}$. Let $s_j \in \{0, 1\}$ represent the outcome of the measurement on qubit j.

The measurement bases that will be used are subject to change up to a sign depending on the results of measurement. To enact the intended rotation, qubit 1 is subjected to a measurement in $\mathcal{B}_1(0)$, with result s_1. Qubit 2 is subjected to a measurement in $\mathcal{B}_2(-\xi(-1)^{s_1})$. Qubit 3 is subjected to a measurement in $\mathcal{B}_3(-\eta(-1)^{s_2})$. Qubit 4 is subjected to a measurement in $\mathcal{B}_4(-\zeta(-1)^{s_1+s_3})$. Qubit 5 will then be in the state $U_{\Sigma,R}U_R(\xi, \eta, \zeta)|\psi\rangle = \sigma_x^{s_2+s_4}\sigma_z^{s_1+s_1}U(\xi, \eta, \zeta)|\psi\rangle$. Qubit 5 is in a state systematically and trivially related to the desired state. $U_{\Sigma,R}$, the byproduct operator, represents the "byproduct" of instituting dynamic changes via remote steering, but its effect is not computationally significant. σ_z operations leave states in the computational basis unchanged, and σ_x operations just flip the computational basis. Depending on what the byproduct operator is, one just interprets the results of measurements on output qubits differently: either flip the resulting outcome or not. The measurement pattern is represented in Figure 4.3.

Other quantum gates can be realized as well. I'll mention as an example how the CNOT gate is realized. This gate requires 15 qubits, arranged as

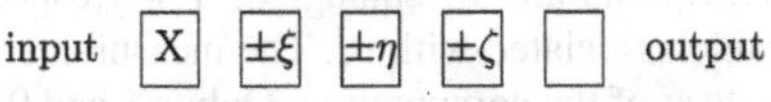

Figure 4.3 The measurement pattern for instituting an arbitrary qubit rotation characterized by the Euler angles ξ, η, ζ. The boxes and their arrangement represent a set of ordered qubits that are entangled. The measurements that are performed on the qubits are indicated inside the boxes. Measurements proceed sequentially from left to right, where the sign of the measurement angles depends on the results of earlier measurements. The boxed X is a measurement of the observable associated with σ_x. No measurements are made on qubit 5, which represents the output qubit

indicated in Figure 4.4. Like the general rotation described above, the outputs of the CNOT gate will have a byproduct operator associated with it – again, some trivial product of σ_x and σ_z operators which can be dealt with at the end of the computation. The same can be said of other common quantum gates, Hadamard (H), $\pi/2$-phase, and X, Y, and Z gates. These gates, as well as the CNOT gate, have measurement patterns that can be measured simultaneously, unlike the general rotation described above.

The fact that, for certain transformations, measurements must be made in a certain order imposes an order on the qubit cluster associated with an MBQC. Some qubits have measurement bases that are fixed. All of these qubits can be measured simultaneously and initially. Call this set of qubits Q_1. There will be a set of qubits whose measurement angles are set on the basis of the results in Q_1. All of these qubits can be measured simultaneously. Call this set Q_2. There will be another set of qubits whose measurement angles are set on the basis of measurement results on $Q_1 \cup Q_2$. Call this Q_3, and so on.

The Clifford group gates – I, X, Y, Z, $\pi/2$-phase, H and CNOT – and any other general rotation not in that group–together form a universal set of quantum gates. That means that any unitary transformation can be made to an arbitrary degree of accuracy using combinations of gates from that set.[6] This fact, combined with the fact that Clifford group gates have fixed measurement patterns, has significant ramifications for the way that an MBQC can simulate a quantum circuit computer. Suppose for conceptual convenience that a quantum circuit computer utilizes the Clifford group gates and an additional

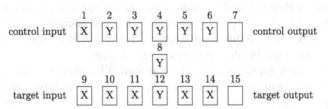

Figure 4.4 The measurement pattern for a CNOT gate, which forms a two-dimensional lattice. Adjacent qubits are all entangled. Y corresponds to a measurement of the observable associated with σ_y. The measurement angles are completely fixed at the outset of the computation. Qubits 1 and 9 play the role of the input target and control qubits, respectively. Qubits 7 and 15 play the role of the output control and target qubits, respectively

---

[6] Any universal set of gates is as good as any other insofar as there will be a fixed computational cost to simulate a universal gate from one universal set with gates using another universal set. Such fixed costs are thought to be computationally irrelevant. See Nielsen and Chuang (2000, pp. 138–141) for a discussion of what constitutes a computationally relevant or significant difference.

rotation as its primitive computing components. The gates of a quantum circuit computer are carefully ordered so that they can institute the intended computation. When an MBQC is simulating a quantum circuit computer, there will be a sequence of measurement patterns on the qubit cluster that corresponds to the sequence of gates in the simulated quantum circuit computer. That said, all of the dynamic changes associated with Clifford group gates will be in Q_1. So, dynamic changes associated with many *different* steps in the operation of a quantum circuit computer will take place *simultaneously* and *initially* in an MBQC that is simulating it. Further, the qubits that correspond to the outputs of a quantum circuit computer are also part of Q_1. The output qubits of a quantum circuit computer will always be in eigenstates of the computational basis and a measurement in that basis will reveal the outputs. On an MBQC, this is just a measurement of the observable associated with σ_z.

One can reasonably wonder how this temporal flexibility in instituting dynamic changes is possible. It is explained quite expediently by focusing on a few facts. Remember that the qubit cluster composing the quantum computer is entangled. Individual measurements on members of the cluster always commute. Any dynamic changes induced on individual qubits in the cluster will have no locally detectable effects. The reduced density operator for individual qubits is always $I/2$. Dynamic changes made on individual qubits are only detectable when we examine *correlations* between entangled systems. The temporal details of how the correlations were instantiated do not matter.

An example will serve to illustrate. Suppose that Alice and Bob each have one member of an entangled pair in the state $|\psi\rangle_{AB} = (|00\rangle_{AB} + |11\rangle_{AB})/\sqrt{2}$. Suppose that Alice wants to rotate her qubit about the x-axis by an angle θ. The corresponding unitary operator for that rotation is $U_x(\theta) = e^{i\theta\sigma_x/2}$. After Alice performs this rotation on her member of the pair, the state becomes

$$U_x(\theta)_A \otimes I_B|\psi\rangle_{AB} = (\cos(\theta/2)|00\rangle_{AB} - i\sin(\theta/2)|01\rangle_{AB}$$
$$- i\sin(\theta/2)|10\rangle_{AB} + \cos(\theta/2)|11\rangle_{AB})/\sqrt{2}. \quad (4.4)$$

But now notice that the pair ends up in the same state if Bob performs the rotation instead of Alice:

$$I_A \otimes U_x(\theta)_B|\psi\rangle_{AB} = (\cos(\theta/2)|00\rangle_{AB} - i\sin(\theta/2)|01\rangle_{AB}$$
$$- i\sin(\theta/2)|10\rangle_{AB} + \cos(\theta/2)|11\rangle_{AB})/\sqrt{2}. \quad (4.5)$$

When performing a single qubit rotation on a fully entangled system, there is no observable local effect, as the equations above indicate. The change can only be revealed via the correlations between the entangled systems. Indeed, the reduced density operators of Alice and Bob's system are $\rho_A = \rho_B = I/2$. That said, if we check the probability of Alice and Bob both measuring 0 when they measure in the computational basis, for example, it is $\cos^2(\theta/2)/2$.

Notice that it makes no difference whether Alice or Bob measures first. Similarly, it makes no difference whether Alice rotates or Bob rotates before they do their measurement, either. The probabilities of their individual measurement results are invariant and so too are the correlations. This is because $[P_{|0\rangle A} U_x(\theta)_A \otimes I_B, I_A \otimes P_{|0\rangle B}] = 0$ and $[P_{|0\rangle A} \otimes I_B, I_A \otimes P_{|0\rangle B} U_x(\theta)_B] = 0$, where $P_{|0\rangle A}$ is a projection operator for the result 0 in a measurement of the computational basis on Alice's qubit, etc.

The upshot of this example is that the temporal flexibility in the time order of measurements in an MBQC that is simulating a quantum circuit computer ought to be completely expected. When making dynamic changes to individual qubits in an entangled system, the effects are nonlocal. What matters is the correlations between systems. As long as one keeps track of the correlations between measurements, one can do useful computational work, which is exactly what MBQCs do.

This explanation of the temporal flexibility in MBQCs might strike one as suspect because it leads one to expect that there should be no temporal order to measurements at all, as all individual measurements on the qubits commute. A temporal structure associated with measurements in a MBQC is imposed because one is using remote steering to implement changes in the cluster. Remote steering is a rather coarse means of implementing these changes. For certain kinds of dynamic changes, like general rotations, several measurements are required, and one needs to make corrections along the way by altering measurement bases in order to get a state that is (computationally trivially) related to the intended state. The commutativity of individual qubit measurements on the cluster preserves correlations regardless of the order of measurements, but one has to chose the right measurements to ensure that the right correlations are generated. One can only come by that information by making measurements, hence the temporal order in MBQCs.

From the point of view of the mechanistic account of computation, MBQCs are quite curious – see Duwell (2017) for details. Unlike quantum circuit computers, one cannot track a string of qubits entering the machine and being sequentially manipulated by primitive computing components, each described by a particular deterministic transition rule. The cbits generated via measurement results are completely random, so they cannot be understood as being sequentially processed according to deterministic rules either, as they would be in a classical circuit computer. A better way to conceptualize how MBQCs fit into the mechanistic account is to think of them as being a hybrid computing system whose vehicles are classical and quantum. Additionally, the entire qubit cluster is best conceived of as an individual vehicle. This requires no stretch of the imagination because the qubit cluster is a nonseparable system due to entanglement. Measurements drive dynamical changes in the quantum vehicle as well as dynamical changes to the classical vehicles. The changes in these vehicles will appear completely random when examined individually, but the correlations between the vehicles are systematic. Given this understanding of MBQCs, it is possible to appreciate Steane's objection to the QPT with respect to them.

4.6   MBQCs and the QPT′

As has been described above, all of the gates in a universal quantum set can be simulated by an MBQC. So, any algorithm involving the quantum parallelism process can be simulated as well. Yet, the quantum state associated with the qubit cluster that in part constitutes the MBQC will not generally instantiate the quantum parallelism process described by Eq. (4.2). It does not do so for several reasons. First, the qubits associated with the cluster begin in an entangled state. Even though the qubit cluster alters its state with each measurement, the qubits that compose the cluster will typically be found neither in the state $\frac{1}{\sqrt{2^n}} \sum_{x=0}^{2^n-1} |x\rangle |0\rangle$, the state associated with the input to the gate associated with f, nor in the output state $\frac{1}{\sqrt{2^n}} \sum_{x=0}^{2^n-1} |x\rangle |f(x)\rangle$. Second, because of the temporal flexibility associated with measurements on an MBQC, some of the dynamic changes associated with the f-gate will take place in a different order than they would have in a quantum circuit computer. So, it certainly seems reasonable to claim that the operation of an MBQC is not readily described in terms of the QPT′, much less that multiple values of a function are computed in a single computational step.

There is something unfair about the comparison between MBQCs and quantum circuits of the kind we are considering. The problem is that the quantum parallelism process, Eq. (4.2), involves an oracle. An oracle, in terms of circuits, is a gate that is deemed to incur the same computational cost as a primitive computing component, but which generally is not a primitive computing component, and would generally take multiple primitive computing components from a universal set to simulate. In this sense, if one were to "open up the oracle," so to speak, one would generally find that multiple primitive computing operations were taking place inside the oracle, but their computational cost was hidden. So, the idea that the quantum parallelism process can take place in a single computational step described by Eq. (4.2) is, strictly speaking, false. The transition certainly occurs, but in steps corresponding to the primitive computing components. MBQCs, in virtue of their construction, invite what might be termed an "open-oracle" view when simulating quantum circuits with an oracle. In order for the comparison between quantum circuit computers and MBQCs to be fair, I will adopt an open-oracle view of quantum circuit computers.[7]

It is a bit of a misnomer to suggest that the explanatory power of the QPT′ in a non-oracle setting needs to be examined because the formulation of the QPT′ is essentially tied to oracle computing. The problematic phrase in the QPT′ is "in a single step." That language is part of the QPT′ because an oracle, by definition, is taken to have the same computational cost as a primitive computing component. We need a formulation of a potentially explanatory principle that is relevantly similar to the QPT′ in order to examine its power with respect to MBQCs.

In the non-oracle case, one must take into consideration the computational cost of computing the function, f, in terms of the primitive computing components. Call the associated primitive computing components the *oracle gates* for f. It is well-known that any computation that a classical computer can perform can be efficiently simulated by a quantum computer. So, however many oracle gates for f there are in the classical case, the number of quantum oracle gates for f will be computationally on par. That said, in order to generate a state correlated to multiple values of a function in the classical case, one will require exponentially more classical oracle gates for f to compute all values of f simultaneously than in the quantum case. So, a suitable statement of QPT′ for oracle-free computation (OF-QPT′) is:

OF-QPT′: A quantum computer can generate a system in a state correlated to multiple values of a function in exponentially fewer steps than a classical computer can.

The question is whether OF-QPT′ is true of MBQCs that simulate the quantum parallelism process.

There are two cases to consider. The first is when one simulates quantum circuit computers, gate by gate, preserving the temporal ordering of gates, using an MBQC. In this case, despite the fact that the qubits associated with gates at a later computational stage *can be* in the first simultaneously measurable set of qubits, Q_1, in this case they *are not* measured simultaneously and initially. The measurements are ordered according to their corresponding gates in a quantum circuit computer. In this case, there is no barrier to the OF-QPT′ applying. The dynamic changes will be applied in the right order, so that the entire state of the MBQC will be correlated with all values of a function when all of the gates of the quantum circuit computer associated with preparation of inputs and evaluation of the oracle have been simulated. The outputs will

---

[7] Alternatively, one can imagine equipping MBQCs with oracles, which would impose a unit computational cost on the measurements associated with the oracle, as well as impose a temporal ordering on the measurements more closely mimicking the steps a quantum circuit computer would undergo. I take it that it is a stronger challenge to the QPT′ when MBQCs enjoy their full temporal flexibility in ordering measurement operations and only focus on that case.

generally have byproduct operators associated with them in order to effect the exact state transition associated with the simulated quantum circuit computer, but these are alterations, not obliterations of the requisite correlations. The OF-QPT' does not indicate anything about the exact nature of the correlation between the state of a computer and the values of a function, just that a correlation exists. So, it can be multiply realizable and is generally realized differently in MBQCs than in quantum circuit computers.

The second case to consider is when one utilizes the temporal flexibility of measurements associated with making changes on entangled systems. In this case, dynamic changes associated with earlier and later stages of an operation of a quantum circuit computer will take place simultaneously and initially with respect to those changes associated with Q_1, and similarly for the other simultaneously measured sets of qubits.

A person who rejects that the OF-QPT' is true of MBQCs simulating the quantum parallelism process faces a tough challenge. The OF-QPT' seems to function perfectly well with respect to the case where an MBQC simulates a quantum circuit computer gate by gate, but, somehow, in the case where we perform measurements in a different order, the state of the computer is not correlated with all values of a function in exponentially fewer steps. There is no debate regarding whether the MBQC does its job in exponentially fewer steps than classical computers, so the only dispute could be with the claim that the state of an MBQC is correlated with all values of a function being computed. That claim seems to be at odds with the quantum mechanical description of the computer.

The very same *correlations*, up to the uncertainty associated with the byproduct operators, are generated no matter what the order of the measurements is *because they commute*. The individual states of the qubits that compose the cluster have a reduced density operator of $I/2$, are maximally mixed, and are individually computationally useless. It is in virtue of the *correlations* generated that an MBQC computes.

One might object using the following reasoning. The reason why MBQCs are able to simulate a quantum circuit computer is because they can simulate it gate by gate. The reason why it is reasonable to think that certain measurement patterns simulate the action of a gate is because the *state* of the final qubit or qubits in the sequence of measurements that simulates a gate is systematically related to the output of the simulated gate. If one changes the order of measurements, then that generally won't remain true. The cluster qubit that corresponds to the output of the gate in a circuit computer will no longer be in a state systematically related to the output of the gate.

This is a crucial objection and if it cannot be met, then OF-QPT' would be in jeopardy for MBQCs. To address it, one must recall how the output of a simulated gate is related to the output qubit in an MBQC that simulates that gate. It is related by a sequence of σ_x and σ_z operators. The action of these operators either flips the qubit in the computational basis (σ_x) or flips the phase of the qubit in the computational basis (σ_z). It would be computationally superfluous to actually perform those rotations on the output qubit to get it into a state that is exactly the same as that of the output of the simulated gate. One can simply measure in the computational basis, getting either a 0 or a 1, and then compute (classically) from the output of previous measurements whether to flip that output to reveal the solution to the computational task. The upshot is that what is computationally salient is to know how the measurement outcome on the output qubit in an MBQC is systematically correlated with measurements on the output of a simulated gate. If it is the correlations that matter, then the objection is neutralized. Changing the order of commuting measurements leaves the correlations alone, up to the uncertainty associated with the byproduct operators. If the state of an MBQC that simulates a circuit computer gate by gate is correlated to all values of the function being computed at some point, then the same will be true of an MBQC that changes the order of commuting measurements.

4.7   Are the QPT′ and the OF-QPT′ Explanatory?

I have argued that in certain cases the QPT′ and the OF-QPT′ are true. That said, one can reasonably wonder whether they can do any explanatory work. In this section I will argue that they can.

The mechanistic account of computation has provided us with a notion of computational explanation which explains the capacities of a computing system in terms of the computations it executes. In explaining quantum speedup, we aren't trying to explain the capacities of a particular computing device, but rather why quantum computers can do some things more efficiently than classical computers seem to be able to do in some cases. In keeping with the general spirit of the notion of computational explanation associated with the mechanistic account of computation, it appears most natural that an explanation would be afforded in terms of differences in the kinds of computations the different models can perform, i.e., in differences in the kinds of transformations in computational vehicles the primitive computing components afford. If there exists a kind of transformation of computational vehicles in one computing device that cannot be efficiently simulated in another, this would pinpoint a source of computational efficiency requisite to effecting the desired explanation. This seems to be exactly what either the QPT′ or OF-QPT′ provides.

The QPT′ and OF-QPT′ indicate that there is a certain kind of process available in quantum computers that is not efficiently simulable in classical computers. In the cases of algorithms that exhibit quantum speedup and of which QPT′ and OF-QPT′ are true, if they would be efficiently simulable in the cases in which quantum speedup occurs, then classical computers would be as efficient as quantum computers. An attractive feature of the OF-QPT′ is that it applies to quantum circuit computers as well as MBQCs, even though the ways that they go about generating states correlated to multiple values of a function are very different.

The explanatory power of the OF-QPT′ is limited in a couple of ways. Neither QPT′ nor OF-QPT′ by themselves provide a computational explanation of the capacities of the computing devices that they are true of. To do so, one would have to provide the details of the computational description of the device and relate it to the computational task at issue. For example, consider the task of deciding whether a Boolean function is constant or balanced. The Deutsch-Jozsa algorithm for quantum circuit computers does utilize the quantum parallelism process and the QPT′ is true of it. In order to complete the computational task, it isn't enough just to have a state correlated to all values of a function: One must have a way of extracting the information regarding whether the function is constant or balanced. A computational explanation would account for this. That said, this is not the explanatory target of the QPT′ or the OF-QPT′. QPT′ and OF-QPT′ are useful for comparison between different kinds of computing devices for certain kinds of computational tasks, those that involve quantum speedup.

The other way in which the explanatory power of the QPT′ or OF-QPT′ is limited is that they will only afford explanations of quantum speedup in cases in which they are true. There is no guarantee that every case of quantum speedup will involve the truth of these theses.[8]

One might also have worries about OF-QPT′ in particular. One of the motivations for the QPT and QPT′ is that one can simply "see" parallel computations by looking at the quantum algorithms that invoke the quantum parallelism process, as Hewitt-Horsman (2009, p. 876) has emphasized. In MBQCs in which the OF-QPT′ is true, one certainly cannot "see" anything of the sort. Might this challenge whether the principle is explanatory?

Certainly not. Whether a principle is explanatory has nothing to do with whether it is easily recognizable as being true or applicable. Whether a principle is easily recognizable as being true or applicable is a psychological property, one that has nothing to do with explaining computational capacities of physical systems. What makes a principle explanatory is whether it is explanatorily relevant, and I have argued that it is.

---

[8] Cuffaro (2013) has argued that the intelligent use of entanglement explains quantum speedup in all cases. I view that claim as completely compatible with the claim that QPT′ and OF-QPT′ can explain quantum speedup. An investigation into the comparative advantages and disadvantages of these accounts is beyond the scope of this paper.

4.8   Conclusions

In this paper I've attempted to make the case that despite what seem like contradictory claims being made in the literature on quantum computing, especially regarding the truth of the quantum parallelism thesis, there is actually a great deal of agreement. The disagreement that exists is largely a result of talking past one another using the language of computation without a suitable shared framework to govern the use of that language. I have argued that the appropriate formulation of the quantum parallelism thesis that involves no controversial claims regarding when a function is computed is:

QPT': A quantum computer can generate a system in a state correlated to multiple values of a function in a single computational step.

I also pointed out that the QPT' is essentially tied to oracle computing. Insofar as MBQCs are not typically conceived of as being equipped with an oracle, it is hard to see the QPT' as being explanatory of their capacities. That said, one can formulate the QPT' without the essential tie to an oracle. The resulting formulation is:

OF-QPT': A quantum computer can generate a system in a state correlated to multiple values of a function in exponentially fewer steps than a classical computer can.

I have argued that MBQCs that simulate the quantum parallelism process satisfy the OF-QPT'.

Finally, I have argued that the mechanistic account of computation suggests that an explanation of quantum speedup should compare the computational processes available to classical and quantum computers. A process that is essential to some of the algorithms that exhibit speedup and cannot be efficiently simulated classically will, at least in part, explain why one is faster than another. I have argued that for quantum algorithms that exhibit speedup and satisfy the QPT' or the OF-QPT', these principles are explanatory of their speedup.

Acknowledgments

Thanks to Mike Cuffaro, Sam Fletcher, and an anonymous referee for their comments on this paper.

5 How is There a Physics of Information? On Characterizing Physical Evolution as Information Processing

Owen J. E. Maroney and Christopher G. Timpson

5.1 Introduction

We have a conundrum. Information theory, whether considered in terms of communication or computational problems, is based upon analyzing information-processing tasks in the abstract, separated from the details of the physical substrate used to instantiate the task. However, in the last several decades there has been significant growth in a field called "The Physics of Information," whose motivating conception is precisely that information processing must be instantiated through physical systems, and that the nature of these systems matters. The key slogan is that "Information is Physical" (Landauer 1991, 1996; Lloyd 2006b; Vedral 2010). As the late Rolf Landauer asserted:

Information is not a disembodied abstract entity; it is always tied to a physical representation. (Landauer 1996, p. 188)

Although the exact sense in which this slogan is to be understood is often left somewhat ambiguous (cf. Timpson 2013), core claims in the field concern the existence of fundamental physical resource costs associated with information-processing tasks, where these costs arise from the nature of the abstract information-processing task itself. For some, this motivates construing information as a bona fide physical entity.

Here we address some of the puzzles concerning these claims in the physics of information. We will begin by reviewing various ways in which physics and information interconnect, and the idea that information-processing tasks have intrinsic physical costs. We will suggest that in order to make good upon the adumbrated resource-cost claims, the physics of information needs criteria for determining when a physical system instantiates a given information-processing task, where these criteria must be independent of the model of information processing involved and of the physical theory describing the systems.

We propose a five-fold criterion which is clearly sufficient – and we argue is also necessary – to underpin resource-cost claims. The final form of this criterion will be seen to emerge from the necessity of responding to the

103

problem of the ambiguity of representation, i.e., the problem of specifying when some physical evolution counts as implementing one specific computation rather than another. We shall suggest that characterizing a given physical evolution as an instance of information processing requires there to be a physically embodied agent for whom the processing task is performed. Identifying the agent lifts the ambiguity of representation enough for a physical resource cost meaningfully to be defined. We conclude that there is a physics of information not because information is physical, but because physically embodied agents can use physical systems to perform information-processing tasks.

5.2   Why is There a Physics of Information?

Ab initio, information theory proceeds by separating the analysis of information-processing tasks from the physical systems that perform those tasks. Both the technical notions of computability and of information come about through a move away from ties to specific physical models.

The definition of a Turing machine (Turing 1936) made it possible to study what is computable, and with what difficulty, without being constrained to the particular details of the physical systems needed to perform the computations. The fact that a Universal Turing Machine can efficiently (i.e., with at most a polynomial time increase) compute any algorithm that can be computed by any particular Turing machine allows one to categorize the complexity of a given computational problem on the basis of how long it would take a universal Turing machine to solve the problem (while bearing in mind that open questions still exist regarding what precisely that categorization is), and makes it possible to group computational problems into distinct families based upon which problems could be solved by conversion (in polynomial time) into some other problem with a known solution strategy. This categorization does not seem to depend on how the universal Turing machine is implemented, nor on whether one might use instead a network of logic gates or some other implementation such as cellular automata, for example.

Similarly, Shannon's coding theorems (Shannon 1948) made it possible to study information-communication problems without being tied to the particular details of how a signal might be physically coded. Instead of looking at examples of specific communication schemes, or the content of the messages themselves, Shannon simply looked at the frequencies f_i with which signals occurred, and showed that the best possible compressibility and rate of transmission of the signals could be given as a function of those frequencies. The result was an operational definition of the information content of a source of signals that was independent of the details of the particular message or medium:

$$H = -\sum_i f_i \ln f_i. \tag{5.1}$$

In both cases it seemed that a universal resource cost for certain tasks could be derived independently of the specific physical means by which the tasks might be instantiated. Call the *Separation Thesis* the idea that minimum resource costs associated with information processing can be wholly separated in this kind of way from the physical problems of building systems to perform the tasks. We now wish to consider a number of challenges to the Separation Thesis.

5.2.1   The Challenge of Landauer's Principle

A key starting point in the move to consider information as physically embodied – or physically embroiled – in some significant manner comes from the thermal properties of computation. Computers typically give off large quantities of heat. The question of whether there is a non-trivial *minimum* quantity of heat that a computer must emit started to receive attention from the 1940s (von Neumann 1966 [1949]; Brillouin 1951; Rothstein 1951; Gabor 1964) before Landauer (1961) proposed what is now generally regarded as being the correct way to understand this:

$$\Delta Q \geq -kT\Delta H. \tag{5.2}$$

The heat (ΔQ) generated by a computation – however implemented – is at least equal to the reduction in Shannon information over the course of the computation (ΔH) multiplied by Boltzmann's constant (k) and the temperature (T) of the environment in which the heat is generated. Given the similarity in form of the Shannon information measure and the Gibbs entropy measure

$$S = -k \int dxdp \, P(x,p) \ln P(x,p), \tag{5.3}$$

it might seem entirely plausible that this should be the case. Given that the Landauer lower bound on the heat cost is independent of the details of physical implementation, Landauer's principle (Eq. (5.2)) is in line with the Separation Thesis. However, controversy has arisen (Earman and Norton 1998, 1999; Maroney 2005, 2009b; Norton 2005, 2011; Ladyman et al. 2007; Ladyman and Robertson 2013, 2014) over what exact assumptions could underlie the claim that *any* physical system – of whatever kind – instantiating a particular computation must bear a thermodynamic resource cost.

Whilst it is fair to say that much of the controversy on this question has, at heart, concerned whether the kinds of probabilities that are used in Eq. (5.1) are really of the same kind as are used in Eq. (5.3), there is also the question (more interesting for our current purposes) of how one can evaluate the generalization "any physical system." As Norton complains about statements of Landauer's principle:

One does not so much learn the general principle, as one gets the hang of the sorts of cases to which it can be applied. (Norton 2005, p. 383)

There have been a number of attempts (Piechocinska 2000; Ladyman et al. 2007; Maroney 2009b; Turgut 2009) to address this complaint and prove Landauer's principle from general physical principles, but there is a remarkable lack of agreement in such attempts as to what kind of assumptions can be taken as valid in constructing such a proof.

5.2.2   The Challenge of Quantum Computing

The next challenge to the Separation Thesis comes from a concrete example of how a change in physics has led to a change in the minimum resource cost associated with an information-processing task. The computational complexity of a given problem is, as we have mentioned, based on how quickly the best possible computer program, running on a universal Turing machine, will solve that problem as the size of the input to the problem increases. Simple cases, such as adding several numbers, increase in direct proportion to size and quantity of numbers being added. More complicated problems increase in time polynomially on the size of the input, and particularly complex problems can scale exponentially or even factorially with the size of the problem.

For a universal Turing machine, the problem of multiplying two numbers together increases polynomially as the size of the numbers increases. However, the problem of taking a number and finding its prime factors increases super-polynomially.

Research in the 1980s and 1990s (Feynman 1982; Deutsch 1985; Shor 1994) discovered that a computer based upon the principles of quantum theory could factorize a number into its constituent primes in only polynomial time. Constructing a physical system that is based upon primitive quantum operations, rather than classical logic gates, seems to lead to the possibility of an exponential increase in the speed by which some problems can be solved. To date, no known algorithm operating on a universal Turing machine can perform this feat. The fundamental resource cost associated with solving the computational problem seems to be dependent upon physics.

But if a universal Turing machine has hidden physical assumptions, then what has happened to the idea of computational complexity as a feature which floats free of the physical substrate? And if it is the non-classical logic instantiated by quantum computers that allows the speed-up, how sure can we be that a classical physical system could not be built which would allow efficient instantiation of equivalent non-classical logic gates? Might we not just be missing a trick on the classical side of things?

5.2.3　The Challenge of Quantum Information

In tandem with the generalization of the universal Turing machine to a universal quantum computer, the second pillar of information theory – the Shannon information measure – has also had to be revised in the face of quantum theory. This is, in some ways, an even more remarkable challenge to the Separation Thesis. While no rigorous proof had ever been presented that a universal Turing machine really was efficiently equivalent to any other computational process (relying more on plausibility arguments that every realistic process had been shown to be so equivalent), in the case of Shannon information, the assumptions seemed much weaker and the conclusion correspondingly stronger. The entire purpose of Shannon's theorem is to characterize the signal capacity of a communication channel independently of the physical means by which the information is transmitted.

However, in the mid-1990s (Jozsa and Schumacher 1994; Schumacher 1995) it was shown that a communication channel sending signals in quantum states could achieve a transmission rate that exceeded that of the Shannon theorem: If a quantum signal is represented by the quantum state ψ_i, then the quantum information capacity of a channel is given by

$$H_Q = -\text{tr}[\rho \ln \rho], \tag{5.4}$$

where $\rho = \sum_i p_i \psi_i$, and $H_Q \leq H$, with equality only occurring when the quantum signal states are all mutually orthogonal. For a given size of channel, that is, many more distinct quantum states could be sent intact from sender to receiver than would be possible classically.

We therefore face the same problem: The Shannon information measure was intended to characterize a fundamental resource cost that was independent of the physical means by which the task was instantiated. But the change in physics, from classical to quantum, seems to change this fundamental cost.

5.2.4　The Challenge of Exotic Models of Computation

Further questions for the Separation Thesis arise, moreover, independently of the subtleties of shifting from classical to quantum physics. How sure ought we to have been in the first place that the universal Turing machine does in fact adequately characterize the best means by which a physical system can instantiate a computation? Might we not be able to do things more cheaply, via *exotic models* of computation?

Analog Computation   Analog computation provides the best-known challenge in this family. Much of the proof of what is computable – and with what complexity – for universal Turing machines rests upon their having only a finite (if arbitrarily large) amount of data-storage space available. An analog computer could potentially have a continuous infinity of information storage available to it, so these proofs would not apply. The standard story is that analog computers should be discounted as realistic possibilities due to the effects of noise: In the presence of noise, the computational state of an analog computer would be knocked into some different state, destroying the reliability of the computation. To be robust against noise, the computational states would have to be represented by extended regions of the physical state space, thus reducing the analog computer back to a finite amount of data storage. (More generally, any computational speed-up due to analog computing seems to require a resource cost of an exponentially increasing precision of the measurement of the computational state.)

This story is prima facie plausible, but there remains a concern about its basis: While we do not doubt the necessity of including noise in practical problems, the question might be raised as to why noise is, in principle, an unavoidable problem. The existence of noise is not written into the laws of classical physics, and quantum computing – for example – generally requires the existence of decoherence-free subspaces to proceed: Whence the proof that "noise-free subspaces" are not possible in classical physics, allowing analog computing to proceed?

Objet Trouvé Computation   A second challenge in this family arises from reflecting on the possibility that there might exist physical systems that simply happen to be able to solve a complex problem in a short period of time. Suppose, for example, that it was possible to come across, or to build, a quantum system with an atom confined in a potential, where the eigenvalues of the Hamiltonian of the system correspond to the sequence of prime numbers (Mussardo 1997). Then if one wished to test whether an integer was prime or not – a computationally hard problem – one could simply shine a laser beam on the system, with a frequency that matches the integer in question. If the system resonates, then the integer is prime. The time taken to perform this test would not seem to have any dependence on the size of the number being tested.

Two responses to this kind of example can readily be discerned. The first would note that the precision of both the laser frequency and the potential would need to be very fine, to avoid a non-prime-numbered frequency resonating a nearby prime-number eigenvalue. (This is akin to the noise objection to analog computing.) The second is a concern as to whether it is genuinely possible to calculate the potential required accurately without already knowing the sequence of prime numbers. In other words, to build the system it might be necessary first already to have solved the problem itself, in which case solving the problem requires simply looking up the previously found answer, rather than deploying a laser. (See Timpson [2013, sec. 6.4] for discussion of a related example of Nielsen's [1997].)

Niagara Falls Computation   The next cases adapt a well-known problem that has been raised in the literature on the philosophy of mind: the problem of whether it is even possible to state unambiguously whether a given physical system is uniquely instantiating a particular computation, and not some other (Putnam 1988; Searle 1992). We will have more to say on this later (Section 5.3). For now, let us start by considering Niagara Falls. Standing and looking at Niagara Falls, one will observe a large amount of matter going through a great number of complex physical transformations. It is reasonable to suppose that, for any given computation you care to think of, somewhere in Niagara Falls there will be some part of the flow of water whose complex transformations match the abstract transformations required to perform the computation. Is Niagara Falls factorizing large numbers into their constituent primes, then? Examples of this kind threaten to trivialize the Separability Thesis: How can we derive interesting resource costs if any old system might be doing any (even very tricky) computation?

The problem can be made more absurd by noting two further points: Any classical computation can be efficiently performed by a reversible computer (Bennett 1982; Fredkin and Toffoli 1982) and any reversible computation corresponds to the action of a permutation on the computational states. Now, we can take a sufficiently large, but stationary system, and after labeling which physical states represent the input states for a computational problem, simply re-label the physical states as representing output states that correspond to the appropriate permutation of their input states. Call this *Stationary Computing*. Our stationary system is now apparently performing any computation whatever simply by sitting there, being stationary. Such a conclusion is clearly absurd. What has gone wrong? Two considerations present themselves. The first is that we cannot read out the answer from the stationary computer unless we already know the relevant permutations, and this would correspond to our already having solved the relevant computational problem ourselves independently. Or second, if instead of this we were handed a code book that allowed us to decode the answer from the output state (or equivalently, encode the input state) then either the code is simply a look-up table that requires us already to have solved the problem as before, or else we will find that the decoding process will in fact be the computation, and all the resource costs have simply been transferred from the system to the decoding.

5.2.5   The Challenge of Different Physics

The quantum turn seems to indicate that both the notions of Shannon information and of Turing computation – despite their air of being fully abstract analyses – must have had hidden physical assumptions buried in them all along. Might this not happen again? How sure can we be that there are not further apparently innocuous physical assumptions in our current analyses of information and computation? And what happens if we take into account other physics, such as relativity, or even the possibility of our current best theories being replaced by entirely new physics?

One example in this line is that of hypercomputation. The notion of time resource associated with the hierarchy of complexity classes is not based upon the passage of physical time, but rather upon the number of elementary computational steps that are required. If the length of time per computational step is constant, then things are straightforward. But various hypercomputational models exploit the possibility that if each successive computational step takes a shorter and shorter length of time, then even an infinite number of computational steps might take place within a finite length of time. Is this physically possible? While quantum theory appears to provide "speed bumps" (Lloyd 2000) that would rule this out, general relativity seems to allow the possibility of space-times in which hypercomputation is possible (Earman and Norton 1993; Hogarth 1994). In a Malament-Hogarth space-time (which is not globally hyperbolic) one observer can, in a finite proper time, see the worldline of a second system pass through an infinite proper time (before hitting a singularity). If the second system is a universal Turing machine, this can be used to solve the halting problem, which is formally unsolvable using a universal Turing machine. Here then, not only would complexity classes differ from the standard case, but so also would the computability class: the set of problems which can be solved at all.

If solutions to general relativity that are not globally hyperbolic are considered physically possible, then this also raises the possibility of computation involving closed-time-like curves, which would again present a very different landscape (Deutsch 1991; Brun 2003; Bacon 2004; Aaronson and Watrous 2009).

Finally, we note that even in the absence of experimental evidence, it is possible to reason about theories that might replace quantum theory in the same manner that quantum theory replaced classical physics. Generalized probabilistic theories (Barrett 2007; Barnum et al. 2010) would allow information processing to take place differently to quantum theory. Once one has accepted that these things can change because quantum theory comes along, what is to stop one believing that things can change further if something else comes along? If, for example, a Popescu-Rohrlich nonlocal box (Popescu and Rohrlich 1994) were discovered to be physically possible, would this have as dramatic an impact upon the definition of information as the discovery of quantum entanglement has done?

5.2.6 What is the Basis of Resource-Cost Claims?

To recap: The initial success of the theories of information and computation was that they seemed to provide a way of analyzing information processing which was independent of particular models for how the processing task was to be physically instantiated. Resource costs could then be defined – such as channel requirements given data compression, the complexity of computation, or the thermal costs expressed in Landauer's principle – on the basis that any possible physical process that instantiates a given information-processing task will have an intrinsic physical resource cost purely as a result of the properties of the (abstract) information-processing task itself. At first blush, the advent of quantum information and quantum computation seemed an exciting new development within this picture, a sympathetic amendment or extension of it: Here there were interesting new physical resource costs to incorporate into a *richer* physics of information. But equally – and catastrophically – one might see quantum information and computation instead as driving nails into the coffin of the physics of information. For once one has recognized that the details of physical instantiation cannot be ignored, then where do things end, as the challenges represented by our other examples illustrate? The Separation Thesis seems in tatters. How can any non-trivial statements about the resource costs of information processing be defined? How can there be a physics of information at all? Why isn't there just physics?

The first aspect of the problem is dealing with the quantification "any possible physical process." Any analysis which leads to a resource-cost claim might have made implicit physical assumptions about the underlying process, but then perhaps there exists some more clever method of instantiating the task which would avoid the cost.

The second aspect to the problem is that the notion of "possible physical process" is itself a shifting terrain. Quantum theory has shown that abstract notions of information and computational complexity change if the fundamental physics changes. Might this not happen again? With the notion of information becoming dependent upon the physical theory used to model its processing, how can we ask what the physical resource cost is of an information-processing task when the physics itself may be changing?

In our view what is required is a perspective on all this which allows one to step a bit further back. To make a claim that an information-processing task has an intrinsic resource cost, one will need:

• An abstract characterization of the information-processing task: What are the inputs, the outputs, and the logical relations between them?

- Necessary (and sufficient) criteria for determining when a physical system instantiates a given information-processing task.
- A proof, within any given physical theory, whether those criteria can be met, and if so, at what resource cost.

And we need to be able to apply these to any information-processing logic, instantiated by any physical theory.

To obtain the required perspective we invoke again a final, familiar, problem.

5.3   The Ambiguity of Representation

Even in the most well-understood case (classical information processing on a classical system), what it might be for a physical system to instantiate a computation turns out to be a somewhat difficult and controversial problem.

Take the simplest suggestion: one defines a logical gate as a map from a set of input states to a set of output states. In the case of the AND gate, for example, the map is shown in Figure 5.1.

Now we design a physical system to instantiate this operation, in the simplest manner imaginable. We pick three physical systems, each with a large number of physical states x_1, x_2, and x_3 respectively, and we associate a region of the physical state space of each with a logical state, so that if the physical state of the first system is in the region $\{x_1\}_0$, that represents A being in logical state 0, and if the physical state of the second system is in the region $\{x_2\}_1$, that represents B being in logical state 1, etc.

Now all that is required of our physical instantiation of the AND gate is to find some physical evolution of the system, f, that takes[1] physical state x into physical state $y = f(x)$, and which ensures that whenever the first two systems initially lie in the joint region $\{x_1\}_A\{x_2\}_B$ then at the end of the process the third system will always lie in the region $\{x_3\}_{(A \cdot B)}$. More generally we can state the requirement as: If the set of physical states $\{x\}_\alpha$ represent the overall logical state α then the physical evolution $y = f(x)$ instantiates the logical map $\beta = M(\alpha)$, if and only if $\forall \alpha, \beta, y = f(x) \in \{x\}_\beta$ whenever $x \in \{x\}_\alpha$.

A	B	$A \cdot B$
0	0	0
0	1	0
1	0	0
1	1	1

Figure 5.1 The AND gate

[1] We are treating the evolution as deterministic in this simple example. Generalizing this formulation to a fundamental stochastic dynamics is straightforward, but adds nothing to the argument.

Now, the "$\forall \alpha, \beta$" is important here when calculating the resource cost of some operation. It is essential that the counterfactual *if some other input had been used, the correct output would still have been reached* holds, if the physical system is truly to be instantiating the operation.

To see this, consider the following example (which we owe to James Ladyman). A colleague suggests that they have a machine which can add two numbers much faster than a normal computer. To demonstrate they take you into a room with racks and racks of boxes, each box labeled with two numbers, and possessed of a button and a display screen. When you suggest two numbers to add, they pick up the box labeled by those numbers and press the button. Immediately, the screen on the box lights up, displaying the sum of those two numbers. When you suggest two different numbers, they do the same but with a correspondingly different box.

Is this a legitimate instantiation of the computation? The problem is that there is no reason to believe that any box is really performing the addition operation at all – for all one can tell, there might as well just be a light inside, with the pre-calculated answer already printed on the screen ready to show when the screen is back-lit.[2] We need one physical process that could have started with *any* valid input, and ends with the corresponding correct output in each case.[3]

Suppose now that we happily build our physical instantiation of an AND gate, satisfying the mapping criterion stated above, and excitedly show it to a colleague. She looks blankly at us and responds, "What are you talking about? That's an OR gate!" What has happened is that the physical states that we took to represent logical state 0, our colleague took to represent logical state 1, and vice versa. And under that different choice of representation, the exact same physical system is indeed representing an OR gate.

Now it might be tempting to seek to establish that the physical system either *really is* instantiating an AND gate, or *really is* instantiating an OR gate, but we suggest that this would be a mistake. Suppose we now want to build a larger operation, out of AND and OR gates; we could quite happily use exactly the same kind of physical process for each of the gates, but just wire up the connections to the inputs and outputs differently in the two cases. So it seems that there just is a fundamental ambiguity in which kind of gate the physical process instantiates.

Unfortunately, things do not stop there. Another colleague looks at the same system, and claims we have built an entirely different device altogether, one

---

[2] For illustrative purposes we have just framed the issue in epistemic terms: "no reason to believe ... for all we can tell." But the point is actually about what it takes for a device to be implementing some logical operation; this is a modal – a counterfactually loaded – notion.

[3] The room-full of numbered boxes would, perhaps, be a valid instantiation of addition, provided that all the pre-processing associated with deciding *which* box to select was included in the evaluation of the resource cost. Of course, even in this case, the presumably finite number of boxes involved would mean this approach would not scale to the addition of large numbers.

that performs a complex matrix multiplication. And when asked to explain how, it turns out that he has chosen a division of the physical states that is entirely different from ours, when representing the logical states. This is the problem that underlies the "Niagara Falls" computation we saw in Section 5.2.4. If we are just using the criterion that the physical evolution must reproduce the logical operation, with complete freedom as to how the physical states represent the logical states, any sufficiently large complicated system could be argued to instantiate any computation whatsoever. And reflecting on the case of Stationary Computation mentioned earlier, this would even apply to a system sitting there doing nothing, since *any* computation can be simulated just by permuting which unevolving physical states are to count as which logical states.

As we have already mentioned there are ways to resist this seemingly pathological prospect of Stationary Computing; but for now, we wish to retain much of the ambiguity of representation, firstly because it may be of practical benefit to be able to use the same physical system to instantiate different operations, and secondly because we should be wary of ruling out some particular exotic computational model too hastily, in case our justification for ruling it out would also rule out another seemingly exotic model which, on development, might have more than a little going for it. Consider:

Stationary Quantum Computing

- Any quantum operation can be implemented by a unitary transformation on a sufficiently large Hilbert space.
- Unitary transformations are just changes of basis.
- So any quantum computation can be performed by just preparing a state and then measuring in a rotated basis.

Stationary quantum computing is, occasionally, a useful way of studying the possibilities of quantum computation, but also, when the complexities of actually performing the measurement through a series of elementary operations is taken it account, this model starts to become very close to *Measurement-Based* Quantum Computing (Raussendorf and Briegel 2001; Raussendorf et al. 2003; Benjamin et al. 2009), which is one of the more viable models for practical quantum computing.

Now: The problem of the ambiguity of representation has been studied extensively, with particular importance for debates around computational theories of cognition (Putnam 1988; Searle 1992; Chalmers 1996; Copeland 1996; Shagrir 1999; Sprevak 2005, 2010; Piccinini 2008). If it is thought that the mind is basically computational in nature, a matter of some interesting software running on some physical (presumably largely neurological) hardware, then it becomes rather important to be able to identify what software is running, and in virtue of what kinds of facts it is the case – if indeed it is the case – that some particular piece of computation is being performed on some occasion as opposed to some other one, or indeed none at all. We do not propose to address this large issue here, interesting and important as it is. For us this problem of the ambiguity of representation is embedded within and quite naturally at home in information theory because – as we said at the start – the whole point about the success of the Church/Turing notions of computability, and of the success of the Shannon coding theorems, was that they analyzed information processing abstracted from the essentially arbitrary ways in which the information might physically be represented and operated upon. And this was a good thing.

5.4   Quin-Criteria

A balance must be struck: On the one hand, some facet of ambiguity of representation may be a good thing; on the other, if our only requirement for a process to count as computational or as information-processing is simply having an evolution of the physical states which reproduces the mapping of the logical states, then this will leave too much ambiguity to be able to analyze the fundamental resource costs of information processing, with it being unclear that it amounts to any restriction at all. More restrictive criteria for computation have been proposed in various contexts, however, and we now turn to examine these, before extracting a common core – the final five criteria. In the following section we shall seek to justify the claim that all five are indeed *necessary* criteria.

5.4.1   Quantum Computing and DiVincenzo

In the 1990s, when the field of quantum computing was really beginning to develop, there were a great many proposals for which kind of physical systems might be viable for a scalable quantum computer. DiVincenzo (1997) gave the canonical statement of what should be required. His criteria were (paraphrasing):

1. A Hilbert space representation of the quantum states. The proposed systems must not simply be classical systems representing quantum states, but actually use a physical structure whose state space was a Hilbert space.
2. Initialization of the physical system into a standard state. In order to get the computation going, the quantum system had to first be put into a "fiducial starting quantum state."
3. Noise tolerance, especially with respect to decoherence effects. To perform quantum operations successfully, quantum coherence must be maintained during the operations, so the system should have a high degree of isolation

from the environment. This proved to be one of the biggest problems for early proposals until the development of quantum error-correction codes and fault tolerance (Shor 1995; Steane 1996; Aharonov and Ben-Or 2008).

4. Ability to perform all the needed unitary operations. By analogy to a classical computer, a network constructed from a small set of elementary, universal logic gates (such as AND and NOT), a quantum computer needed to be a network constructed from a set of elementary, universal quantum operations. The development of more exotic models of quantum computation, such as quantum annealing or measurement-based quantum computation, led to this requirement being relaxed slightly: the evolution of a quantum computer had only to be unitarily equivalent to a network model of a universal quantum computer.

5. Ability to perform a strong projective measurement to read the output. While in a classical system the end of the computation simply leaves the device holding the answer, in a quantum system it is necessary to perform a quantum measurement, meaning something rather different than simply a further unitary evolution, so that the result of the computation could be read out.

5.4.2   *Landauer's Principle and Maroney*

As described earlier (Section 5.2.1), one of the complaints raised against the resource claim associated with Landauer's principle is that no proof is offered that any physical system must be of the kind to which Landauer's principle applies. In attempting to clarify the basis of Landauer's principle, one of us put forward necessary criteria for a physical system to implement a classical logical operation (Maroney 2005, sec. 1.2; Maroney 2009b, sec. II.D–II.E). While the criteria varied slightly between these two papers, they shared a number of features in common (with the numbering below following the criteria as stated in Maroney [2005]):

1. Distinct regions of the physical state space are to be used to represent the distinct logical states (2009b, sec. II.D nos. 1–5).

2. When the physical system is placed in a state in one of the regions, it does not leave that region unless acted on by an operation (2009b, sec. II.D no. 6).

3. It should be possible to set the system reliably into a physical state within one of the regions, even if it is not possible to set it into a precise physical state. (This criterion was not explicitly mentioned in Maroney [2009b].)

4. It must be possible to determine the region of the physical state space in which the system lies. (This criterion is found only implicitly in the requirements of Maroney [2009b, sec. II.D nos. 3–4].)

5. There must be physical interactions that allow the logical operations to be performed (2009b, sec. II.E).

Beyond the second requirement, of the stability of the representation of the logical states, noise tolerance was not explicitly referred to for the operations themselves.

5.4.3   Unconventional Computation and Horsman et al.

In parallel to our own thinking on these matters, Horsman et al. (2014) were independently seeking to address the question of when exotic computational models (which they refer to as "unconventional computation") could be properly understood as performing a computation. While their analysis was expressed in different terms, exploring an analogy between confidence in unconventional computation and confidence in physical theories, the criteria they identify matched the criteria we were developing:

1. There exists a representation relation between the input and output states of the physical system and the abstract states of the computational model.
2. This representation results in a commutative diagram when the evolution of the physical states, according to a theory of the physical system in which there was a high degree of confidence, is compared to the abstract computation in the model.
3. The physical state of the system at the start is encoded with the input logical state through an initialization process.
4. The logical output state is read off from the physical state of the system at the end, through a decoding process.
5. The high degree of confidence in the physical theory used to predict the evolution of the system is based upon its having been experimentally tested to within a good error bound.

5.4.4   The Final Five Criteria

The various criteria listed above display some variation as each was developed to analyze a particular kind of problem. However, one can see clear similarity across these different cases:

1. **Representation of Logical States by Physical States**
   The physical system has to have the right kind of physical states available to represent all the required logical states. This representation will need faithfully to preserve the relevant properties of the logical states. In the case of classical computational states, for example, they need to be distinct, so the physical representations must correspond to distinguishable regions of the physical state space. Quantum computational states, by contrast, will need to be represented by physical states which include the phase relationships between the logical states.

The question of what the relevant structural properties are will, of course, vary between different kinds of logics of information processing, but provided the relevant properties are there in the physical representation, there is the full freedom to choose how to implement this, which the ambiguity of representation allows.

2. **Initialization of physical states**

   There must be a controllable physical process which can reliably set the system to be in each of the initial representative states, and is in fact used to do so on any occasion on which a computation is understood to take place. The resource costs (such as time, space, or work) that this initialization process uses must be counted as part of the resource cost of the information-processing task. So all pre-processing must be included, otherwise it is possible simply to transfer the costs of performing the task to an initial encoding operation, such as in the case of the "Stationary Computing Model," with an initial encoding that actually performs the permutation part of the computation.

3. **Equivalence of evolution**

   The information-processing task will involve a set of operations which map[4] input logical states to output logical states. The physical processes must evolve the representative physical states equivalently to the mapping of the logical states. It is important that the physical system does not just instantiate a particular mapping of input to output state. All the counterfactual claims must also be true of the system: if it had been presented with a representation of a different input state, the physical process would have still produced the representation of the correct corresponding output state. The resource costs (time, space, work) associated with this are part of the resource cost of the information-processing task.

4. **Readout of results**

   It is important to understand this not just as specifying the physical representation of the final logical states. It is the requirement that the final output of the information-processing task must be in a form that is directly accessible to anyone who needs the information, in the sense not only that the output be of a *kind* which is readable, but that it also be fixed exactly what logical output is represented.

   In classical information processing, it might be assumed that the internal logical states can be read off from the physical states without any problems. However, for quantum computation, this is not necessarily the case. A quantum computation would generally involve non-orthogonal states, which cannot be distinguished, and cannot be measured without destroying phase coherence and disrupting the computation. Here, then, a final measurement

---

[4] This includes stochastic maps, so that probabilistic operations and machines are included.

stage is needed at the end of the computation. For models of analog computation, the readout stage also cannot be ignored: the need to measure the physical state with an exponentially increasing precision is of course one of the principal limitations which makes analog computing unfeasible.

Once again, the resource cost (time, space, work) of this final stage must be included in the overall resource cost of the information-processing task. All post-processing must be included, otherwise we have the problem of the "Stationary Computing Model," where the resource costs are simply shifted to a final decoding step.

5. **Error tolerance**

As the physical system is initialized, evolves, and is measured, errors will inevitably occur. The errors depend upon both the kind of physical system being used and the kind of information-processing task. Error tolerance is a significant hurdle for quantum computers and for analog computing is the main reason it is generally considered not to be feasible. Although error tolerance does not attract much attention in studies of classical computation, that is simply because the problem has been practically solved, and highly fault-tolerant classical computers exist. That it can still be a problem becomes much clearer when computers overheat: in modern computing centers, the air-conditioning is one of the significant constraints upon computing power. In fact, error tolerance was understood to be a potential problem in the early development of classical computing, leading von Neumann to prove that classical computing could be made error tolerant (von Neumann 1956).

Criteria 1 and 3 are just those we noted in Section 5.3. Criterion 5 adds error tolerance, but this does not affect the problems that were raised by the ambiguity of representation. If our final five are to resolve all the problems we have encountered, it must be due to the addition of Criteria 2 and 4: the initialization and readout stages.

5.5   Information and the Physically Embodied Agent

Our Criteria 1, 3, and 5 seem uncontroversially to be necessary for an information-processing task to be performed, but they do not seem to be sufficient. Our contention is that Criteria 2 and 4 are also necessary conditions, and that together 1–5 constitute a sufficient condition.

But how are 2 and 4 to be justified? It is not obvious to all that they do indeed constitute necessary conditions. Consider the following countervailing contention of Ladyman:

In practice of course it is only possible to use a system as a computer if: (a) the relevant physical states are distinguishable by us (with our measurement devices); and (b) it is possible for us to put the system into a chosen initial state so as to compute the function in question for it. However, in principle some reason must be given as to why a system with respect to which these constraints are not met cannot be considered to be computing nonetheless. (Ladyman 2009, p. 382)

Given our argument so far, it seems there would be trouble indeed if some principled reason could not be offered here, or trouble at least for the physics of information. For we have tried to argue that physical resource-cost claims cannot reasonably be made absent conditions like our full range 1–5 above, specifically including Criteria 2 and 4, which latter pair Ladyman denies to be essential. So much the worse for the physics of information, it might be maintained! But we wish to preserve a domain for the physics of information – properly understood – and we think that a principled reason can be offered for why Criteria 2 and 4 should be conceived to be necessary conditions for computation to be happening at all, not just for its being useful.

To develop this, let us clarify further what we understand by an information-processing task.

We began our discussion with the general problem of trying to understand how information-processing tasks could come with physical resource costs, purely by virtue of the identity of the information-processing task itself. One solution to this which we have not pressed so far would be to propose that information itself should be conceived of as an independent physical property, such as energy or charge, or perhaps as some new kind of stuff, a new material or quasi-material substance, perhaps. This might make the idea that there is a physical cost to information processing seem perfectly natural. However, it would also seem to require there to be an unambiguous matter of physical fact that a given physical system is instantiating a specific information-processing task, and that seems to be problematic.

Here we take an alternative view of the relationship between information and physics, drawing on the ideas developed in Timpson (2013). It is more than a little useful, in our view, to reflect that the term "information" is an abstract noun, and correlatively to note that:

Very often, abstract nouns arise as nominalizations of various adjectival or verbal forms ... their function may be explained in terms of the conceptually simpler adjectives or verbs from which they derive; ... 'information' is to be explained in terms of the verb 'to inform'. Information ... is what is provided when somebody is informed of something. (Timpson 2013, p. 11)

Phrases such as "information is gathered," "information was communicated," and "information will be stored" often seem to be used as if information were some object or property that we can search for, pick up, move around, and put somewhere. But there are also situations where it is not clear how that conception could make sense: For example, if some data are encrypted, and the encrypted data are stored separately from the encryption keys, the information seems only to be present in the correlations between the encrypted data and the encryption key. Destroy either, and the information appears lost (thereby also raising questions like "where did the information go?").

We do well to step back and think about information not *as* a physical object or property, but rather as pertaining to a process or an activity for informing: it is "what is provided when someone is informed of something." So in each case, rather than use the abstract noun, we could more fundamentally refer instead to the process of informing:

- "information is gathered" vs. "an information-gathering task is performed,"
- "information is communicated" vs. "an information-communication task is performed,"
- "information is stored" vs. "an information-storage task is performed,"
- "information is processed" vs. "an information-processing task is performed," etc.

On this conception, the different kinds of information-processing tasks are different means by which someone is actually or potentially informed of something (and by someone). These tasks might be very complicated: the someone who is being informed may also be the someone who is doing the informing, for example if they are using a computer to perform a calculation; a communication task may involve multiple partners trying to communicate different things to each other, whilst concealing things from other people; some tasks, such as information storage, may be defined by the need indefinitely to defer the informing stage.

The important point we wish to draw from this for our purposes in this chapter is that when seen in this way, an information-processing task is defined by the question of who is being informed of what, and by whom, and the question of how that task may be physically instantiated is going to need to include a specification of how the informers and informees – as physically embodied systems themselves – interact with the information-processing system. It is this which enforces and underwrites Criteria 2 and 4 as necessary requirements (*pace* Ladyman), in our view.

It is key that these *someones* are physically embodied agents. To inform, they must be able physically to start the process of initialization, and they must do so. If they cannot do this, the information-processing task cannot begin. To be informed, the information must be presented to them in a form that they can directly perceive. The readout stage is required for this, and until the output is in readable form, the information-processing task cannot be said to have been completed. If either of those processes fail, the information-processing task has not succeeded. Equally, if both those processes succeed, and the remaining criteria are also met, then the information-processing task has certainly taken place. Thus we argue, on this view of information, the five criteria are individually necessary and jointly sufficient for a physical system to instantiate a well-defined information-processing task.

Let us look again at the humble example of the AND gate. The same physical process could be used to instantiate an OR gate. The difference between the two cases seemed to be how the physical states were chosen to represent the logical states, which seemed arbitrary. But given that we have said information processing is to be understood in terms of informing, what our criteria now add – resolving the arbitrariness – is the requirement that when the system is actually being used to instantiate a particular information-processing task, the initialization and readout stages will fix exactly *how* the physical system is being used. Specifying those stages as an integral part of the task lifts the ambiguity of representation. The question of whether the physical process is an AND or an OR gate cannot be determined simply by looking at its physical evolution alone, but must involve a broader context. The question will be determined instead by how, in a given instance, the physical system is actually being used to inform, in virtue of its relation to *someones* as informers and informees.

Once the initialization and readout stages are included as necessary components in any information-processing task, the answers to many of the challenges we outlined earlier become unified. We have already seen how including the initialization and readout naturally answers the problem of Stationary Computing, where the actual computation is shifted into a calculation of how physical states must be permuted to represent either the input or output logical states. It also answers the problem of the Niagara Falls computer: While there may be complex physical evolutions taking place within the waterfall, there is no one who is initializing the state of the falling water, nor anyone able to read out the supposed result.[5] No one is being informed of any computation by such physical processes, so they are not information-processing tasks.

We also wish to note that the need to include the physical process of initialization and readout within the definition of the information-processing task still leaves a great deal of freedom to exploit the ambiguity of representation in the way the task is instantiated between input and output. The definition of the information-processing task must refer to the logical inputs and outputs as they are perceived by the physically embodied agents, but the equivalence-of-evolution criterion just refers to the overall mapping of these inputs and outputs and does not need to specify a particular method by which it is achieved: There may be many different algorithms for achieving the

---

[5] We may also note that it is unclear if the counterfactual condition on the equivalence of evolution holds in this model: If, somehow, someone had been able to "initialize" a portion of the water to a different state, would the subsequent evolution still have produced the correct outcome?

same information-processing task. So we need not identify the information-processing task with any particular algorithm or computational model. It may even be the case that a given physical process for achieving the task could be interpreted in more than one way, in the same manner that the same physical process could be interpreted as an AND gate or an OR gate. Provided the initialization and readout stages define the manner in which the physically embodied agents are informed, however, the instantiation of the information-processing task itself will not suffer from this ambiguity.

5.5.1 Boltzmann's Laptop Falls into a Black Hole

To emphasize some consequences of our view, consider the following somewhat unlikely scenario. A large cloud of interstellar dust is going through thermal fluctuations when, by chance, these fluctuations lead to a chunk of matter coalescing into a physical object that happens to be in exactly the same state as the laptop on which this chapter is being written. The keyboard of this fluctuation-born laptop gets bounced around, leading to this very sentence appearing on its screen. And then the laptop falls into a nearby black hole and no one ever sees it.

Now it might seem reasonable to say that the fluctuation-born laptop did exactly the same computation and exactly the same information-processing tasks as the laptop on which this paper was written. But it is a consequence of our point of view that this is not the case: No one was informing nor was anyone being informed of anything, so no information-processing task was taking place. It just happened to be a random thing that happened in space.

But note: Had a passing spaceship come across the laptop before it fell into the black hole, and scooped it up, then they would still have a perfectly good laptop which they could then use to perform computations, including writing about computations. The fact that it had come into existence by a random nonintentional process would not matter.

Yet there is a difference between having the capacity to perform a task and actually performing the task. While the laptop was floating through the dust cloud, randomly looking – if there were anyone to look – like it was typing up a chapter, it can be considered to have shown the physical capacity to perform an information-processing task, but it was not actually performing any such task. Only were the laptop to be scooped up and put to use would it indeed be using that capacity to inform the passing space-travelers.

Similarly, the physical system instantiating the AND gate has the capacity to be used as an AND gate or an OR gate. Left to itself, it is doing neither. Only when it is placed in the context of being used by someone in an information-processing task can it be said actually to be operating as one gate or another.

5.6   What Makes a Someone?

Our proposal is that the question of how information-processing tasks can be physically instantiated cannot be separated from the question of the existence of physically embodied agents who are informing and who might be being informed by the task. We have argued that this means that initialization and readout stages must be included within the definition of the task, and that this also provides a principled answer to problems involving the ambiguities of representation when calculating the resource costs associated with information processing. However, the question still arises, what kind of physically embodied system is needed to fulfill the role of a *someone* to inform or be informed? There may be danger of a lurking circularity or a descent into vacuity here.

One might seek to adopt quite a liberal conception of what could count as a *someone* who might inform and be informed, which might include non-human animals, artifacts, simple automata, and various kinds of things in general. But too liberal and one is returned to an empty position and the problem of ambiguity of representation. One might wish to define a *someone* in terms of the network of causal relationships they are part of, or in terms of agency – their capacities to interact with and respond to states of the world external to them – but this might invite a return to the problem of the ambiguity of representation, applied now one level up, to the entire system which includes the physically embodied *someones* being informed. One might, therefore, dig in one's heels and maintain that proper informers and informees are creatures pretty much like ourselves: Other cases, in which we are inclined to use information-talk when there are not familiar cognitive agents such as ourselves to inform or be informed, are to be understood by way of extension or analogy. This is a stable position, but perhaps seems a little overly restrictive.

5.6.1   Epistemic Communities of Agents

This restrictiveness might to some considerable extent be mollified, however, if one could identify clear senses in which different classes of agent would count as equivalent or interchangeable. One of the key features of physically embodied agents as we are conceiving them is that they must be able to initialize the process and receive the readout. It is their physical attributes and abilities that ultimately fix the starting and ending points of the process. This need not mean that any physical resource cost associated with an information-processing task will be different for different *someones*, however.

In practice, initialization and readout procedures will be relatively insensitive to many – perhaps the majority – of the details of the individual physical agent being informed, as agents will tend to form what we may term (with a nod to van Fraassen) *shared epistemic communities*. A shared epistemic community means a group of people, entities, or agents that can share and access the same information – that is, they have the ability to interact with the same kinds of devices through the same kind of physical means. Put another way, we can think in terms of initialization and readout being defined relative to a set of physically well-defined observers that share roughly the same kind of capacities to respond to and change the world surrounding them. It will be an objective physical fact what the initialization and readout capacities of a given epistemic community are. We now conjecture: The resource costs of initialization and readout will not be very sensitive to variation between different epistemic communities, and we would therefore expect that there could be efficient translation between the initialization and readout needs of these different epistemic communities.

What if we move further afield? What if one were to encounter aliens, with radically different perceptual capacities? First let us note: If the perceptions of the aliens depend upon a different physics to the physics of our perceptions, then we are just talking about the cost of information processing evaluated by a different physical theory. So let us assume that the aliens' perceptions come from the same physics, but use radically different physical processes to ours.

In this case it should be possible to analyze and understand those physical processes. We therefore conjecture again: If aliens could perceive something, a talented experimenter could build initialization and readout systems to use those perceptual mechanisms for us (and vice versa). So, assuming we had not unnecessarily restricted ourselves in some way when analyzing fundamental resource costs, then a different shared epistemic community, based on alien perceptions but using the same underlying physics, should still yield the same fundamental physical resource costs.

5.7   Conclusion

So where does this leave our original problem? How can there be a physics of information whilst still holding on to the Separation Thesis that information theory succeeds by abstracting the definition of information-processing tasks away from the details of the physical substrates used to instantiate the tasks?

We have approached the physics of information from the perspective of establishing physical resource costs associated with performing information-processing tasks. To try and quantify these costs, we defined an information-processing task in terms of "someone being informed of something by someone." While acknowledging that there is a great deal of complexity that can be hidden in those terms, the *someones* involved are themselves physically embodied agents, and the physical means by which they start and end the process of informing must be included within the specification of how the information-processing task is to be instantiated. These are the initialization and readout stages of information processing, which provide physical connections between the physically embodied agents and the information-processing devices.

The physical means by which the task is performed does not need to be specified at this point: Indeed, if the objective is to derive a fundamental resource cost, then specific models should be avoided. However, a proof of a physical resource cost cannot be wholly independent of the fundamental physics used to evaluate that cost. So any resource-cost claim has to be made relative to a theory of physics. The greatest practical importance should be attached to the assumptions that correspond to our current best physical theories, but the ever-present possibility that these theories may need revision should make us cautious about ruling out exotic models of computation in principle. There is an interesting corollary here, though: Actually building a system to perform some computational task with a resource cost below a minimum threshold could be taken as prima facie experimental evidence against a physical theory that sets that threshold.

Relative to a given physical theory, there may be some particular tasks that can be efficiently implemented, and may be efficiently combined as building blocks out of which all other tasks can be constructed. These building blocks may be used to define a logic of information processing that is well suited to the physical theory: Thus Boolean logic gates are a basis for classical information processing, and quantum operations are a basis for quantum information processing. Equally, some resource costs may turn out to be particularly useful or ubiquitous within a given physical theory, and, like the Shannon and Schumacher measures, become regarded as a measure of information within that theory.

It should be noted that the consistency of a given abstract logic of information processing is not dependent upon physics: the only question is whether that logic can be efficiently implemented within a given physical theory. So information processing can continue to be studied in the abstract, independent of the means of physical instantiation. However, the interplay with physics means that sometimes physics will suggest new ways of analyzing information-processing tasks, enriching both fields, and logics which best match our current best physics will be the ones best suited for understanding the physical resource costs of real information-processing devices.

Acknowledgments

We thank the editors for the invitation to contribute to this volume, for their patience during the process, and for their comments. We thank also an anonymous referee. This work was supported by a grant from the Templeton World Charity Foundation.

6 Abstraction/Representation Theory and the Natural Science of Computation

Dominic Horsman, Viv Kendon, and Susan Stepney

6.1 Introduction

Is computation an intrinsic property of physical systems, or is there a distinction between a computer and other objects in the universe? Computer science as a theoretical discipline, traditionally dealing only with issues of abstract computation, has tended to ignore this question. Amongst more philosophical approaches to computing, the first view, pancomputationalism (Piccinini 2017), has been argued for in various guises. Assertions such as "the universe is a computer" (Ball 2002) are indeed superficially appealing, given the great success of modern computing theory and technology. Yet they lose their apparent content as we look at them more closely. If everything computes merely by virtue of existence, then what more do we say about an object when we call it a computer? How are novel and unconventional computing devices to be characterized, if computing occurs universally and intrinsically in physical objects?

We present here a more discriminating view, in which the use of a physical system to carry out an abstract process (a computation) depends on a number of specific properties that both the physical device (the computer) and physical process (it computing) must have (Section 6.3). Not all physical processes constitute computing: A key element in physical computation is the use of a physical system to manipulate the representation of abstractly encoded data in specific processes. In presenting our framework, *Abstraction/Representation (AR) theory* (Horsman et al. 2014; Horsman 2015), we show that physically carrying out computation and doing science are closely related activities (Section 6.2). Both involve *representational activity*. By looking at how computers are developed from both fundamental science and then engineering and technology, we show the crucial physical nature of computing.

Key to AR theory is the *representation relation* between physical objects and objects in the abstract mathematico-logical domain. AR theory takes as primary a realist physical domain; we discuss other potentially compatible views in Section 6.2.1. The representation relation is structured and directed, from physical to abstract: Scientific modeling is fundamental in AR theory. We consider representation functionally, in terms of its use and properties within the physical sciences. We are not primarily concerned with it as semantics or meaning, or, relatedly, as knowledge and information. AR theory's construction of computation is not of symbol manipulations, nor are there notions of representational states (Putnam 1960; Fodor 1975). If science is to happen in a physical external world, and if there is any part of reality that corresponds to what is considered as an abstract world (where mathematics, logic, computations, etc. live), then there must be a map from the physical to that abstract world. Representation is that map.

AR theory separates out (i) physical systems and processes, (ii) abstract objects, and (iii) the representation relation that maps between them. Abstract entities do not mirror physical ones (Putnam 1960; Rorty 1979), but stand in quasi-functional relations to them. By analyzing the role and specific functionality of representation in science and technology, we give the AR framework for computing in terms of the interrelations of these three elements. In the specific context of computing, the representation relation allows abstract computations to be instantiated in physical computing systems, and the effects of physical processes to be represented as abstract computational results. There are not some physical states that are computational states; rather, there are computing *cycles* that require all elements to be present in specific ways before computing can be said to occur.

AR theory's analysis of physical computation captures computing in commuting-diagrammatic form: Physical computers use representation to act as predicting devices for the results of abstract computations. This acts as the converse situation to that of experimental science, where a physical theory functions as an abstract predictor for the behavior of physical systems. AR theory shows us the deep structural connections between computing and natural science. Experimental science and physical computing both require the interplay of abstract and physical objects. This is mediated via representation in such a way that the diagrams of AR theory commute: The same result is gained through either physical or abstract space-time evolution.

The central role of representation leads to the requirement for a *representational entity*. This is the entity that supports the representation relation between physical and abstract. For the natural sciences, this is the human experimenter or theorist. Within a computational process this is usually the programmer, designer, and/or end-user. However, AR theory does not require the representational entity to be human, or even conscious (Section 6.2.5).

AR theory treats both physical and abstract objects, mediated by representation, within the same framework. This allows us to discuss the connection between the scientific description of a system or device (as a theory of physics or biology, say) and its computational description as two distinct representations. We can use this distinction in the search for novel systems whose physical properties allow us to compute in new and interesting ways. Computer science has previously lacked a formal connection between physical device and abstract theory. AR theory now provides this, distinguishing the physical system from its abstract scientific and computational representations. Conflation of this three-way separation lies at the root of much of the confusion that surrounds both the development of unconventional computing and the relationship between scientific theory and computation (Section 6.4).

Not everything that supports a scientific representation supports a computational one. In this way, pancomputationalism is taken out of the picture by AR theory (Section 6.4.1). The connection between the physical system and an abstract representation is not in general coextensive: The physical system will have properties not captured in the abstract representation, and the abstract representation may have properties not realized in the physical system. Failing to take these differences into account gives rise to problems both of the existence of "side-channels" (Section 6.4.2) and the more extravagant claims of hypercomputing abilities (Section 6.4.3).

AR theory allows us to show the fundamental relationship between scientific and computational models, whilst preserving their necessary distinctiveness. Without this, a physical theory of computation tends to flounder. AR theory forms the backbone of a new framework for the foundations of computer science, treating this mutual relation between distinct models as fundamental: a natural science of computing.

6.2   Introduction to AR Theory

6.2.1   Science and Ontology

Key Role of Representation   Natural science can be viewed, at its basic level, as concerning objects from two distinct domains. Objects, processes, and systems within the physical world are the subject of scientific theories scrutinized during experimental observations, and may be manipulated during experimental tests. Abstract objects are used to model these physical systems within the domain of mathematics, logic, or any other language of the sciences. The fundamental operation of science is this modeling of a physical object by an abstract one: the process of *representation* (Frigg 2006; van Fraassen 2008). This conception of science, and the crucial role of representation, is the starting point for the framework of AR theory.

Figure 6.1 illustrates a physical system represented as an abstract model. This is a fundamental use of the *representation relation* $\mathcal{R}_T$, which quasi-functionally relates physical objects to abstract ones; it is not a mathematical function, as that would require both domain and range to be abstract. In Figure 6.1, a glass bead is represented using $\mathcal{R}_T$ as a volume V. In general, we can talk about a domain **P** of physical objects and a domain M of abstract objects. (We use boldface for physical objects, italics for abstract ones,

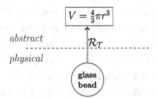

Figure 6.1 A spherical glass bead represented by its volume V. The representation relation $\mathcal{R}_{\mathcal{T}}$ relates the physical object to its abstract model, within the theory $\mathcal{T}$

and script for representations.) Representation is theory-dependent, which we denote by the subscript $\mathcal{T}$ in $\mathcal{R}_{\mathcal{T}}$. In this case, the theory is that of the classical mechanics of solid objects. If we were instead interested in the refractive properties of a glass sphere, we would represent it by its refractive index and surface geometry, and the theory would be classical optics.

There is no unique theory for a given physical system: Multiple theories can have different forms of validity, depending on the properties of the physical system in question, the degree to which they need to conform to experiment, and the domain for which they are required to be applicable.

Representation, Ontology, and Realism   The field of study of representation in philosophy is vast. Since Rorty's demolition of the notion of representation as a simple mirroring of abstract and physical (Rorty 1979), the nature and structure of different types of representation have been extensively explored. AR theory's representation is primarily that of scientific representation: The physical world is fundamental, and representation is structure given abstractly on top of it. As used here, it does not come pre-loaded with implications of intentionality or meaning; it is, at its most fundamental, a mapping from physical to abstract. We almost never talk about "information" or "knowledge" or "meaning" in using AR theory. AR theory draws strongly on the framework for scientific representation explored by van Fraassen (2008), and with many similarities (as a more developed framework) to the basic modeling theory of Hughes (1997). For a philosophical overview of models in science, see, for example, Suppes (1960) and Frigg and Hartmann (2016).

Views of computation that regard it as the manipulation of representation have historically come from a semantic account of computation; see, for example, Putnam (1960) and Fodor (1975). This comes out of an emphasis on the similarities between computation and cognition, thought, and language use. The abstract – thoughts, representation, language – is considered primary. The set of all objects is the set of abstractions: "The facts in logical space are the world" (Wittgenstein 1922, prop. 1.13). In such a view, if anything is to be called into question, then it is the existence of the physical world. At best this is idealism; at worst, hard-line logical positivism.

This concept of computation, as closely allied with cognition and a philosophy of logic and language, is the opposite of our starting-point for AR theory: A computation is abstract; a computer is physical. The question is not "what types of abstract states and processes form computations?" but now "which physical systems are computing?" We assume the existence of a physical world. This is the space of all physical objects, **P**. Idealist-leaning, nominalist, or verificationalist projects fail ultimately in their inability to abstract out ongoing and necessary interaction with an external world. Carnap's project for a basic sense-data language (Carnap 1950a) – which can be viewed in computer science terms as analogous to a universal concrete semantics – failed in ways that make it clear it cannot succeed. Following Quine (1971), our epistemology is naturalized: how abstract content interfaces with physical reality is found by interrogating our best scientific theories. In AR theory this also includes representation: Representation is always within the framework of a physical theory.

We do not use a cognitive starting-point for representation and computation in AR theory, but we do not reject outright the applicability of questions around intention and mental representation to notions of computing. The relationship between abstract computation and physical computer is the mind/body problem *du jour*, and questions of computation and cognition have a lot to say to each other. AR theory draws on many aspects of the long tradition of work around the abstract/physical divide concerning both mind/body and theory/reality in science, prioritizing the latter pair rather than the former.

Such a notion of "naturalized representation" gives the starting-point for AR theory: how representation functions within scientific theories. It is through scientific representation that we come to computing. The physical world and device are basic, and representation is structure in addition. This inversion of the usual conception of representation is shown in the direction of its basic mapping; $\mathcal{R}_\mathcal{T} : \mathbf{P} \rightarrow M$, rather than an abstract object being represented as a physical one, $M \rightarrow \mathbf{P}$.

AR theory starts from scientific realism, but need not end there. In terms of realism, the AR framework gives a way to talk about the physical world **P** without adding a structural commitment to one "foundational" or "fundamental" representation. It is not meant to solve all the issues of realist theories, but to give a structure compatible with a solution. The level of basic physical ontology in AR theory is very stripped down. The term "the physical world" can be viewed as a placeholder for whatever the physical world is, not identical with a representation of it. The aim here is to give a framework in which ontology and representation are considered separately, and the threads of multiple representations – physical, mathematical, computational, amongst others – can be teased apart. This avoids identifying fundamental ontology with whatever our favorite physical theory says it is this week. It also avoids having to choose one description as "more" fundamental than another (particles, fields, qubits, membranes, etc.). This removes the question-begging presumption of a relationship between different representations ("how can a collection of Fock spaces *be* a computer?") where there may be no such relationship. It also removes the issue in the foundations of computer science where a physical computer is frequently, and incorrectly, considered as identical with its concrete semantics.

By stripping the physical world of its representational structure, we are left with what might be termed the "fundamental problem of representation." **P** is supposed to be the set of physical objects without representational structure, yet to write down "**P**" is to violate this by giving it a representation. One response is to accept this necessary use of representation to perform communication as a Wittgensteinian ladder that is cheerfully kicked away after use. A less glib response is to take the notion of a naturalized representation and apply it to the whole system here: ontology is given by our best scientific theories, not just in what it is represented as, but in what is considered to exist physically.

This identification of that which is being represented as basic ontology necessarily opens us up to criticism by pathological case. For example, empty space is not represented in many physical theories but becomes a thing-in-itself in quantum field theories. A frequent response to these issues of realism is scientific empiricism: What is being represented are events of observation. It may well be that AR theory is compatible with such a view. In this case then the domain **P** changes depending on context: It is, as it were, representational turtles all the way down. This is the empiricist way van Fraassen (2008) views scientific representation. It is also worth noting the fit between the commuting-diagrammatic structure of AR theory, below, and Quinean notions of the "field of force" at the edges of scientific theories where they make contact with empirical reality (Quine 1951); the reader is invited there to view "the physical world" as a convenient shorthand for empirical observations, if this better fits their basic metaphysics. Otherwise, we note that any sane realist theory contains a physical domain, and we continue to consider in AR theory how this interfaces with the abstract representations that we use for it.

6.2.2   Prediction in Science

With the context established, let us see where such a framework can take us. First, let us consider a physics experiment on a system **p** represented by abstract model $m_\mathbf{p}$. The system evolves under some dynamics **H** as **H(p)** to become **p**′. We have a theory of this process **H**; call it C_T. Applying C_T to $m_\mathbf{p}$ we obtain $m'_\mathbf{p}$. We would like to test our theory by finding out whether $m'_\mathbf{p}$ corresponds to the result of the experiment, **p**′. We cannot compare $m'_\mathbf{p}$ to **p**′

directly; we first have to use the representation relation $\mathcal{R}_\mathcal{T}$ on $\mathbf{p}'$ to obtain $m_{\mathbf{p}'}$. This includes the process of measuring the outcome of the experiment. We can then ask, does $m'_\mathbf{p} = m_{\mathbf{p}'}$, and if not, what is the difference between them?

We do not need equality for $C_\mathcal{T}$ to be a good theory of the dynamics of $\mathbf{p}$. No theory is perfect, and we have limited precision for our experimental measurements. For $C_\mathcal{T}$ to be a good theory, we just need "close enough," say $|m'_\mathbf{p} - m| < \varepsilon$ for some suitable measure $|.|$ and suitably small ε. This process is illustrated in Figure 6.2, an example of an ε-*commuting diagram*, where theory and experiment agree to within some parameter ε.

Of course, this is a vastly simplified picture of the scientific process, in which many such diagrams interlock and underpin each other to build confidence in theories through many different and repeated experiments. We are most emphatically not claiming to have solved the philosophical questions of how science is done, only that something like this must be part of the story. All we need for our purposes here is that a "good theory," however established, can be described by a diagram like Figure 6.2. There are similarities between AR diagrams and others used for describing physical computation, especially those given by Ladyman et al. (2007), Maroney (private communication), and, in computer science, Abstract Interpretation (Cousot and Cousot 1977). However, AR representation (i) goes between physical and abstract, not functionally from abstract to abstract, (ii) is directed, and (iii) includes ε-closeness conditions.

Once we have a sufficiently good theory, we can use it to predict the behavior of physical systems. Figure 6.3(a) shows an ε-commuting diagram with both the physical system and the abstract model of its time evolution providing their outcome of the process. Figure 6.3(b) shows the use of the theory to *predict* the behavior of the physical system, without actually carrying out the physical process. From this, we can see that not only do experiments guide

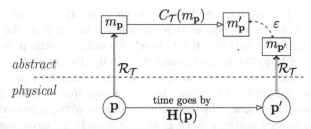

Figure 6.2 A simple experiment in which a physical system $\mathbf{p}$ evolves under some dynamics $\mathbf{H}$ to $\mathbf{p}'$. The corresponding abstract representations $m_\mathbf{p}$ and $m_{\mathbf{p}'}$ are obtained using $\mathcal{R}_\mathcal{T}$. The theory corresponding to the experiment $C_\mathcal{T}$ is used to calculate the expected outcome $m'_\mathbf{p}$, which is then compared with $m_{\mathbf{p}'}$, with resulting difference ε

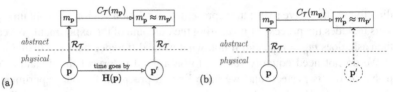

Figure 6.3 (a) A "good theory" has $m'_p \approx m_{p'}$; (b) this allows the outcome of a physical process to be *predicted*, without having to check by running the experiment

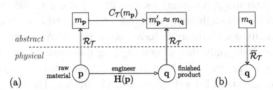

Figure 6.4 (a) The engineering process: making a **q** from a **p** using process **H** with theory C_T; (b) the instantiation relation $\tilde{\mathcal{R}}_T$ is a shorthand for the engineering process

the development of theories to explain their results, but theories can predict the outcome of experiments that have not yet been carried out, suggesting directions for new experiments. Both of these are observed as part of scientific practice. Moreover, good theories are not only used to guide further scientific research, they are also used to underpin new technology.

6.2.3 Technology and Engineering

Instantiation: Engineering an Artifact A good theory as illustrated in Figure 6.3 tells us the outcome of **H(p)** without our having to do the experiment. Furthermore, it tells us that we can reliably make a system **p**′ from a **p** using a process **H(p)** that we understand well through the theory C_T. This is *engineering*; again, we are not claiming to have solved the philosophical questions around engineering. Our ability to make things from detailed designs is unquestionable, evidenced by the multitude of high-tech gadgets available for purchase in their millions. What is most relevant to the development of our framework is that engineering effectively reverses the representation arrow $\mathcal{R}_T$, allowing us to instantiate theoretical objects in certain well-defined circumstances. This is illustrated in Figure 6.4(a), in which a product **q** is made from raw material **p**. The theory provides the method for the engineering process $H(p)$, and the abstract comparison $m'_p \simeq m_q$ verifies that the finished product q is sufficiently close to the theoretical specification. We can abbreviate this by the instantiation relation $\widetilde{\mathcal{R}}_T$, Figure 6.4(b), in which the abstract model m_q is instantiated as a physical object q. It is important to note that representation and instantiation are not symmetric processes: Making models that represent physical systems is easier than making physical objects that instantiate abstract models. In particular, it is possible to devise *unphysical* abstract models that have no possible real-world physical instantiation.

Using an Engineered Physical Artifact   Given a "good theory," we can use it to *engineer* systems, in concert with instantiation as described above, and then put them to use. For example, we probably want to test that our artifact does in fact conform to the engineering specification for its intended use. Figure 6.5 illustrates this process.

Figure 6.5(a) shows the ε-commuting diagram with both use and theory providing their outcomes of the process, allowing its suitability for the task to be checked. Figure 6.5(b) shows the artifact used to *predict* the outcome of the theory without actually carrying out the abstract calculations. This is what normal use of an engineered artifact corresponds to: Our confidence in the theory behind the engineering allows us to use the artifact without having to check it will do what we want it to do.

While the full diagram for engineering in Figure 6.5(a) looks superficially similar to the full diagram for science in Figure 6.3(a), there are fundamental distinctions. The first difference is the starting point of the process: note the instantiation arrow in Figure 6.5(a). For science, the starting point is a physical system to be modeled (represented) and understood. For engineering, the starting point is a problem encoded into an abstract model (engineering specification), to be engineered (instantiated) as the desired physical artifact. Science starts with the *physical* systems, engineering starts with the *abstract* models. The second difference is in the desired endpoint, or goal, of the scientific or

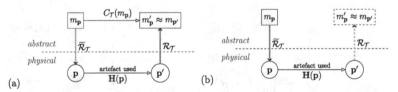

Figure 6.5 (a) A well-engineered system has $m'_p \approx m_{p'}$; (b) this allows the outcome of an abstract calculation (of the designed behavior of the physical system) to be *predicted* without having to check by doing the abstract design calculation

engineering process. For science, the goal is a theory that describes reality sufficiently well. For engineering, the goal is a physical artifact that meets the specified design. Scientific goals are abstract, engineering goals are physical.

As a consequence of these different goals, the response to an insufficiently small ε is different in science and engineering (once the possibility of a faulty experiment or incorrect specification has been eliminated). In science, when ε is too large, it means that the theory fails to adequately describe physical reality, and so the theory needs to be improved. In engineering, when ε is too large, it means that the engineered product fails to meet the theoretical specification, and so the physical object needs to be improved.

6.2.4   Computing Technology

Abstract Prediction   Computers are one type of physical system among the many and varied things that we engineer. However, they differ in one fundamental way from the engineered artifacts described above. For computing, the goal is to carry out an abstract computation that is (in general) unrelated to the details of the physical computer. We can now use our framework to address our original question about when a physical system computes. Figure 6.6 shows a physical system **p** carrying out an abstract computation C_T. Note that we are not here addressing what explicitly characterizes this abstract *computation* (for which see, for example, Piccinini [2015, 2017] on semantic accounts of physical computation); rather, we are addressing the question of when we can say that a physical system is *computing* an abstract *computation*.

In Figure 6.6, $m_\mathbf{p}$ is the encoding of our problem into a suitable abstract computational model; we discuss the encoding stage in more detail shortly.

Given this abstractly encoded problem $m_\mathbf{p}$, we instantiate it in the physical computer **p** and let it run. If successful, the result **p**$'$ can be inspected to obtain the abstract answer.

While the full diagram for using an engineered artifact in Figure 6.5(a) is identical to the full diagram for computation in Figure 6.6(a), there is again a

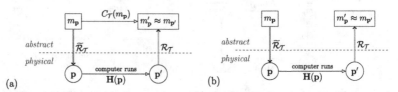

Figure 6.6 (a) A well-engineered instantiated computer has $m'_\mathbf{p} \approx m_{\mathbf{p}'}$ for computations it is capable of performing; (b) this allows the outcome of an abstract computation (the desired goal of computation) to be *predicted*

Table 6.1 *The essential differences between science, engineering, and computation: start: whether the starting point is a physical system* **p**, *or an abstract specification* m_p; *goal: whether the desired goal is a physical system* **p**, *or an abstract result* m_p; ε *too large: what part must be changed to make* ε *sufficiently small*

	start	goal	ε too large
science:	physical	abstract	change m_p
engineering:	abstract	physical	change **p**
computation:	abstract	abstract	change **p**

fundamental difference, and again this is in what is considered the goal of the process. For engineering, the goal is a physical object that meets the specified design. For computing, the goal is an abstract result of a computation. Engineering results are physical, while computational results are abstract. These differences are summarized in Table 6.1.

Performing a computation is a form of engineering where the desired result is an abstract representation rather than the physical system itself.

Encoding and Decoding Our initial discussion of physical computation above skips over some important details. Unlike the science and engineering diagrams so far, where the physical system **p** has been directly represented by the model m_p, we now have an abstract calculation that is initially unrelated to the physical computer or our model of the computer. The calculation problem may be the reason that the computer has been engineered in the first place (as with the earliest computers, or with modern specific-use devices), or it may be a new problem that the user has reason to believe is amenable to being solved on existing hardware. In either case, there is though no a priori connection between the abstract specification of the problem, c, and the abstract specification of the computer, m_p. This connection is to be found in the process of encoding. Figure 6.7 shows a physical computer **p** being used to do calculation c. Since c is in general unrelated to the computer we want to use, the first step is to map the calculation onto the model of the computer m_p. This *encoding* step includes checking that the computer is capable of representing the calculation (e.g., has enough memory and a suitable set of operations).

A modern programmable computer has a lot of existing programs (software, apps) to assist with the process of encoding a new problem, each of them running their own computations, resulting in many nested computational processes. It is easier to see how the basic encoding process works on a simpler computer, such as a pocket calculator. Suppose we have a restaurant bill for £93.47 that we need to divide equally between seven guests (thereby ignoring

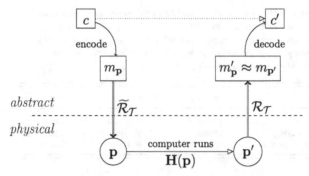

Figure 6.7   Using physical system **p** to carry out the abstract computation *c*

complications of who had the rice). The sequence of button presses 9 3 . 4
7 ÷ 7 = will, on most calculators, result in it displaying the answer. That
is an almost trivial encoding where the calculation can be entered straight into
the calculator.

In general the result will need to be *decoded* from the abstract representation
$m_{\mathbf{p}'}$ of the final computer state $\mathbf{p}'$. In the case of the restaurant bill, the decoding
is again straightforward. Reading the display you announce that each person
owes £13.36. The decoding you do is to interpret the array of seven segment
displays as a number, then add the "£" to that number to interpret it as an
amount of money. And you also rounded the amount up to a full penny, because
amounts of money smaller than this are not useful for settling a restaurant bill.

Encoding and decoding are important steps in physical computation. Indeed,
one of the characteristics of the associated representational processes is that
there is an element of choice about the encoding. The possibility of different choices implies both that the same abstract computation can be carried
out on different physical computers, and that the same physical computer
can in general carry out different computations. What is important is that the
computational semantics of the operations is given by the information that is
processed by them, and that there are different choices of the physical process
by which that happens. The possibility of different choices is one of the ways
by which we can see that computation is indeed happening.

6.2.5   Representational Entities for Computation

AR theory enables us to specify the elements of representation and the interplay of physical and abstract that happen during the use of computers by human
beings. It is not, however, restricted to this. A common, if unsatisfactory,
counter to pancomputationalism is to restrict the definition of a computing system to one used by conscious users via a definition of information (processing)

that is strongly semantic and intentional; see, for example, Bar-Hillel and Carnap (1953) and Mackay (1969).

AR theory does not do this. Within AR theory, the requirement is not for a human being, or even an entity that can think or communicate; the requirement is for representation. We thus do not need to take a position in the ongoing discussions of whether representation requires a thinking or human entity, and we refer to the compute cycle requirement as a *representational entity*. As we discuss in Section 6.2.1, we take a realist view: Whatever the ultimate ontology of "abstract" objects is, there is only ever access to represented objects through physical systems capable of supporting representation. The paradigm example is the human brain: We use representation to model physical objects in the world around us (including ourselves) as abstract notions. This representation happens, though, in the physical brain, for example, when a human uses a laptop computer. In such a case, we are the physical entities using the representation relation: We are the representational entities.

Thus, we have two conceptually distinct, but not necessarily physically distinct, physical entities. First, there is the physical object $\mathbf{p}$, as labeled in the diagrams in Section 6.2, that participates in representational activity (be it science, engineering, computing). Second, there is the representational entity $\mathbf{e}$ (denoted with bold font as, *ex hypothesi*, the representational entity must be physical) that supports the representation relation $\mathcal{R}_T$ it is using for $\mathbf{p}$. We say that the system comprising $\mathbf{p}$, $\mathbf{e}$, and $\mathcal{R}_T$ forms a *closed representational system*. If the cycle is a compute cycle, then this system forms a *closed computational system*: The *system* is computing.

In AR theory, the locatedness of the representational entity is important for determining the type of representational activity happening in a system. If the system comprising $\mathbf{p}$ and $\mathcal{R}_T$ does *not* include the physical representational entity $\mathbf{e}$, that is, if $\mathbf{p}$ and $\mathbf{e}$ are physically as well as conceptually distinct, then we say that the system is *open under representation*. In all the examples of human-designed computer use given above, the steps that go across the divide between physical and abstract (the representational and instantiational steps) all rely on a human representational entity. This entity is separate from the system that is the computer. So in human-designed computing, the computer alone (laptop, Difference Engine, slime mold, etc.) does not form a closed representational system: The representational entity is separate from the computing device, and not even necessarily co-located with it. Human-designed computers are open under representation, and require a human representational entity to close them.

Identifying the steps in a computation (or other representational activity) will identify the representational entity. Horsman et al. (2017) argue that it is possible to have a closed computational system without a human representational entity, analyzing the example of a bacterium performing chemotaxis, that is, changing its direction of motion towards a source of food. In such a closed computational system there are additional challenges to identifying the computational steps: A non-human representational entity cannot communicate to us that it is using representation. It therefore falls to us to determine if the system under consideration is itself using representation. That is, can we represent it as a closed representational system? We do not assign ourselves as the representational entities here: Rather, we determine whether we can describe the entire system as using the representational aspect of parts of itself in certain processes. A physical system that becomes representational only when an $\mathcal{R}_T$ is given by a human observer is *not* a closed representational system.

Thus the AR framework is able to discriminate between computing behavior and non-computing processes even in the absence of intelligent users or designers. The representation relation is always used by some entity, though, and it is that entity which is using the interfacing between abstract and physical that is a key part of physical computing.

6.3   Identifying Computing with the AR Framework

With the AR framework on board, we can use it to generate criteria that distinguish "computers" from other elements of physical reality. One of our original motivations for developing the theory is to provide a critical evaluation of proposed unconventional computational devices. Here we apply it to slime mold computation, which has received significant attention in the past decade. First, we show how the framework describes our familiar digital computers, and also one of the earliest computers, the Babbage Difference Engine. By identifying the components that make up processes that are known to be computing, we can then extend this into the territory of novel and unconventional computing.

Figure 6.7 shows the six essential components to an AR framework description of a computation: **theory**, **encoding**, **instantiation**, **physical process**, **representation**, and **decoding**. For each example, we first list what each of those components consist of, then discuss any issues that arise in identifying the components.

6.3.1   *Classical Digital Computing*

By classical digital computing, we mean the technology that underpins the computers we use on our desks, as laptops, in our smartphones, running the internet, and in many other types of technology, such as modern cars. We take this to be computing, uncontroversially. By showing how the components fit together to produce a compute cycle in this technology that is nowadays the paradigm example of computing, we pave the way for demonstrating it in non-standard devices.

Theory: The theory of classical computing covers the hardware (including how the transistors implement Boolean logic, and how the architecture implements the von Neumann model) and the software (including programming language semantics, refinement, compilers, testing, and debugging).

Encode: The problem is encoded as a computational problem by making design decisions and casting it in an appropriate formal representation.

Instantiate: Instantiation covers the hardware (building the physical computer) and the software (downloading the program and instantiating it with input data).

Run: The program executes on the physical hardware: The laws of physics describe how the transistors, exquisitely arranged as processing units and memory, and instantiated into a particular initial state, act to produce the system's final state when execution halts.

Represent: The final state of the physical system is represented as the abstract result, for example, as the relevant numbers or characters.

Decode: The represented computational result is decoded into the problem's answer.

Despite the complexity of today's computers, the underlying theory is highly developed and well-understood, the result of years of development and testing as each advance in functionality is introduced.

6.3.2   Babbage's Difference Engine

There is ongoing debate about the first "true" human-designed computer (with Stonehenge and the Antikythera mechanism, amongst others, vying for the title). The Babbage Difference Engine is one of the first recognizably modern computing devices. In particular, it was the forerunner of the Analytical Engine, the first proposed programmable machine with associated programming, given by Lovelace (1843) as she laid the foundations of modern computer science. The Difference Engine was, unlike the Analytical Engine, actually built. It computes tables of logarithmic functions by approximating them as sums of polynomials. It therefore needs to find the value of a function, e.g., $f(x) = x - \frac{1}{2}x^2$, for a whole set of values $x = 1, 2, 3, \ldots$ It uses the "method of differences" to change the problem into one of addition and subtraction, finding first the difference between subsequent function values (the "first difference"), and then the difference between subsequent values of the first difference (the "second difference"), and so on, a degree-n polynomial having n differences. Essentially it mechanizes a difficult calculation (a logarithm) by turning it into a relatively straightforward one (first into polynomials and then into addition). The addition is performed physically by combining the rotations from different cogs in the device, each of which is set to the required input.

Theory: The theory of the Difference Engine as a device comprises the theory of how a set of interacting gears can generate an addition function. Crudely, this can be thought of as combining the rotation of separate cogs (the inputs) onto other cogs (the output). The Difference Engine theory also includes how the addition operations for each difference combine to form the correct addition function to generate the next value of $f(x)$.

Encode: The problem (calculate a logarithm) is encoded as an approximation of sums of polynomial functions, which are in turn reduced to addition functions.

Instantiate: The computation is instantiated firstly in the engineering of the device itself: The hardware is not programmable. The input is then given by turning the input dials to the settings that correspond to the first n values (usually around four) of the function $f(x)$ to be determined.

Run: The Engine runs (powered by turning a crank handle). As the gears interact, differences are calculated and then added.

Represent: The final position of the output gears is coupled to a "printer," whose written numerical output depends on the position of the cogs.

Decode: The value of $f(x)$ for the specific x computed by the Engine is used to calculate the required logarithm (by hand).

The Difference Engine demonstrates how computation can occur without full programmability. The Engine can perform only specific addition tasks, given by its construction. Addition is "hard-coded" into the design of the physical device. The Engine performs a range of calculations by virtue of the fact that it can take different inputs. This is one way we can see that the device is processing information through its physical operation. The theory of the device was originally developed to design clockwork. However, small differences between the physical system and the model originally made the operation faulty, as errors cascaded through the system. While these are resolvable with modern precision-engineering (for example, the Difference Engine in the London Science Museum works with negligible errors), the much more sensitive Analytic Engine has still not been constructed with a physical gearing that matches the necessary precision of its theory.

6.3.3   Slime Mold Maze Solver

Adamatzky (2010) documents a range of computational uses for the slime mold *Physarum polycephalum*. Nakagaki et al. (2000) describe the observation of the slime mold finding the shortest path through a maze. That work is described in experimental terms: testing a theory of slime mold behavior in the presence of food.

The maze to be solved is implemented as a small physical structure, a few centimeters across, suitable for a slime mold to inhabit (Figure 6.8). The tested maze has multiple possible routes and several dead ends. Initially, the slime mold is grown to fill the whole of the maze. Then blocks of slime mold food, oat flakes in agar, are placed at the entrance and exit of the maze. The behavior of the slime mold is then observed over the next few hours. Having discovered the food, it withdraws from the dead ends of the maze and concentrates along the shortest path(s) between the two food blocks. After four hours, the slime mold has withdrawn from all the maze dead ends, but still exists along parallel alternative paths. After eight hours, the slime mold has withdrawn from the longer parallel routes and has found the shortest path.

Now having evidence for that theory, we can exploit the same process to *compute* the shortest path.

Theory: Slime mold forms a minimum-length body between food sources, as a consequence of the way its contraction frequency changes in the presence of food (for which see references cited in Nakagaki et al. [2000]).

Encode: If the abstract problem is to compute the shortest path through the maze, or simply any path through the maze, the encoding is essentially trivial: $c = m_p$. If the maze abstraction is a more indirect analog of some other problem, the encoding would be more complex. Analogs tend to be fairly direct encodings, exploiting a clear analogy.

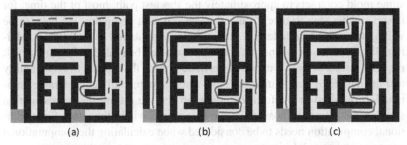

(a)   (b)   (c)

Figure 6.8 The slime mold *Physarum polycephalum* physically computing a path through a maze: diagram after Nakagaki et al. (2000). The dark areas are the maze walls; the pale regions are the maze. The gray squares indicate the position of the oat-flake-containing agar food blocks at the entrance and exit of the maze. (a) Multiple paths: On the left ("α routes") the solid path fragment is 20 percent shorter than the dashed path; on the right ("β routes"), the solid path is 2 percent shorter than the dashed path. At the initial state of the physical system **p**, the maze is filled with slime mold. (b) Half way through execution, after four hours: The slime mold (marked as a line) has shrunk and moved out of all the maze dead ends. (c) The final state of the system **p′**: After eight hours, the slime mold has found the shortest path

Instantiate: This has three main parts: build a physical maze that instantiates the abstract maze problem from materials supporting slime mold reconfiguration; place food sources (oat flakes) at the entry and exit positions; cover the maze with pieces of slime mold.

Run: Allow the slime mold pieces to coalesce into a single organism, and then wait for it to contract to the shortest path through the maze. The reported system took approximately eight hours to run, on a maze of approximately 4×4 cm.

Represent: Read off the final position of the slime mold in the maze, which requires the use of image processing to detect, and represent this as the position in the abstract maze.

Decode: Decode the abstract slime mold position into a route through the maze.

In this example of unconventional computation, we can see how the AR framework allows us to analyze the claims of computation.

There is first an issue at the level of the theory: The minimization is approximate. Nakagaki et al. (2000) report results from 19 experiments. In two cases no path formed. In three cases the slime mold did not fully contract, occupying all branches. When it did contract to a single route, for the β routes differing by only 2 percent, it chose the shorter route five times and the longer route six times. When it contracted to a single α route, which differ by 20 percent, it always chose the shorter route. So the theory would be better stated that the slime mold contracts to approximately the shortest path, most of the time, for mazes of this size. Hence, to use this system as a computer, we have to be willing to accept a quite large ε, and run the computation several times. Additionally, there is no evidence that the approach can be scaled to large mazes. It requires further scientific experiments to determine the domain of applicability and the degree of approximation. Note also the potentially considerable computation required in the form of image processing during the representation stage for detecting the position of the slime mold within the maze. Such additional computation needs to be considered when calculating the computational power of a physical device.

In this slime mold example, there is no clear distinction between computer (hardware), program (software) and input data (configured run of software). Moreover, the construction and programming effort can no longer be amortized over a potentially unbounded number of runs. Zauner and Conrad (1996) argue that one-shot "instance machines," which can solve only a single instance of a problem, have their advantages for certain substrates. Such machines avoid the need for resetting to some initial configuration, and so the compute step can irreversibly alter the state of the physical system, which is often a necessity when using a complex biological substrate. However, the cost of each instantiation needs to be low for this to be a viable strategy, significantly lower than the cost of a reusable computer for the same problem.

6.4   Distinguishing Representation(s) and Reality

We have seen in Section 6.3 how AR theory allows us to analyze specific systems for their computational activities. Separating out physical system, physical theory, and computational representations enables us to identify when all these elements, and the necessary connections between them, are present. This separation, and AR theory in general, can also be used to help design unconventional computer architectures, and to address some issues and claims about computation.

Both the abstract model m_p and the physical system **p** are essential components of AR theory, as are the instantiation and representation relations between them. Failing to properly distinguish which claims are about the abstract model and which are about the physical system can cause confusion. Confusion also arises by failing to differentiate claims about different abstract representations.

6.4.1   Pancomputational Rocks Aren't

Pancomputationalism is the view that everything – rocks, hurricanes, planetary systems, galaxies – are computing systems (Piccinini 2017).

The weak form of pancomputationalism holds that every system is computing (at least) itself. Consider the claim of a rock computing "itself" in the framework of AR theory, Figure 6.6. The rock is the physical system **p**. But in this context there is, importantly, no representational entity **e** and hence no representation relation $\mathcal{R}_T$. So there is no encoding of a computational problem, no relevant abstract model m_p representing **p**, no instantiation of that model as the rock, no representation of the rock's final state back to an abstract result, and no decoding of that abstract result into the solution of the problem. There is only the physical rock **p**, which does not form a *closed computational system* (Section 6.2.5).

One might try to argue that the rock is its *own* representational entity, that **p** = **e**. As we argue elsewhere (Horsman et al. 2017), in AR theory such a claim requires us to demonstrate representational activity occurring. As we show in that paper, this is a highly non-trivial process requiring the active and explicit use of representation intrinsic to a system's processes. Even for organisms such as bacteria it is controversial to claim representational activity. It is highly implausible that the criteria can be established for rocks. In the AR theory definition, systems do *not* compute themselves *for* themselves, because they do not represent themselves.

The strong form of pancomputationalism holds that every physical system performs a combinatorially vast number of computations, of every finite automaton that its microstate can encode through some tabular representation (Putnam 1988). Again, consider the claim of a rock computing one of these automata in the framework of AR theory, Figure 6.6. Again, the rock is the physical system **p**. Now there is a representational entity **e**: a person pointing at the rock, allegedly encoding their problem, and decoding the result, by using a representation of the rock's relevant microstate, establishing the relevant table defining the automaton.

Despite the existence of all the components in the theory in this case, there is nevertheless an issue. All the "computation" of establishing the mapping from rock states to table entries is being done in the representation stage: The rock *itself* has computed none of this. The representational entity could equally well have pointed to *any* rock; a different representational mapping would be needed, and would need to be computed in its entirety without the aid of the indicated rock. One might equally say that a broken clock is measuring the time. We observe the final state of the broken clock, but then we must use another clock to establish the correct representation of its broken state: *All* the measurement of the amount of time that has passed is being done in the representation (using a second clock); the broken clock has measured none of this. Again, the identified components **p**, **e**, and $\mathcal{R}_T$ do not form a *closed computational system*: A further computer is needed, and performs the totality of the purported computation. This contrasts with the case of a clock that is known to be, say, five minutes slow. Its physical state can be represented abstractly by **e** as one that is five minutes later in time than the standard representation would suggest. Alternatively, **e** could use the standard representation, and then decode the resulting abstract state by adding the five minutes. Here **e** is responsible for only a small, and well-determined, amount of computation to perform the representation and decoding.

So according to AR theory, there is no pancomputationalism, either weak or strong. For computation to occur at all, we require a representational entity, instantiation, and representation, in addition to the physical processes. For these identified representational entity, physical system, and representation relation to be sufficient to be performing the purported computation, they must additionally form a closed computational system. Rocks don't.

6.4.2   *The Representation is Less than Physical Reality: Side Channels*

The abstract model is just that: a model. It necessarily omits details about the physical system it is modeling, and may make simplifying assumptions. Other models, and their corresponding representations, are possible. Mathematical proofs are performed at the level of the model, and so concern only things in the model. If the physical system is richer, it can exhibit behavior not represented by the model, or the proofs. In particular, if a different model is used for a given physical system, with its own representation, different properties may hold than in the original model.

For example, a particular system, such as a crypto system, may be proved secure. Such a proof is performed at the abstract level and may depend on assumptions about the physical system, particularly assumptions about what can be *observed* (and so represented) about the physical system. Different models support different observations; hence these may break the assumptions underlying the mathematical proofs.

Such alternative observations in the case of security systems, for example timing observations (Kocher 1996), are called *side channels*, and many kinds exist (Clark et al. 2005). The original analysis of kinds of side-channel in Clark et al. (2005) was performed purely in the context of abstract-level refinement concepts, although physical issues were considered. AR theory can augment such analyses by exploiting its clear distinction between physical-to-abstract representation relations and abstract-to-abstract mathematical refinement relations, helping to expose where properties can be subverted and attacked.

6.4.3   The Representation is More Than Physical Reality: Hypercomputation

The AR theory diagram only "ε-commutes"; there may be small differences between the desired computational result and the physically computed result. Digital systems are designed to commute exactly: Because a physical gap is engineered between the instantiation of abstract 0 and 1, small errors in the physical system either do not lead to errors in the representation and decoding, or can be identified and corrected. In contrast, continuous analog systems are not exact and have no such gap, hence errors can propagate.

When deducing properties of a computational system, it is important to realize that the abstract model can have different properties from the physical system. In particular, abstract models of "continuous" systems often model state variables using real numbers. This does not mean that the physical system somehow "implements" such real numbers. That this is so can be readily seen in some cases. For example, Lotka-Volterra-style predator-prey models (Wangersky 1978) use a real-valued variable to model the population size using a continuum approximation. The population size is in reality a discrete quantity, and such models break down when the continuum approximation is no longer valid. For another example, the Banach-Tarski paradox (Wagon 1985; Wapner 2005) is a theorem that states that it is possible to take a sphere, partition it into a finite number of pieces, and reassemble those pieces into two spheres each the same size as the original; the proof relies on properties of the reals that cannot be exploited to double a physical ball of material made of discrete atoms.

An (abstract) real number potentially has infinite information content. Claims of hypercomputing (systems that can compute some non-Turing computable functions) and super-Turing computing (systems that can compute some Turing-computable functions with exponential speedup over Turing Machines) that rely on this infinite information content being physically accessible appear to be confusing the mathematical real number power of the abstract model with the physical capabilities of the modeled physical system (Broersma et al. in press). If that abstract content could be exploited physically, it would lead to these claimed forms of hypercomputing and super-Turing computing. However, there is no evidence that physical systems can do this: infinite-precision real-valued variables cannot be *instantiated* in a physical material system. Physical variables such as position and momentum are classically *modeled* using real values, yet according to our current best physical theories, in particular quantum theory, the physical world is ultimately discrete, and so the set of values these variables can take is ultimately countable, and do not form a continuum. The real numbers used in the abstract model are just that: a model.

AR theory, with its careful distinction between the physical system and its abstract model, helps us to analyze the claimed computational power of various systems, determining whether the power is part of the abstract model, the physical device, or the representation relation.

6.5   Summary

Landauer (1996) famously claimed that "information is physical." The physical nature of computing has been acknowledged by the unconventional computing community, but computer science in general has hitherto viewed its subject matter to be one of mathematics and logic, relegating the physical details of computing devices to engineering. As various forms of non-standard computing come to prominence, in particular quantum and internet-based technologies, this division has become increasingly untenable. Various fields from physics to biology to the social sciences have begun to import the language of information processing to describe their model systems in ways that are at odds with the usual foundations of computer science. AR theory allows us to bridge this divide, and to put the physical nature of computing devices back into the core of computer science, while also preserving its specific domain of applicability. Demonstrating the foundational part physical devices play in computing is not to extend the definition of "computer" to encompass every physical object. Computing is physical, but not everything that is physical computes.

The natural scientific description of a physical system and its computational description share many important properties. Both are model representations of the underlying physical system. They can relate to each other, and to the system being modeled, in a number of different ways. Carefully distinguishing these allow us to find when a physical system can support a computational representation, and when it is, in fact, being used as a computer. This is a complex process: A number of important criteria must be met, with, fundamentally, representation occurring and being used in specific ways in the physical system. Differences between physical system, physics-based models, and computational representations give rise to many of the problematic behaviors of computing systems: side-channel attacks, over-ambitious claims of computing power, and lack of clarity about what in a computer is computing and when. AR theory gives a framework in which all these claims can be defined and then analyzed. This opens the field for a formal computer science of unconventional devices, and for a new foundational understanding of the relationship between computation and the physical sciences.

Acknowledgments

DH and VK are supported by EPSRC under Grant EP/L022303/1. SS acknowledges partial funding by the EU FP7 FET Coordination Activity TRUCE (Training and Research in Unconventional Computation in Europe), project reference number 318235. We thank the various referees for their detailed comments, which have helped improve this chapter.

Part III

Physical Perspectives on Computer Science

7   Physics-like Models of Computation

Klaus Sutner

7.1   Introduction

The formal theory of computation, as developed by Church, Kleene, Herbrand, Gödel, Turing, and Post, stands isolated from physics, in part since mathematical theories of physics rely heavily on second-order or set-theoretic machinery such as the real numbers, Hilbert spaces, and the like. As it turns out, *classical recursion theory* (CRT), to use the old-fashioned term, even stands isolated from much of mathematics, including its discrete parts, in that there is little transfer of results and techniques. An extreme example in this context is Fenstad's (1980) axiomatization of computability. To be sure, there are some obvious connections to other areas of logic and in particular to proof theory. For example, a key notion in CRT is recursive enumerability: a set of discrete objects is recursively enumerable if there is an algorithm that generates exactly all the elements of the set (running forever if the set is infinite). The collection of all theorems of any axiomatizable theory is recursively enumerable and Feferman (1957) has shown that recursively enumerable sets of all possible degrees of complexity arise in this fashion. Also, the arithmetization in Gödel's incompleteness theorems requires some basic machinery from recursion theory. Modern computational complexity theory naturally derives from CRT, though it appears that many of the techniques break down. For example, on the one hand, one can check that all the standard results in CRT are invariant under oracles: one can add a hypothetical database to a computing device without affecting the basic theory. As a case in point, it will still be true in the presence of an oracle that there are recursively enumerable sets that fail to be decidable. On the other hand, it was shown by Baker, Gill, and Solovay (1975) that the infamous question of whether $\mathbb{P} = \mathbb{NP}$ does not relativize with respect to oracles: one can construct oracles where this identity holds, and others where it fails. However, none of these applications seems to engage anything like the full power of the methods developed in CRT over the last century. The only noteworthy exception to this claim appears to be Soare's (2004) work on differential geometry that relies on sophisticated results concerning the structure of the recursively enumerable degrees.

On the other hand, there are some undeniable connections to physics: there is a clear intuitive sense in which (some) computations are physically realizable in the sense than one can construct physical systems whose evolutions correspond to a given computation. Several models that are often more popular with theoretical computer scientists and complexity theorists, such as Turing machines, register machines, and cellular automata, are physics-like in this sense. Of course, all these models and the more purely logical ones are equivalent in a certain precise technical sense, but implementations of the former class are far more direct and they avoid tedious coding issues. The digital computer on which this paper was written is a perfect example of a physical system that is capable of executing a wide variety of computations. Alas, it is also of dubious use in our context, since it is hopelessly complicated to establish a precise semantics for the programming languages, compilers, run-time systems, and underlying hardware involved. More precisely, some physical systems can be construed to represent a particular data type, and the evolution of these systems under the standard laws of physics represents the manipulation of these data types associated with a computation. The point here is that the translations back and forth between the mathematical and physical system are fairly straightforward; in particular they do not carry the computational burden of the simulation.

Landauer (1991, p. 23) has made the following assertion:

Computation is inevitably done with real physical degrees of freedom, obeying the laws of physics, and using parts available in our actual physical universe. How does this restrict the process?

As he also points out (Landauer 1996, p. 188), this "amounts to an assertion that mathematics and computer science are a part of physics." As we will argue below, this view of computation appears to be rather too restrictive; indeed, in a sense, only a vanishingly small fragment of all logically possible computations can be realized in actual physics. It is difficult to see why all the rest should simply be dismissed. On the other hand, the idea to use physical computation as a source of inspiration for the logical study thereof is indeed potentially quite fruitful, and deserves more attention than it has received so far.

In order to bring some weight-bearing structure to the discussion, we will introduce a simple axiomatization of computability in a first-order system in Section 7.2. Unfortunately, at present this axiomatization has no counterpart in the realm of physics, so the notion of "physically realizable computation" remains somewhat vague, as discussed in Section 7.3. To be more precise, we can interpret computability in some subtheory of physics as long as this subtheory is consistent with the whole theory. Of course, this requirement causes problems with, say, classical Newtonian mechanics, which fails to account for relativistic effects. For the opposite direction there is no scaling back: If we want to interpret physics in terms of computability, we need a comprehensive theory of physics. To be sure, it is still of interest to work with fragments of the full theory, but ultimately there is no way around the great challenge that was first articulated by Hilbert (1900) in his famous address in 1900 at the International Congress of Mathematicians in Paris: Axiomatize physics. In Section 7.4 we propose two areas of computability theory that could ostensibly benefit greatly from some intuition derived from physics, small universal systems and intermediate degrees.

7.2   Models of Computation

The standard models of computation were developed starting in the 1930s: Church's λ-calculus, Herbrand-Gödel equations, Turing machines, Kleene's μ-recursion, Post's canonical systems, and Shepherdson-Sturgis register machines (Post 1943; Shepherdson and Sturgis 1963; Rogers 1967; Soare 1987; Odifreddi 1989, 1999). As far as their definitional power is concerned, they are all equivalent to each other in that they produce exactly the same class of computable functions, given any kind of reasonable input/output and coding conventions. Church's thesis captures this equivalence and points to the fact that computability is a rather robust notion. In fact, there are easily computable transformations that convert the representation of a computable function from any one of those systems to any other one. Bowing to history and Kleene's considerable effort to develop the theory of computability, we will refer to these maps as *partial recursive functions* and use the notion of computability in an informal sense; see Soare (1987) for a complementary perspective. While the models just mentioned are all ingenious, they are also ad hoc: the λ-calculus, for example, is a general theory of function application and abstraction, for which data types and computability appear almost as an afterthought.

Following Hilbert's suggestion, the axiomatization of mathematical theories has become the de facto gold standard during the twentieth century. Bourbaki's (1949) work is a colossal and all-encompassing effort in this direction, its bizarre logical framework notwithstanding. A rather general axiomatization of computability has been proposed by Fenstad (1980): in a nutshell, one considers a set Θ of triples (e, x, y) that are intended to mean that the partial recursive function with index e on input x returns output y. By imposing suitable conditions on Θ, one obtains a general framework for computability. For our purposes, this is a bit too ambitious; we are here not interested in generalized computability on, say, ordinals or other exotic set-theoretic structures, such as initial segments of the constructible hierarchy. Instead, we want to focus solely on CRT. One should mention that Fenstad's approach, and the formalization described below, are not without critics. Feferman (1977) in particular complained about the use of "coordinates" implicit in index-based descriptions. With a view towards physical realization this seems a non-issue, as any device implementing computable functions will produce some kind of coordinate, much like a program being executed on a digital computer.

Very concise axiomatizations that are more directly aimed at CRT were proposed by Blum (1967), Wagner (1969), Strong (1970), and Pippenger (1997). The idea is to consider *clones* of partial functions over the naturals, where a clone is any family of partial functions that contains all projections and is closed under composition in the following sense. A projection is a total function

$$P_i^n : \mathbb{N}^n \to \mathbb{N}, \qquad P_i^n(x_1, \ldots, x_n) = x_i,$$

where $1 \le i \le n$; these are needed solely for bureaucratic reasons. Given partial functions $g_i \colon \mathbb{N}^m \hookrightarrow \mathbb{N}$ for $i = 1, \ldots, n$, $h \colon \mathbb{N}^n \hookrightarrow \mathbb{N}$, we define a new function $f \colon \mathbb{N}^m \hookrightarrow \mathbb{N}$ by composition as follows:

$$f(\boldsymbol{x}) = h(g_1(\boldsymbol{x}), \ldots, g_n(\boldsymbol{x})).$$

For example, the projections themselves form a clone. To properly represent computability, one has to include certain basic functions in the clone. In this case, it is sufficient to include a monadic zero function, the monadic successor function, a dyadic pairing function π and an if-then-else construct such as the tetradic $\mathrm{ifte}(x, y, u, v) = u$ if $x = y$, and v otherwise. It is easy to see that simple subclasses of the partial recursive functions such as the Kalmar-elementary ones (Odifreddi 1989) or the primitive recursive ones satisfy the axioms so far. To reach full computability, one needs one additional requirement: the existence of a *universal* function, first introduced in Turing's seminal paper (Turing 1936). Following Kleene, universality can be expressed as an index theorem: There has to be a monadic function $\mathcal{U}$ such that, for each k-adic function f in the clone, there is some index $e \in \mathbb{N}$ such that for all arguments $\boldsymbol{x} \in \mathbb{N}^k$ we have

$$f(\boldsymbol{x}) \simeq \mathcal{U}(\langle e, \boldsymbol{x} \rangle).$$

Here $\langle . \rangle$ is a polyadic coding function on $\mathbb{N}^*$, the collection of all finite sequences of natural numbers, that is obtained from π by recursion.[1] Thus, the universal function can simulate the computation of any function in the clone, given a program for the function and the actual inputs. No such functions exist in the clones of elementary or primitive recursive functions, but Turing showed that they do exist in the realm of functions computable by Turing machines, and thus for partial recursive functions.

---

[1] This is a simplified version of the axiom. The actual technical definitions are slightly more complicated and avoid direct use of the polyadic map $\langle . \rangle$.

It takes some effort to show that the least clone defined by these conditions is indeed identical to the computable functions produced by all the other models. In fact, Kleene's recursion theorem, one of the more surprising results in computability theory, plays a major role in these arguments (Pippenger 1997). As is often the case, though, concise axioms are easier to interpret in other theories than more elaborate ones. In the context of physical realizability, the condition of being a clone is clearly satisfied: Projections are obviously realizable as physical devices, as is closure under composition: All that is needed is to collect the outputs of several devices and package them as input to another device. Owing to their great simplicity, the basic functions are similarly realizable, so the only problem is universality. But in the physical context, a universal device simply needs to be capable of feeding a given input to a given device and controlling the actual execution. Commercial digital computers could not exist without this feature.

In the axiomatic setting, universality is guaranteed by fiat. Of course, in concrete models, it may require a lengthy argument to establish the existence of a universal device (Turing 1936). One standard application of universality is undecidability: A Cantor-style diagonalization argument can be used to show that halting of a universal device is undecidable. This should be referred to as *strong universality*, as we exploit the fact that the device can effectively compute all partial recursive functions. In a sense, this is a bit of overkill, as was first pointed out by Davis in the 1950s (Davis 1956, 1957; Hooper 1966). For simplicity, let us keep the setting of Turing machines from the references, as the ideas easily carry over to other models. Davis suggests that a Turing machine is *weakly universal* if the set of all configurations of the machine on which the machine halts is recursively enumerable complete, and thus as complicated as the standard Halting Set. It is easy to see that every strongly universal machine is also weakly universal, but the converse implication fails to hold in general (as the machine may simply erase the tape, write 0, and halt, an example of information hiding; see also Section 7.4.2). Still, one can compute arbitrary partial recursive functions given the halting configurations as an oracle; Davis has shown that the mediating functions are very simple.

The conciseness of the axioms has another, less desirable side-effect: It entirely obscures resource bounds such as time or space. To be sure, by establishing effective equivalence to, say, the Turing model, one can introduce these bounds ex post facto, but it is preferable to have resource bounds appear as part of the axiomatization itself. Indeed, Pippenger has suggested to add an axiom that anchors Blum's (1967) approach to resource bounds directly in the axioms. More precisely, one insists that there is a dyadic complexity function Φ in the clone such that, for any k-adic function f in the clone, there is an index e such that the domain of f and $\lambda x.\Phi(e, \langle x \rangle)$ agree and the predicate $\Phi(e, \langle x \rangle) = t$ is decidable. The intent here is that $\Phi(e, \langle x \rangle) = t$ expresses the assertion that the computation of f on input x requires a certain amount t of resources. There is considerable leeway in what exactly is meant by a resource; the most obvious choices are time and space requirements for Turing machines, but we could also consider the length of a derivation of a Herbrand-Gödel system of equations. In the physical realm, energy consumption would be a plausible example.

Note, though, that there is substantial friction between axiomatic complexity and physical realizability. As a case in point, consider extremely rapidly growing functions that are, from the perspective of CRT, quite easy to compute. A classical example is Ackermann's function $A : \mathbb{N}^2 \to \mathbb{N}$ (or, more precisely, Rózsa Péter's version of the original, triadic function). As we will see, this function is easily computable in a physical sense, but only for very, very small arguments. The function is defined by recursion like so:

$$A(0, y) = y^+,$$
$$A(x^+, 0) = A(x, 1),$$
$$A(x^+, y^+) = A(x, A(x^+, y)).$$

Here a term of the form u^+ is an abbreviation for $u + 1$. Proof-theoretically, this function is exceedingly tame; e.g., a simple induction over ω^2 establishes totality, an induction that is easily carried out in Peano arithmetic. As far as intuitive computability is concerned, a rather too glib argument would be to point to the following program fragment, written in standard C: it properly implements A, at least if we ignore limitations of machine-sized integers.

```
int acker(int x, int y) {
  return( x ? (acker( x-1, y ? acker( x, y-1 )
    : 1 )): y+1 );
}
```

This is essentially just an application of the if-then-else operator together with a general form of recursion, easily justified abstractly in our axiom system, but much more problematic in the physical setting: quite a bit of bookkeeping is required to implement general recursions, a fact obscured in modern programming languages by the existence of powerful compilers. Here is an alternative algorithm that is more easily realized. A computation of $A(2, 1)$ might start like $A(2, 1) = A(1, A(2, 0)) = A(1, A(1, 1)) = A(1, A(0, A(1, 0))) = \ldots$. Note that the function name A and parentheses are just syntactic sugar. A more concise description of the computational process would be

$$2, 1 \rightsquigarrow 1, 2, 0 \rightsquigarrow 1, 1, 1 \rightsquigarrow 1, 0, 1, 0 \rightsquigarrow 1, 0, 0, 1 \rightsquigarrow 1, 0, 2 \rightsquigarrow 1, 3 \rightsquigarrow 0, 1, 2$$
$$\rightsquigarrow 0, 0, 1, 1 \rightsquigarrow 0, 0, 0, 1, 0 \rightsquigarrow 0, 0, 0, 0, 1 \rightsquigarrow 0, 0, 0, 2 \rightsquigarrow 0, 0, 3 \rightsquigarrow 0, 4 \rightsquigarrow 5.$$

We can recover the original expressions by applying A like a right-associative binary operator to these argument lists. Hence, we can model these steps by a function Δ defined on sequences of natural numbers.

$$\Delta(\ldots, 0, y) = (\ldots, y^+),$$
$$\Delta(\ldots, x^+, 0) = (\ldots, x, 1),$$
$$\Delta(\ldots, x^+, y^+) = (\ldots, x, x^+, y).$$

We know that A is total, so for any a and b there is some $t \geq 1$ such that

$$A(a, b) = \text{fst}(\Delta^t(a, b)).$$

Here fst extracts the first element of a list, and the halting time t is characterized by the fact that the length of the list after t steps is just 1. The one-step operation comes down to simple copy, delete, and edit operations and is clearly implementable. For example, one could exploit recent efforts to achieve very high density of information storage by placing chlorine atoms on a copper grid (Kalff et al. 2016). Under the proper circumstances, one can store, write, and read information in this setting, albeit with the help of a scanning electron microscope.

It is often convenient to think of the Ackermann function as a hierarchy of monadic functions defined by $A_x(y) = A(x, y)$. The first few levels of this hierarchy are easily understood: A_0 is the successor function, A_1 is "plus 2," and

$$A_2(y) = 2y + 3,$$
$$A_3(y) = 2^{y+3} - 3,$$
$$A_4(y) = 2^{2^{\cdot^{\cdot^{2}}}} - 3.$$

In A_4, the stack of 2's in the exponentiation has height $y + 3$; this operation is often referred to as super-exponentiation. But beyond level 4, things become quite murky and there is little one can say about A_{10}. Yet higher levels of the Ackermann hierarchy do occur in computability theory. One well-known example is the bound on the complexity of the union-find algorithm (Tarjan and van Leeuwen 1984) that differs from linear by an exceedingly slow-growing multiplicative value that can be described in terms of the inverse of the function $A_x(x)$.

Perhaps more compelling is the appearance of very fast-growing functions in elementary combinatorics that are associated with very simple algorithms, yet are entirely unrealizable in any physical sense. Here is an example due to

H. Friedman (2001). Consider the subsequence order on words, where $u = u_1 \ldots u_n$ precedes $v = v_1 v_2 \ldots v_m$ if there exists a strictly increasing sequence $1 \leq i_1 < i_2 < \ldots < i_n \leq m$ of positions such that $u = v_{i_1} v_{i_2} \ldots v_{i_n}$. In other words, one can erase a few letters in v to obtain u. By a famous result of Higman (1952), the subsequence order does not admit infinite antichains: Any collection of pairwise incomparable words must already be finite. This can be exploited as follows: For a 1-indexed word x, write $x[i]$ for the block $x_i, x_{i+1}, \ldots, x_{2i}$. A word is *self-avoiding* if, for all positions $i < j$, the block $x[i]$ is not a subsequence of $x[j]$. It is an easy consequence of Higman's theorem that every self-avoiding word is finite: Otherwise, the collection of its blocks would produce an infinite antichain.

Fix a digit alphabet $\Sigma_k = \{0, 1, \ldots, k - 1\}$ and think of the collection Σ_k^* of all finite words over Σ_k as the infinite k-ary tree: The empty word ε is the root, and every node x has k children xi, $0 \leq i < k$, obtained by appending another digit to x. Now consider the collection $T \subseteq \Sigma_k^*$ of all self-avoiding words. A prefix of any self-avoiding word is also self-avoiding, so T forms a subtree. Moreover, every node in T has at most k children, and T has only branches of finite length. By a combinatorial principle, known as König's lemma, T itself must be finite and thus have a longest branch. Write $\alpha(k)$ for the length of any such branch. Then $\alpha(k)$ is the maximum length of any self-avoiding word over any alphabet of size k, and we have a straightforward algorithm to compute this function. It is easy to show that $\alpha(1) = 3$; the algorithm shows that $\alpha(2) = 11$, as witnessed by the word 01110000000. But a lower bound for $\alpha(3)$ appears at level 7190 of the Ackermann hierarchy, thousands of levels away from anything that could be considered physically meaningful. Of course, this is but the tip of an iceberg: See Harrington et al. (1985) for many other examples of extremely fast-growing functions, all computable in the logical sense, but entirely unrealizable in any physical setting. Any kind of physical bound, be it space, time, mass, or energy, is easily dwarfed by the computations associated to these functions.

Computability theory has a number of speed-up results that coexist uneasily with physical realizability. A simple, machine-dependent version says that computations on a Turing machine can be accelerated by an arbitrary constant factor: In essence, instead of reading a single bit on the tape, one reads a whole block of bits. This is perfectly fine as long as the number of bits is 8 or 64, but becomes rather problematic when we try to read, say, $\alpha(3)$ bits at a time. Worse, a famous result known as Blum's speed-up theorem (Blum 1967) shows that some computations can be accelerated arbitrarily. Informally, the theorem says that there is a decidable set so that, for any algorithm E that determines membership in the set, there is another algorithm E' doing the same, but with an exponential speed-up over E. The actual theorem is significantly stronger: For any computable function f,

$$f(x, \Phi(e', x)) \leq \Phi(e, x)$$

for all but finitely many inputs x. If we choose f to be a very rapidly growing function like Friedman's α from above, it becomes clear that speed-up is not compatible with physical realizability – except if one exploits the hedge that finitely many inputs are exempt to rule out any actually realizable computations.

7.3   Physical Interpretation

To take the step towards a physical interpretation of computability, one needs to be a bit careful in choosing the right logical model as a starting point. On one side, the objects in Herbrand-Gödel, μ-recursion, λ-calculus, and Post systems can all be represented by the data type "string": We can ignore the semantics of, say, an equation and consider it to be simply a sequence of characters. Alas, a computation in any of these systems then requires some kind of rewrite system, which in turn is based on pattern matching and fairly sophisticated string manipulations. For example, in the Herbrand-Gödel setting we need to implement substitution (replace a free variable everywhere in an equation by a numeral) and replacement (replace a term $f(\underline{a_1}, \dots, \underline{a_k})$ by a numeral $\underline{b}$ if we have already derived $f(\underline{a_1}, \dots, \underline{a_k}) = \underline{b}$). This is similar to the list operation used above to evaluate the Ackermann function, but considerably more complicated in the general setting of unconstrained recursive equations that require more sophisticated bookkeeping.

It seems preferable to resort to machine-based models, either Turing machines or register machines. Some 80 years after their introduction by Turing, his eponymous machines are still the coin of the realm when it comes to computability and computational complexity. (Incidentally, it was Turing's work that convinced the ever-hesitant Gödel that the problem of defining computability was finally solved [Gödel 1995].) A configuration of a Turing machine is but a string

$$x_n x_{n-1} \dots x_1 x_0 \, p \, y_0 y_1 \dots y_{m-1} y_m$$

in $\mathcal{C}$, the collection of all words over the tape alphabet, plus one extra symbol p representing a state as well as the head position. The update operation (or relation in the nondeterministic case) is exceedingly simple, a so-called synchronous transduction that can be computed by a finite-state machine (Sakarovitch 2009). As a consequence, the first-order theory of this operation over $\mathcal{C}$ is decidable by automata-theoretic machinery. For example, we can check whether every configuration evolves in at most 5 steps to a configuration that has exactly 2 predecessor configurations. First-order logic turns out not to be expressive enough to deal with the transitive closure of the updates, and thus with the long-term evolution of a machine configuration, so all the usual questions regarding global questions remain undecidable. A physical realization of a Turing machine requires storage of a configuration in $\mathcal{C}$, presumably on some one-dimensional grid, and an implementation of the update operation. The latter is based on a finite state control, which is easily implemented, but one needs to be careful with communication between the control and the configuration; in general, they will become unboundedly separated in space (unless one wishes to move the control about, which seems less attractive).

Register machines, also known as counter machines (Minsky 1961; Shepherdson and Sturgis 1963), seem more suited to the task of physical realization. A register machine consists of a finite collection $R_1, \ldots, R_m$ of registers, each capable of storing a natural number. We write $[R]$ for the content of register R. The machine operates on these registers using a very spartan set of instructions:

- `inc r k`
  Increment register R_r, go to instruction k.
- `dec r k l`
  If $[R_r] > 0$ decrement register R_r and go to instruction k, otherwise go to instruction l.
- `halt`

Here we assume that the instructions are numbered from 1 to n, providing targets for the "go to" instructions. In fact, some computational hardness results do not even require the presence of a halt instruction, as explained below.

It is well-known that register machines with just two registers are weakly universal in the sense discussed in Section 7.2, though they fail to be able to compute the exponential function. Three registers suffice to produce a strongly universal, albeit somewhat cryptic machine (Minsky 1961). It is a straightforward exercise in program design to construct surveyable universal register machines with just a handful of registers and a few dozen instructions.

To implement a register machine, we again need to construct a finite state control that is responsible for executing the instructions; this can clearly be accomplished in a small volume of space for the universal machine from above. To represent the registers, think of a one-dimensional grid that is subdivided into m tracks. One end of the grid abuts the control, but the other end can be extended arbitrarily far. A pebble on each track indicates the current content of the corresponding register. To execute an increment instruction, it suffices to send a signal down the appropriate track, and to move the pebble accordingly;

lastly, an acknowledgment is sent back to the control. Of course, this may necessitate an extension of the supporting grid. To test whether a register is zero is a strictly local problem and can be handled by the control; if the register is positive, the decrement operation proceeds very much like increment. This is the unary version; one could similarly obtain an exponentially more compact representation by storing binary numbers on the grid.

There is an interesting variation due to H. Friedman (2008) of the idea of using the spatial location of pebbles to code memory content as well as state that makes it possible to decentralize the control unit. Friedman's automata also avoid explicit coordinates and are arguably more robust than the last construction. Formally, an *ordered partition automaton* (OPA) is defined as follows. By an ordered partition of length k of a set A we mean a list $P = (A_1, A_2, \ldots, A_k)$ of non-empty, pairwise disjoint subsets A_i of A whose union is A. The cardinality of A will be referred to as the rank of P. Thus, the list is ordered but its components are unordered sets. It is easy to see that the number of ordered partitions of rank n is $\sum_{k=1}^{n} k! \, S(n, k)$ where S indicates the Sterling number of the second kind. Hence, these numbers grow rather quickly, e.g., for $n = 12$ we have 28 091 567 595 partitions.

A configuration of an ordered partition automaton of rank n is a vector $C \in \mathbb{Z}^n$, indicating the positions of n pebbles on a discrete, one-dimensional grid. Several, and in fact all, pebbles may occupy the same position; this will not present a problem since the number of pebbles will be quite small. We can interpret a configuration vector as an ordered partition of $[n]$. For example, $C = (3, 1, 5, 3)$ gives rise to the partition $P = (\{2\}, \{1, 4\}, \{3\})$: The leftmost pebble is on track 2, next there are two pebbles on tracks 1 and 4, and the rightmost pebble is on track 3. Let us suppose that every pebble is associated with a lookup table $\sigma : \text{OP} \to \{-1, 0, 1\}$ that maps every ordered partition to an integer -1, 0 or 1; we will interpret $\Delta(P)$ as an instruction to move left, stay put, or move right. By collecting all these instructions, we can associate a configuration C with a vector $\Delta(C) \in \mathbb{Z}^n$, and we can update the configuration by adding this vector, that is to say, by moving all the pebbles as required by their tables. For example, the rank-2 system

$$
\begin{array}{lrr}
(\{1\}, \{2\}) & 1 & -1 \\
(\{2\}, \{1\}) & 0 & 1 \\
(\{1, 2\}) & 1 & -1
\end{array}
$$

generates simple, ultimately periodic orbits as seen in Figure 7.1 for the initial configuration $(1, 8)$. It is not difficult to classify the behavior of all rank-2 systems; even ignoring symmetries there are only 729 of them. A simple calculation shows that, for $n = 3$, there are already $3^{39} \approx 4.05 \times 10^{18}$ systems, and, for $n = 4$, this number increases to $3^{300} \approx 1.37 \times 10^{143}$.

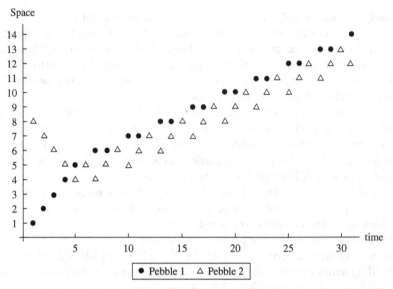

Figure 7.1 The evolution of an ordered partition system of rank 2. The movement of two pebbles for 30 steps is shown

To show that these devices easily lead to undecidable propositions, it is convenient to consider a *hybrid ordered partition automaton* that is augmented with a finite state control. The pebbles are now constrained to locations corresponding to nonnegative integers only. These hybrid automata can be shown to simulate Turing machines or register machines, with a small number of additional control states. Indeed, at the cost of a few additional tracks and pebbles, one can then eliminate the coordinate system as well as the states. Moreover, most of the pebbles can be required to remain in a spatially bounded area for all computations; it suffices to have only three free pebbles that can move arbitrarily far. In the resulting OPA, pebble positions are updated solely according to their relative positions and we obtain a more uniform system that is quite robust when it comes to synchronization and spatial arrangement. Friedman (2008) has shown that despite the simplicity of these systems, universality in the weak sense already appears at rank 12: specifically, it is undecidable whether the distances between all pebbles in an orbit remain bounded (and to show this, a halting state in the simulated register machine is not required). One interesting feature of Friedman's argument is that the description of the universal system is much smaller than the formal definition from above might suggest; the total size of the movement tables in direct form is approximately 3.37×10^{11}, but all the relevant rules can easily be specified in just a few pages. Pushing a bit further, Friedman even conjectures that for a mere 20 pebbles the question of whether an orbit is spatially bounded might be independent from ZFC. More precisely, the conjecture is that there is an OPA of rank 20 such that the question of whether the one-point initial configuration (which is the one in which all pebbles are in the same location) becomes unbounded is neither provable nor refutable in ZFC. Moreover, Friedman suggests that the description of such a system can be very simple, thus placing it close to physical realizability. Given the enormous power of ZFC, currently the standard formalization of mathematics (Bourbaki 1949), this may seem surprising, but note that computability theory can produce apparently bizarre results that wreak havoc on the intrinsic power of set theory. A prime example is based on Chaitin's Ω, the probability that a randomly constructed Turing machine will halt, subject to some technical conditions (Li and Vitányi 1997). Ω is a real number between 0 and 1, but it is exceedingly difficult to determine even a few digits. For example, knowledge of the first 10,000 digits would suffice to settle major open problems in number theory such as the Riemann hypothesis or the Goldbach conjecture. Solovay (2000) has shown that there is a set-theoretic universe that cannot determine even a single bit of Chaitin's Ω, one of the most counterintuitive results in computability theory. The argument depends on a very clever application of Kleene's recursion theorem.

As far as a physical realization of an OPA is concerned, Friedman's original proposal was to think of the pebbles as emitting some kind of "radiation." This seems dubious since the spatial excursions of some of the pebbles will be unbounded and it will be exceedingly difficult to discern their relative positions from an arbitrarily weak signal. In a more realistic setting, there is a fundamental choice to be made regarding the global control mechanism. On the one hand, one can keep the pebbles simple and resort to an external mechanism that scans the arrangement, determines the current ordered partition and then moves the pebbles accordingly. An alternate solution is to distribute the control by using "smart pebbles" that determine the current partition on their own. This requires a bit of global communication between the pebbles and some degree of local control, including synchronization. By Friedman's construction, the amount of local control required to achieve universality is quite small; there are no obstructions in the form of unachievably large local state sets.

7.4   Connecting the Fields

In 1900, Hilbert posed his famous twenty-three problems at the International Congress of Mathematicians in Paris (Hilbert 1900). It is difficult to overestimate the importance of Hilbert's challenge for the development of mathematics throughout the twentieth century. Problem number six reads like so:

6. Mathematical Treatment of the Axioms of Physics. The investigations on the foundations of geometry suggest the problem: To treat in the same manner, by means of

axioms, those physical sciences in which already today mathematics plays an important part; in the first rank are the theory of probabilities and mechanics.

Hilbert himself gave an excellent example of the type of construction he had in mind in the domain of planar geometry in 1899 (Hilbert 1950). As Hilbert points out, the problem is to conduct a "logical analysis of our intuition of space" by choosing simple, complete, and independent axioms. Frege considered Hilbert's methods substantially flawed, but they are nowadays universally accepted (Blanchette 2014). Kolmogorov (1933) solved the purely mathematical problem of axiomatizing probability theory very elegantly and concisely, though it is less clear how probability can be formalized in a physical context. Similarly, tremendous progress has been made on the physical part of Hilbert's sixth problem throughout the twentieth century. In fact, there are even first attempts to explore the logical consequences of some partial physical theories in a strictly formal setting, i.e., by verifying theorems in a proof assistant. Stannett and Németi (2014) have shown how to axiomatize relativity theory in the context of first-order logic and how to verify certain claims ("no inertial observer can travel faster than light") using the Isabelle proof assistant (Wenzel 2012). However, there is currently no clear way forward towards the unification of general relativity and the standard model of particle physics in a single logical framework. Both are heavily supported by experimental evidence and enjoy substantial predictive power, yet they appear to be strangely orthogonal. Moreover, unlike with computability theory, the mathematical machinery required to deal with physics is fairly complicated. As a case in point, E. Witten received the Fields Medal in 1990 for work on the interface between differential geometry and physics. Unless the proponents of a digital physics, such as Zuse (1967), Fredkin (1990), and Wolfram (2002) are right, it is clear that a careful formalization of physics will be significantly more complicated than a formalization of computability. And yet, to ignore physical computation as a source of insights and potentially fruitful methods for the study of pure computation would be a "Missed Opportunity" in the sense of Dyson (1972). We consider two such opportunities.

7.4.1   *Discovered Universality*

For particular computational models, such as Turing machines, we now have a fairly complete understanding of the frontier of undecidability (Neary and Woods 2006a,b). And yet, it was a physics-like model of computation that produced the first example of a universal system that was discovered, rather than designed: the elementary cellular automaton number 110 (Wolfram 2002; Cook 2004). The automaton is given by the triadic Boolean function $\rho(x, y, z)$ shown in Table 7.1.

Table 7.1 *The triadic Boolean function* $\rho(x,y,z)$

x	y	z	ρ	x	y	z	ρ
0	0	0	0	1	0	0	0
0	0	1	1	1	0	1	1
0	1	0	1	1	1	0	1
0	1	1	1	1	1	1	0

We can think of ρ as acting on bi-infinite sequences of bits by applying it synchronously to all overlapping blocks of length 3. Thus, if xyz is one such block, y is replaced by $\rho(x,y,z)$. (See Hedlund [1968] for the fundamental theory of these devices and Wolfram [2002] for a more recent account.) One can speculate about the reasons why this automaton is universal: interpret the second argument y as a control bit, and x and z as the actual inputs. For $y = 0$ we obtain a left-shift, whereas $y = 1$ yields the standard NAND operation. This is mildly promising since NAND is a basis for all Boolean functions. Of course, since we are dealing with a cellular automaton rather than a clone, the interconnections between the synchronous action of ρ on all cells are heavily constrained by geometry.

It is interesting that the argument requires a relaxation of the standard finiteness requirement imposed on configurations $C \in 2^{\mathbb{Z}}$, the space of all bi-infinite sequences, in the context of computability arguments: only finitely many bits are 1, so we are dealing with a finitary object. On these configurations rule 110 behaves in an entirely decidable fashion. Instead we consider configurations of the form $^{\omega}uwv^{\omega} = \ldots uuuwvvv\ldots$, where u, w, and v are all finite words, so-called almost periodic configurations. These configurations may seem somewhat artificial, but they turn out to be entirely natural: They provide the least elementary substructure of the full configuration space containing all finite configurations (Sutner 2013). Informally, some first-order statement about ρ is true in the full uncountable space $2^{\mathbb{Z}}$ if, and only if, it is true in the countable subspace of almost periodic configurations. Given the dynamics of rule 110, these configurations barely suffice to encode combinatorial structures called cyclic tag systems (Post 1943; Cook 2004) and thus establish universality. At present, the proof by Cook is based on the notion of particles and their interactions, the particles being represented by clusters of bits in the automaton, and the interactions being determined by rule 110. The geometric ideas of the proof are quite compelling, but it would be most reassuring to construct a more formal argument in a proof assistant. After all, the utter lack of continuity leaves geometric arguments a bit fragile.

7.4.2   Intermediate Degrees

Semi-decidable sets appear to come in just two types: those that ultimately turn out to be decidable, and those that turn out to be complete, i.e., as complicated as all other semi-decidable sets. Needless to say, the existence of the latter type can easily be established via universality. In particular the Halting Set is easily seen to be complete. Post asked whether there might be other, intermediate semi-decidable sets: too complicated to be decidable, yet not as complicated as Halting (Soare 1987). The question is natural enough and has become known as *Post's Problem*; it was answered in the affirmative by Friedberg and Muchnik a half century ago (Muchnik 1956; Friedberg 1957; Soare 1972). Yet the only direct construction known to produce an intermediate set, a so-called *priority argument*, is highly technical and, in a sense, artificial; worse, it produces a solution that appears to have no connection to any other object in computability theory, as pointed out by Davis: There is no established and known computational problem whose complexity has turned out to be intermediate; all natural examples are either decidable or complete (though the proof may take seven decades, as in the case of Diophantine equations). From a proof-theoretic perspective, the priority argument establishing the existence of intermediate sets can be carried out in a rather weak system of arithmetic: essentially just a bit of elementary arithmetic plus induction is needed to justify the argument. And yet, despite the logical simplicity, the results are artificial; the sets so constructed serve no other purpose than to establish a positive solution to Post's Problem.

In a nutshell, the construction uses the well-known equivalence between semi-decidable sets and recursively enumerable sets (Soare 1987). To effectively enumerate an intermediate set, it is convenient to enumerate two sets A and B in parallel, subject to the following constraints, which are referred to as *requirements* in this context:

$$(R_e) \quad A \neq \{e\}^B, \qquad (R'_e) \quad B \neq \{e\}^A.$$

Here $\{e\}^B$ is the eth partial recursive function relative to oracle B and we abuse notation slightly by identifying the sets A and B with their characteristic functions. In other words, we add the characteristic function of B to the basic functions in our description of the clone of B-computable functions. This idea of computation relative to an oracle was first promoted by Turing (1939), in a paper on ordinal-based systems of logic that appears to have caused much confusion among those interested in hypercomputation (Davis 2006; Soare 2009). Alas, Turing never explored his oracle machines in any depth and it took several years before Kleene and Post (1954) showed how to use his framework to introduce a partial order on decision problems.

Note that there are infinitely many requirements and that they are all based on undecidable propositions as stated. To overcome this problem, the construction is organized in ω-many stages σ and uses finite approximations A_σ to A and B_σ to B at each stage σ. Moreover, the computations in the requirements are also constrained to at most σ steps, a standard device in computability theory (Rogers 1967; Soare 1987). This renders the approximate conditions trivially decidable. The only permissible action is to place a new element into A or B, so the resulting sets are indeed recursively enumerable. Unfortunately, the requirements may well clash: We may want to place an element into B to achieve some $R_{e'}$, thus violating another R_e. Resolving these conflicts so that the constructions succeeds in the limit (i.e., ultimately all requirements are satisfied) requires a fairly delicate and combinatorially sophisticated construction.

Now consider the priority argument from the perspective of an actual physical implementation. Lerman (1973) proposed a pin-ball-like device with the intent of making the construction a bit more intuitive. Let us stick instead with the ordered partition model from above. It is straightforward to adapt the machinery to the construction of a recursively enumerable set: One binary pebble can indicate the times when a new element has been produced, and another pebble then indicates the corresponding numeral. The geometric setup can easily be described using Hilbert's (1950) axiomatization, and in particular the Archimedean axiom 8.V which says in essence that a line segment of any finite length can be covered by a sufficiently large number of line segments of unit length. Note, though, that a physical realization of the grid underlying Friedman's device does not require equidistance as in the axiom, as a fairly rough approximation thereto will be entirely sufficient.

An external observer of this OPA, focusing solely on the two critical pebbles, would indeed conclude that knowledge of their movements in the ω limit would be insufficient to decide the Halting Set, yet there would be no decision algorithm for the membership problem of its constructed set. As it turns out, we have to willfully blindfold the observer to achieve this effect: If the observer had knowledge of the movements of all pebbles, he would indeed be able to decide the Halting Set. In other words, the physical process that implements the priority construction uses the full power of computational universality, but then simply hides part of the information generated to create the illusion of intermediate complexity. The idea of including the observer in the evaluation of a computational process can help to pin down this problem. Specifically, we would like to obtain an intermediate computational process in the sense that it admits an observer who concludes that the process is undecidable; yet there is no observer who can extract the Halting Set from the process (Sutner 2011). For this to make sense, the observers have to be computationally weak – say, finite-state machines. It is entirely plausible that in this setting intermediate processes indeed fail to exist, though it is currently unclear how one might go about proving this assertion.

7.5   Conclusion

The logical aspects of computation are easily formalized, leading to axiom systems of complexity comparable to, say, Hilbert's formalization of planar geometry, or Peano's formalization of arithmetic. No such claim can be made for physical computability, and it is difficult to see how this could change in the near future. As Tegmark has quipped, "It's not enough to be right in physics. One must be right for the right reasons" (Sarfatti 2004, p. 75); at present it is not really clear what the right reasons might be. The lack of a truly solid foundation on one end of the bridge leads to a surprisingly high level of controversy; consider, for example, the endless debate about hypercomputation and its physical realizability, or lack thereof. The best we can do at present is to consider a fragment Γ_0 of the ultimate theory Γ, such as Newtonian mechanics, relativity theory, and so on. One can then attempt to construct interpretations between computability and Γ_0, an endeavor that is of interest in both directions. The notion of a fragment here is not meant in the technical sense of a subtheory; rather, it is supposed to suggest a description of some aspect of physical reality, which description may well be inconsistent with Γ in the logical sense. Similar caveats apply to the notion of interpretation. Of course, some choices of Γ_0 may lead to perfectly absurd results from the perspective of realizable computation; for example, Beggs and Tucker (2007) show that in a strictly Newtonian system all functions are "computable."

On the other hand, the abundance of very rapidly growing computable functions suggests strongly that physically realizable computation is bound to cover only a small portion of the abstract theory. There is no harm in this, and it is fair to say that this particular portion is precisely the one that is most relevant to applications such as theoretical computer science and complexity theory, as evidenced, for example, by the tremendous interest in quantum computation in the last two decades. Exponential speed-up is irrelevant for general computable functions such as the fast-growing ones from Section 7.2, but of fundamental interest for more mundane tasks such as the factoring of integers. However, we would argue that even the general theory of computation can benefit from a study of physics-like computation. One case in point is reversible computation (a topic we could not address in this space), another, the discovery of small universal systems. The universality proof of the elementary cellular automaton number 110 is based on the interpretation of clusters of bits as "particles" that move about and interact in a fairly complicated manner, an idea clearly borrowed from physics. It would be interesting to construct a strictly formal proof of this result, one that could be verified by a theorem prover. The further study of highly stylized and simplified models such as Friedman's automata may also help to further clarify the boundary between computability and the lack thereof. If one thinks of a computation as a physical process, it is quite natural to introduce an observer that tries to extract information from this process. As indicated in Section 7.4.2, this perspective may help to explain the disconcerting absence of natural problems of intermediate degree.

8   Feasible Computation: Methodological Contributions from Computational Science

Robert H. C. Moir

8.1   Introduction

The focus of the standard model of computability is effective calculability, a mechanical method for computing values of a function on the natural numbers or solving a symbolic problem. The Church-Turing thesis then states that this informal concept is equivalent to, or at least correctly rendered as, computability by recursive functions (Copeland 2015). It is this notion of computability that underlies the standard definition of a formal system as a recursive axiomatic system, viz., a system of expressions whose formulas (e.g., theorems) and proofs are effectively decidable. Indeed, this proof-theoretic conception of an axiomatic system was viewed by Gödel as the only permissible interpretation of "formal system" (Sieg 1999). Thus, this formal understanding of mathematical theories represents mathematical proof as a form of computation. Viewed in relation to the common view in philosophy that non-ampliative scientific inference can be faithfully represented as logical inference, this common view regards much of scientific inference as computational.

Whether computability is modeled in terms of recursive functions on the natural numbers or in terms of manipulations of symbols or words, we can view computation as involving a map between an input and an output, so that computation in general can be regarded in terms of function evaluation. Since mathematical problems can be regarded in the same way, as maps between certain input data and output solutions, problem-solving can also be regarded in terms of function evaluation. Given, further, the connection between inference and formal systems, scientific inference can also be regarded as function evaluation. Accordingly, underlying computation, problem-solving and inference is always a form of function evaluation. Consequently, we will move between these perspectives throughout this chapter when it is useful to do so, keeping in mind that computing values of a function underlies each of the notions.

It is traditional to think of computation (and the associated concepts of problem-solving and inference) in exact terms, so that a computational problem is concerned with computing the exact value of a function. In contexts where it is known that exact solutions to a problem exist, the question becomes one of whether there are effective (viz., finitely algorithmically computable) methods to solve them, if so, whether efficient algorithms exist (in the sense of computable in polynomial time/space), what the complexity of available algorithms is, the minimum-complexity among these, etc.

A consideration of computation in scientific practice raises somewhat different complexity concerns, and what counts as an acceptable complexity can vary depending on the dominant constraints in a given context, which can include mathematical, software, hardware, energy, and financial constraints, among others. Thus, there is a subtler condition on computational complexity concerning the ability to compute solutions to a problem in a manner that fits the constraints imposed by the context in which those solutions are to be used in practice. We will call an algorithm that has this property *feasible* and problems for which such algorithms exist *feasibly computable*. A problem is *feasibly uncomputable* when no feasible algorithm exists.[1] Feasibility involves not just finding very low-complexity algorithms, since it is also an epistemological concept, having to do with methods we have epistemic access to. A feasibly uncomputable problem might become feasibly computable in the future if new analytic methods, better algorithms, or improved technology are developed. For example, the Berlekamp-Zassenhaus algorithm for factoring polynomials over the integers was not feasible for computing factorizations when it was first developed in the 1960s, but became feasible with subsequent advances in computing technology.[2] Contrariwise, a problem might be feasibly uncomputable yet can be feasibly approximated.

The field of computational science is concerned with generating feasible algorithms to solve mathematical problems, usually those that are important in scientific applications. An important difference between such feasible algorithms from traditional algorithms considered in computability theory is that, in general, they involve various forms of approximation. We will see that there is a common strategy in computational science that can take a problem that is not feasibly computable, or not sufficiently feasible in the sense that the computational complexity is too high, and then generate a (more) feasible algorithm to a slightly modified problem. This strategy to produce feasible algorithms itself has an algorithmic structure, involving a recursive process of complexity-reducing transformations of the problem into a feasibly computable problem, followed by a recursive back-interpretation of the solution, a process that can itself require problem-solving. We will see how one, more general, version of this strategy underlies numerical computing, which

[1] The concept of "feasibility" intended here is an informal concept, like "effective calculability." Given the inherent contextuality of the concept, it is likely to be resistant to the kind of global formalization that "effective calculability" admits, though useful piecemeal formalizations may be possible.
[2] Thanks to M. Moreno Maza for this example. For details on the algorithm, see Geddes et al. (1992).

uses approximations, and how a more restricted version underlies symbolic computing, which is exact. The nature of this feasible computation strategy in these two branches of computational science has some consequences for computability theory, which we consider in the final section.

Before we introduce this feasible computing strategy, we will consider its roots in the history of science. Various methods of problem-solving and theory-building in science involve reformulations of mathematical problems, reformulations used either to make solutions (more) accessible or to make a successful (more) general theory available. Where these reformulations lead to equivalent or nearly equivalent problems, they present possibilities for methods to make solutions (more) feasibly computable. Indeed, many of the methods of computational science utilize historically established exact or near-exact equivalences of problems. Though the exact version of the feasible computing method arguably has its roots in pure mathematics, we trace the historical origins of the more general, approximate form of the method to techniques developed by physicists to overcome the computational limitations of the mathematical formulation of theories and models of natural phenomena. We find in this process reasons why the feasible computing strategy is likely to be replicated widely in scientific practice, even outside of the mathematical sciences, specifically because the motivation for the method is epistemological: Since scientists are constantly confronted with the computational limitations of their own conceptual tools, they have to find reformulations of problems that make reliable inference feasible.

It emerges, therefore, that feasible computing is a fundamental part of a great deal of scientific inference, as well as at the core of advanced algorithms for solving problems in computational science. Moreover, by revealing an algorithmic strategy of converting feasibly uncomputable problems into feasibly computable ones, we show how computational science extends beyond the traditional theory of computation and is revealing new aspects of computability, aspects of fundamental importance to scientific inference and practical scientific computation.

8.2   Approximate Problem-Solving in Physics

The ability to calculate is a fundamental part of the scientific process as a result of the need to work out the consequences of our theories for real-world phenomena. Much of the time there is a tension between minimizing the complexity of theoretical inference and minimizing the complexity of application of a theory. A simple example of this contrast is the situation in logic, where for (meta)theoretical purposes we work with the most limited possible set of axioms and inference rules – often just *modus ponens* – but for the purpose of doing proofs within a system we want as rich a system of inference rules as possible to simplify the proofs.

In the context of scientific practice, this need to minimize the complexity of theories can lead to a gulf between theory and the ability to describe the phenomena. An extreme example of this occurred in the history of fluid mechanics. Although the standard equations of fluid motion were developed in the eighteenth and early nineteenth centuries by Euler and Navier, based on work by many others, the ability to draw consequences from these theories for realistic fluids was extraordinarily limited. This led to statements like d'Alembert's paradox, which stated that perfect fluids in steady motion exerted no force on fully immersed bodies (Darrigol 2005). It took an extensive development over almost two centuries, involving both theoretical innovations and deep investigations into physical fluid behavior, to resolve this gap. The theoretical development was driven by attempts to apply the theory (Darrigol 2005, ch. 8), and this required novel use of approximations in the construction of the theory itself. Examples of this are modifications of the equations of motion (e.g., adding higher-order terms), allowing discontinuities (e.g., shock waves), and asymptotic behavior (e.g., Prandtl's boundary-layer theory). Thus, approximation is not only important in terms of getting numbers out of known theories or models, but is part of theoretical inference and the process of theory development itself.

Determining the consequences of a theory for the behavior of the phenomena it describes requires generating solutions of its equations in realistic situations. Though exact solutions are extremely valuable, they are only available in very rare cases, and typically only for highly idealized or controlled situations. Since it is a matter of great priority to be able to know what a theory says about the phenomena, it becomes essential to use approximation methods to be able to gain information about the consequences of theoretical models. Given that applying a theory generally requires the construction of models, which introduces forms of error, it is usually justifiable to solve modeling problems using approximation methods provided they introduce less severe forms of error than those introduced in model construction. For these reasons, the history of physics is full of strategies for effective approximation. Three important general classes of approximation methods, namely, numerical methods, asymptotic methods, and perturbation expansions, all have origins in the need to complete inferences approximately that cannot be completed exactly. Moreover, they involve a strategy of *modifying the problem* into one where inference becomes feasible.

To illustrate the nature of these methods as simply as possible, we will consider how they emerge out of series expansions of functions. This actually captures the nature of these methods very well and is indeed how they emerged, since the method to develop an expression into an infinite series occurred very early. A convenient starting point is Taylor's theorem, announced by Taylor in 1712, though it appeared in a manuscript of Gregory in 1671, which was developed out of Newton's forward difference interpolation formula (Struik 1969, p. 332). Taylor's theorem allows one to expand a (suitably differentiable) function $f(x)$ about a point a into a power series

$$f(a + h) = f(a) + f'(a)h + \frac{f''(a)}{2!}h^2 + \frac{f'''(a)}{3!}h^3 + \cdots,$$

where $h = x - a$. This is useful theoretically, since, among other things, it can allow one to compute integrals of functions that are not known by expanding the integrand into a series, integrating term by term, and summing the result. But because h^n goes rapidly to zero as $x \to a$, for well-behaved functions it also provides a very useful algorithm for approximating the value of $f(x)$ near a, i.e., when $|x - a| \ll 1$. Thus, by truncating the series up to a certain order (power of h), we can obtain local approximations of $f(x)$. In so doing, we substitute the function $f(x)$ with a nearby function equal to the truncated series.

It is precisely this sort of strategy that underlies the use of numerical methods to solve differential equations. The simplest such numerical method, attributed to work of Euler in 1768–1769 (Goldstine 1977, p. 141), is motivated simply on the basis of a Taylor expansion of the solution truncated after first order. Consider a differential equation with an initial condition (called an initial value problem) of the form

$$y'(t) = f(y, t), \quad y(0) = y_0, \tag{8.1}$$

where we interpret the independent variable t as the time. We can "take a time step" by approximating the value of the solution at $t = h$, for $0 < h \ll 1$, in terms of known quantities by truncating the Taylor expansion of $y(t)$ about the point $t = 0$, yielding

$$y(h) = y(0) + y'(0)h = y_0 + f(y_0, 0)h,$$

where we have substituted Eq. (8.1) for $y(0)$ and $y'(0)$. Since the initial time is $t_0 = 0$, the values of t and y after the first step are $t_1 = h$ and $y_1 = y(h)$, and we can iterate this procedure, computing the value of $y_{n+1} = y(t_{n+1}) = y((n+1)h)$ in terms of the recurrence

$$y_{n+1} = y_n + f(y_n, t_n)h, \tag{8.2}$$

which is Euler's method for a constant step size h. If we imagine connecting the points (t_k, y_k) by straight lines, then we generate an approximate solution to Eq. (8.1) as a polygonal arc.

There are two important features to note about this procedure. The first is that, under mild conditions on the function f in a region around $(0, y_0)$, as $h \to 0$ the polygonal arc converges to the solution of Eq. (8.1), as was shown by Cauchy prior to 1830, although Euler had the essence of the method more than fifty years earlier (Kline 1972, p. 717). Thus, numerical methods provided an early tool for demonstrating the local existence of solutions to differential equations. The second is that it provides a very simple way of estimating the behavior described by a differential equation, even if the nonlinearity of the function f makes analytic solution impossible. To effectively control the error, however, the size of h needs to be very small, making calculation by hand tedious. Numerical methods were nevertheless sometimes used, particularly to avoid tedious perturbation methods (Linton 2004, p. 405). Although prior to the development of numerical methods for differential equations, in 1757–1758 Clairaut, with the aid of two associates, took months to perform effectively the first large-scale numerical integration to predict the return of what thereafter became known as Halley's comet (Linton 2004, p. 305). The subsequent nineteenth- and early twentieth-century development of numerical methods for differential equations by Adams, Heun, Runge, and Kutta was motivated by the desire to have more computationally efficient methods for applications in celestial mechanics, thermodynamics, and ballistics calculations (Goldstine 1977, p. 286).

The methodological structure of this kind of technique is important for us to note. Given the difficulty in solving differential equations analytically, the need to obtain information about the consequences of differential equations for the behavior of real-world phenomena motivated the development of accurate and efficient computational procedures (numerical methods) to compute approximate solutions. The computational procedure involves a modification of the problem, which here requires moving from a differential equation (8.1) to a difference equation (8.2), for which solutions could be computed (or approximated)[3] feasibly. The (approximate) solution of this modified problem then provides an approximation of the solution of the differential equation, giving insight into its consequences for behavior. Thus, an inferential obstacle at the level of theory is overcome by clever use of approximation.

Even before Brook Taylor proved the famous theorem bearing his name, scientists developed functions into series expansions, many of which were worked out by Newton and Johann Bernoulli. The methods for expanding into series emerged out of the theory of finite differences and interpolation, themselves developed for the purposes of computing numerical values for logarithmic and trigonometric tables (Goldstine 1977). Since such methods provide

---

[3] As we will see in the next section, the difference equations of numerical methods often cannot be computed exactly but give way to an iterative strategy for approximate solution.

means for generating *infinite* series expansions, which can convert mathematical problems involving transcendental functions to equivalent calculations on polynomials, infinite series were an important tool from the inception of the differential calculus. Indeed, Newton relied heavily on such methods.

A number of difficulties arise from this approach, however, and principal among them is the question of when, and in what sense, one can identify a function with its power series expansion. For example, Newton in 1665 and Mercator in 1668 (Kline 1972, p. 354) both obtained the series

$$\log(1 + x) = x - \frac{x^2}{2} + \frac{x^3}{3} - \cdots ,$$

which was observed to have an infinite value for $x = 2$, even though the function value is $\log(3)$ (Kline 1972, p. 437). It was therefore recognized early on that series expansions needed to be treated with care. Responses to the risks of divergent series were varied. Some, such as d'Alembert, singled out convergent series and were suspicious of the use of nonconvergent series, and the first tests of convergence were introduced in the eighteenth century by Maclaurin, d'Alembert, and others. Others, however, and Euler in particular, recognized that nonconvergent series could be useful as representations of functions, so that expressions such as

$$\tfrac{1}{2} = 1 - 1 + 1 - 1 + \cdots$$

could be not only meaningful but useful in algebraic manipulations even though the sum of the series lacked an arithmetic meaning in the sense of the terms adding up to the correct value of the function. Nonconvergent series were not necessarily devoid of arithmetic content, however, and specifically divergent series often proved useful for computing accurate approximate values of functions.

In a 1754/1755 work on divergent series (Kline 1972, pp. 450–451), Euler considered the differential equation

$$t^2 y'(t) + y(t) = t,$$

which has the solution

$$y = e^{1/t} \int_0^t \frac{e^{-1/x}}{x} dx. \tag{8.3}$$

Euler observed that the divergent series

$$y = t - (1!)t^2 + (2!)t^3 - (3!)t^4 + \cdots$$

also satisfies the differential equation *and* that the sum of a small number of terms of this series gave a good approximation to the value of the integral, even though the series diverges. This peculiar property has to do with a particular relationship between partial sums of the series and the value of the integral. It turns out that this relationship can be understood in terms of approximation, which explains why divergent series can nevertheless be useful for approximation.

This behavior can be understood in terms of the remainder between the function being expanded into a series and partial sums of the expansion. Taylor's theorem, as it is presented now, includes not only a prescription for computing a series expansion of a function from its derivatives, but also an expression for the remainder R_n when the series is truncated after n terms, i.e.,

$$f(a + h) = f(a) + f'(a)h + \frac{f''(a)}{2!}h^2 + \cdots + \frac{f^{(n)}(a)}{n!}h^n + R_n,$$

where Lagrange's expression for the remainder term

$$R_n = f^{(n+1)}(a + \theta h)\frac{h^{n+1}}{(n+1)!}, \quad \theta \in (0, 1),$$

is sometimes given. What one may observe from this is that, as h goes to zero, the remainder term R_n vanishes faster than the order-n term of the truncated series. More formally, we may observe that if $T_n = \frac{f^{(n)}(a)}{n!}h^n$ is the order-n term of the expansion, then $\lim_{h \to 0} R_n/T_n = 0$. Stated specifically in terms of h^n, the powers used to expand the function, $\lim_{h \to 0} R_n/h^n = 0$. This latter property is written as $R_n = o(h^n)$ as $h \to 0$, which can be read as saying R_n is *asymptotically dominated by*, or *infinitesimal compared to*, h^n as $h \to 0$.[4] The significance of this is that by knowing that the last term in an approximation dominates the remainder, we can be assured that, sufficiently close to $h = 0$, the partial sum will provide a good approximation. The important point is that this is the case *whether or not the series converges*.

To illustrate this, consider the case of the integral of $\int_1^x e^t/t \, dt$. Through repeated use of integration by parts, this integral can be developed into a series as

$$\int_1^x \frac{e^t}{t}dt = e^x\left(\frac{1}{x} + \frac{1}{x^2} + \frac{2!}{x^3} + \frac{3!}{x^4} + \cdots\right).$$

---

[4] Strictly speaking, the order notation should be written as $R_n \in o(h^n)$ since $o(h^n)$ is a set of functions, but I follow the standard abuse of notation by writing it as an identity.

There are two important features of this series for us to note. First, for no value of x does the series in the parentheses converge. Thus, there is no hope to compute an arithmetic sum of the series to compute the value of the function this way. Nevertheless, the series has the property that the remainder R_n after n terms is $o(1/x^n)$ as $x \to \infty$; thus, by taking a sufficiently large x, partial sums of the series can provide good approximations of the value of the integral. Series that have this property that the remainder is asymptotically infinitesimal compared to the last term of the truncated series are called *asymptotic series*. They generally have the property that, for a given value of x, there is an optimal number of terms to take to obtain the best approximation before the approximation begins to get worse. They also tend to be very advantageous for the purposes of computation because often only a few terms are needed to get very accurate approximations.

To illustrate, by keeping 4 terms of this series, the approximation is correct to 3 decimal places by $x = 12$, and to 4 decimal places by $x = 20$, getting better the larger x becomes. On the other hand if we fix, say, $x = 12$, then though keeping more terms initially yields a better approximation, giving a result correct to 5 decimal places by keeping 9 terms, after that adding more terms makes the result worse. By 13 terms the approximation is only correct to 3 decimal places again, and by 30 terms the approximation is not correct to any decimal place.

A major advantage of asymptotic series, given their tendency to converge rapidly taking only a few terms, is that, provided one is considering phenomena in an appropriate asymptotic regime so that truncated series are accurate, one can transform an unknown or difficult-to-manipulate function into functions that are known or more easily manipulated, which represents an effective reduction in computational complexity. In the series considered above, we can usefully approximate a non-elementary function, essentially the exponential integral $\mathrm{Ei}(x)$, by elementary functions. Thus, by strategic use of truncated asymptotic series, we can complete inferences that would not otherwise be possible if we were to restrict to exact values or solutions, or can complete inferences more easily. This is therefore another case of using approximation to render inferences (more) feasible, and it is once again the difficulty in obtaining exact solutions that motivates it.

The strategy of using asymptotic approximations can also be understood in terms of a modification of the problem. In the case of approximation of integrals, we can see that the result does not differentiate to the integrand we are trying to integrate (i.e., $e^x/x \neq \frac{d}{dx} e^x(1/x + \cdots + (n-1)!/x^n)$) for any n). Thus, we have actually solved a slightly different problem, but it is close enough in the appropriate asymptotic regime to give us useful information.[5] Although subtle in this case, the character of asymptotic methods as a modification of the problem becomes more explicit when they are used to solve mathematical problems, such as differential equations. At least as far back as Euler, but more fully developed in the work of nineteenth-century mathematicians such as Jacobi, Liouville, and Stokes, divergent series expansions were used to solve differential equations approximately, in some cases explicitly involving an asymptotically valid modification of the differential equation (Kline 1972, p. 1100n). In this latter case, the equation itself is considered in an asymptotic regime where it is more easily solved, so that the problem is modified directly. The solution thus obtained then carries information about the solutions to the original problem, and the behavior of phenomena accurately described by it, in the same asymptotic regime. Once again, an inferential obstacle at the level of theory is overcome by a clever use of approximation.

As was stated earlier, exact solutions are extremely valuable when they are available, but not only because they do not introduce error. They are also important for enabling the computation of approximate solutions to problems that differ by a small amount, or the approximate characterization of phenomena that differ only slightly from phenomena that can be described exactly. In this case we have a problem that can be regarded as a small perturbation of a problem that can be solved exactly. As such, the known exact solution can be used to compute approximate solutions to the perturbed problem. The mathematical techniques developed to make this approach work became perturbation theory, and have their origins in astronomy.

Although Newton introduced in the second edition of the *Principia* the idea of computing corrections to the two-body problem of the Moon's orbit around the Earth due to the effect of the gravity of the Sun (Linton 2004, p. 277), the analytic methods of perturbation theory find their roots in Euler. In his work on treating three-body problems as a small perturbation of Keplerian two-body motion, he introduced the technique of giving solutions in terms of trigonometric series, i.e., series in powers of $g \cos \theta$ for small g, and even

---

[5] This approach to analyzing approximation error in terms of modified problems underlies the method of backward error analysis in numerical computing (Corless and Fillion 2013; Fillion and Corless 2014). The essence of the method is to regard approximation error in a computed solution as an equivalent modification of the problem, so that the computed solution is an exact solution of a modified problem. Since the modification of the problem (the backward error) can typically be feasibly computed or estimated, the method has an important epistemic virtue: It gives one an exact solution to a known or approximated modified problem, without needing to know the solution to the original problem. Moreover, since the size of the modification is typically very small (according to a relevant metric or norm), and the computational error can be considered alongside other modifications of the problem, such as those arising from modeling error, we obtain sound reasons to trust the validity of the computational result. Although backward error analysis is very relevant to our discussion here, limitations of space prevent its further consideration. For more details on the assessment of backward error in a modeling context, see Moir (2010, ch. 5).

introduced methods for accelerating their convergence.[6] These methods were developed initially for specific three-body problems by Clairaut, d'Alembert, and Euler himself, and later developed into a more general approach, beginning with Laplace and Lagrange. The perturbative approach became a standard method for constructing analytic theories of planetary and satellite motion, as well as for computing numerical tables for use in astronomical study and navigation.

Much later these methods were developed into a more general method for solving equations that could be regarded as small perturbations of problems with known solutions, an approach now widely used in contemporary physics. The basic idea of the method is the following. Suppose we have a differential equation of the form

$$y' = f(y, \varepsilon),$$

such that when $\varepsilon = 0$ the solution $Y_0(t)$ is known. The strategy is then to expand the solution when $\varepsilon > 0$ in powers of ε so that

$$y(t) = Y_0(t) + Y_1(t)\varepsilon + Y_2(t)\varepsilon^2 + \cdots, \tag{8.4}$$

which is usually truncated after a certain number of terms, determining the order of the perturbations one is considering (e.g., second order if we truncate after the ε^2 term). By substituting this equation into both sides of the differential equation, possibly also developing $f(y, \varepsilon)$ into a power series in ε, and collecting terms with the same power of ε onto one side of the equation so that the other side is zero, one obtains a series of terms multiplied by successively higher powers of ε. Then the coefficient of each power of ε is a differential equation, each of which must be set to zero (since the other side of the equation is zero), yielding a sequence of differential equations, beginning with the one we already have a solution for.[7] Solving these equations for $Y_1(t)$, $Y_2(t)$, etc., then provides first, second, and so on, corrections to the exact solution according to Eq. (8.4). Thus, provided ε is sufficiently smaller than 1, one can obtain a good approximate solution to the perturbed equation. This process as described rarely works in a straightforward way, and can itself involve laborious calculations, but it gives a sense of the method.

Just as for numerical methods and asymptotic expansions, perturbation theory involves developing functions into series, which are typically asymptotic, as was recognized early on but only properly emphasized and treated after Poincaré (Kline 1972, p. 1103n; Linton 2004, p. 422). And as in those cases,

---

[6] Such methods can sometimes convert divergent series into convergent ones, revealing more clearly how a divergent series can still represent a function in an algebraic sense.

[7] The first equation in the sequence is generally nonlinear, where the rest are linear and increasingly tedious to solve.

perturbation methods work by solving a modified problem. Rather than directly approximating or replacing the problem with a nearby one, the approach here is to attempt to stretch an exactly solvable problem into the problem one wishes to solve, or an approximation to it, since the problem one wishes to solve is too difficult to solve directly. The perturbation solutions, particularly as rapidly converging asymptotic series, then convey information about the solutions to the insoluble problem, possibly in some asymptotic regime. Once again, an inferential obstacle at the level of theory is overcome by a clever use of approximation to make inference feasible.

It is evident even from this limited consideration of approximation methods in physics that approximation is about far more than simply calculating numbers or determining behavior in particular situations. Approximation methods are part of the theoretical process and play an important role in the development of theory. This is very clear in the case of perturbation methods in the history of astronomy, which led to better-developed theories of planetary and Moon motion, as well as the discovery of Neptune. Even numerical methods allowed early demonstrations of the existence of solutions to differential equations. This role for approximation methods has only grown, with asymptotic methods becoming essential tools of analysis throughout physics, applied mathematics, and computer science. Perturbation methods also have wide application, underlying the phenomenally successful perturbative approach to quantum field theory and modern fluid mechanics, as well as being extended to many other classes of mathematical problems, finding applications throughout applied mathematics and computer science (Avrachenkov et al. 2013). And numerical methods, far from simply being tools for calculating values, have developed into sophisticated tools of scientific inference, such as the tools of geometric numerical methods (Hairer et al. 2006), and new branches of science, such as computational fluid mechanics (Wesseling 2009).

We have also seen in this section how approximation methods in physics enable one to extract information from theoretical models when exact solutions are not available. There is a common pattern of using approximation to modify the original problem one wished to solve into one from which information or solutions can be more feasibly attained, thereby giving approximate information about behavior of solutions to the original problem. As will be made clear in the next section, this is the kernel of an algorithmic process of feasible problem-solving that underlies the success of methods in scientific computing for rapidly solving difficult mathematical problems.

8.3 Feasible Computation: Algorithmic Problem-Solving

The ability to solve mathematical problems is essential to scientific inference. In the previous section we saw how approximation methods have become an essential tool for making scientific inference feasible given the rarity of exact solutions, and how such methods involve a modification of the problem into one that is solvable. The strategy of modifying a problem to make a solution feasible is not restricted to approximation methods, however, and underlies problem-solving methods in analytic geometry, introduced by Fermat and Descartes in the seventeenth century (Kline 1972, p. 302n), Galois theory and algebraic geometry developed in the nineteenth century, as well as algebraic topology and the modern algebraic geometry developed in the twentieth century, which all solve problems in their original form by converting them into equivalent algebraic ones.[8] Though such methods are exact, they nonetheless involve transforming a problem into a more feasible one, so that solutions obtained not only carry information about the original problem but actually yield exact solutions. In such cases, the problems are equivalent, so that reasoning can be performed in whichever context is preferable.[9]

Whether or not approximation is used to solve an infeasible problem, it is generally necessary to make more than one transformation of the problem to reach a feasibly solvable problem. Such a strategy underlies the common method of reducing a mathematical problem to another class of problem that is simpler or regarded as already solved. A nice example of this approach occurs in the analytic solution of partial differential equations (PDE), which can be "solved" by reducing them to a system of ordinary differential equations (ODE), reducing an infinite-dimensional problem to a finite-dimensional one, a reduction in complexity. Van Dyke (1975) points out that such solutions are generally regarded as "exact" even when the ODE must be solved numerically. This is therefore an example where an exact transformation of a problem (PDE to ODE) is combined with approximation methods (numerics) to render a solution feasible. To see that this sort of scenario is very general, notice that even for uncontroversial cases of exact – in the sense of closed-form – solutions, numerical methods are generally required to evaluate the functions used to express the solutions. Consider for example the difference in feasibility of information for the solution of the hydrogen atom problem in quantum mechanics as expressed in terms of spherical

---

[8] It should be stressed that the motivation was not always a search for equivalent problems, however, but often an increase in rigor or conceptual precision and to discover or construct more general systems from which motivating problems could be recovered as a special case. An important example of this is the role of the Weil conjectures motivating Grothendieck's substantial generalization of algebraic geometry (Grothendieck 1960). Nevertheless, once equivalent formulations have been discovered that make inference more feasible, they can be used to facilitate problem-solving by reducing complexity.

[9] Noting such a situation in his laying the foundations of the calculus of variations, Euler wrote: "It is thus possible to reduce problems of the theory of curves to problems belonging to pure analysis. And conversely, every problem of this kind proposed in pure analysis can be considered and solved as a problem in the theory of curves" (Struik 1969, p. 399). Euler preferred a geometric approach, but as a result of Lagrange's analytic formulation that proceeded directly from Euler's work, the analytic approach has become standard.

harmonics and (associated) Laguerre polynomials versus computed plots of the three-dimensional probability distributions.[10]

The process of making problem-solving feasible, therefore, involves in general a *sequence* of transformations of the problem, each of which makes a solution more feasible, reducing the complexity of computing a solution. Feasibility requires that this process terminates after a small number of steps, at which point one actually obtains a solution to one of the simplified problems. Now, since the purpose of transforming the problem is to solve the original problem, the final stage of the process is to back-interpret the computed solution through the sequence of simplified problems so that it yields a (perhaps approximate) solution to the original problem, or a problem sufficiently close to it. This recursive process of problem simplification to a feasibly solvable problem, followed by back-interpretation of the result, is the characteristic pattern of what I call feasible computation.

We have seen several examples now of this pattern of feasible computation, although they have generally focused on a single step of the iterative process or have left the sequence of steps implicit. To make the structural pattern clear, consider the following illustrative but still simple example of the process. This is the use of logarithmic tables to render arithmetic calculations feasible. This method of calculation was developed in the seventeenth century by John Napier, Joost Bürgi, and Henry Briggs, and used logarithms to convert multiplication, division, exponentiation, and root extraction problems, respectively, into addition, subtraction, multiplication, and division (Goldstine 1977). This approach to simplifying arithmetic underlay the use of the slide rule for arithmetical calculations until the development of inexpensive pocket scientific calculators supplanted its use in the 1970s. This is an interesting example because an exact equivalence of problems is the basis of the method, but it is nevertheless a tool for accurate approximate calculation because of how the tables were constructed and used.

The transformation of an arithmetic problem is based on the basic property of the logarithm that it converts products of its arguments into sums of their logarithms, i.e.,

$$\log(xy) = \log(x) + \log(y),$$

from which it follows that

$$\log\left(\frac{x}{y}\right) = \log(x) - \log(y), \quad \log(x^n) = n\log(x),$$

$$\log(\sqrt[n]{x}) = \frac{\log(x)}{n}. \tag{8.5}$$

---

[10] The hydrogen atom problem is a standard problem in non-relativistic quantum mechanics involving computation of the electron orbitals around a stationary proton; see, e.g., Liboff (2003).

Thus, one makes multiplication or division of quantities more feasible by computing their logarithms, adding or subtracting the result, and then finding the quantity whose logarithm equals the sum or difference. If one had infinite precision, then this would be an exact calculation. Logarithmic tables, however, can only have so many values in them, and it was extraordinarily tedious to compute them – so much so, that only certain values were computed directly, and the intermediate values were computed by interpolation. In fact, methods of interpolation developed a great deal through the need to compute more accurate tables (Goldstine 1977). Thus, one in fact replaced an arithmetic problem with an approximately equivalent one. Further approximation is involved in the use of the tables because the result one computes may not equal exactly an entry in the table, so one finds instead an approximate intermediate result by some convenient method that is sufficiently accurate for the given context.

We see from this process all of the features of the general pattern of feasible computation. First of all, there is the mapping of a problem that is difficult to solve directly – computing products or quotients – to an approximately equivalent problem of summing or subtracting their logarithmic table values, and then back-interpreting the result to the approximate product or quotient of the quantities. We even find the iterative pattern when a table is used for exponentiation and root extraction. In this case, the problem maps to a product or a quotient of quantities according to Eqs. (8.5), which can then reduce to a sum or difference by applying the table a second time, followed by two back-interpretations to yield the result. Thus, the use of logarithmic tables provides an early example of the pattern of feasible computation.

Logarithmic tables are such a nice example because the pattern of transforming the problem iteratively, computing of the solution, and back-interpreting the result is so clear. Moreover, the motivation of the method, viz., to reduce the complexity of arithmetical calculations, typically for the purposes of applying theoretical models to solve practical problems, such as astronomical prediction and navigation, is so clearly tied to feasibility. In most cases the feasible computation pattern is implicit, but nevertheless present, in strategies to simplify problem-solving. It does not always involve back-interpretation to the original problem, if some simpler problem or model suits the scientific purposes at hand. A nice example of this is post-Newtonian celestial mechanics, which uses corrections to Newtonian gravity from general relativity (GR) to model the motion of bodies in weak gravitational fields (Kopeikin et al. 2011). This is a simplification from GR, but the results are not mapped back to the framework of GR since it is easier to model things in terms of vector mechanics on a fixed background space rather than using the tensor framework needed for GR.

We can see at this point that features of the feasible computation strategy are present wherever a problem is modified to simplify its process of solution. In the examples considered so far, with the exception of logarithmic tables, the full iterative aspect of repeated problem simplifications is typically not performed by a single individual. In most of the cases we have considered, one person is doing one kind of reduction, leaving any further reduction to be performed by someone else, or different levels of simplification are divided between theoretical and practical work. This situation has changed dramatically as a result of the development of advanced computing machines, which have made it possible to offload tedious arithmetical calculations to microprocessors, allowing scientists to focus on scientific inference. It has also led to new branches of science, in particular computational science, a multidisciplinary field concerned with developing algorithms to solve difficult mathematical problems as efficiently, accurately, and reliably as possible, using a combination of exact and approximate methods. An important point for us is that computational science uses the recursive feasible computation process to construct algorithms that can solve mathematical problems *automatically*.

Modern scientific computing uses the feasible computation strategy to great effect by reducing difficult mathematical problems to combinations of simpler problems that can be solved rapidly and accurately with well-understood error behavior.[11] A large proportion of mathematical problems can be reduced to problems in linear algebra and polynomial algebra. Breaking down problems in this way then allows highly optimized low-level algorithms to be applied to these simplified problems, so that the results can be combined (back-interpreted) to generate fast, accurate solutions to the original problem. One important such package is the basic linear algebra subprograms (BLAS), an optimized package for low-level linear algebra computations originally written in Fortran in the 1970s, which still operates (sometimes in a further optimized form) under the hood of most systems that do numerical calculations, including the mathematical computing environments MATLAB, MAPLE, and MATHEMATICA.

To make it more clear how advanced algorithms for solving mathematical problems do indeed follow the feasible computing pattern, let us consider an example of how one would go about solving an ordinary differential equation in MATLAB. For concreteness, suppose that we wish to solve the van der Pol equation

$$y'' - \mu(1 - y^2)y' + y = 0, \quad y(0) = y_0, \ y'(0) = y'_0, \tag{8.6}$$

---

[11] It is worth noting in passing that users of computational software still need to do their own error analysis, and cannot blithely accept a computer-generated result as correct, even though most do just that. The reasons for this are varied, but in many cases flow from the fact that the assumptions of the author and user of software often differ and so the user must verify that the computational method being used is appropriate for their problem-solving needs. More specifically, where approximation is involved, one must understand the sensitivity of the problem one is solving to small changes to know that a computational result is valid; where results are exact, they can be incomplete or inapplicable in a case the user is interested in.

for the case $\mu = 1$. Due to the nonlinearity (the $\mu y^2 y'$ term) we do not have analytic solutions for this problem, which forces us to use an approximate method. For examining arbitrary initial conditions we need to simulate the solution, hence we seek a numerical solution. Solving this equation in MAT-LAB is made extremely simple. With a standard reformulation of the problem as a coupled pair of differential equations,[12] we can write four lines of code (easily reduced to two) to solve the equation:

```
f = @(t,y)([y(1)-y(1)^3/3-y(2);y(1)]);
tf = 40;
init = [4;0];
sol = ode45(f,[0,tf],init);
```

For these values of `tf` and `init`, the final time and initial condition,[13] this code executes in less than one hundredth of a second, the results appearing in Figure 8.1a. With two more lines of code, within another hundredth of a second, we can interpolate the sequence of values y_n produced by the numerical method, as is shown in Figure 8.1b. With this simple procedure, we find a stable limit cycle solution of the oscillator. Imagine trying to do the same calculation by hand using numerical methods and logarithmic tables.

This simple procedure masks the recursive feasible computation structure, which is all packaged into the MATLAB routine `ode45`. This procedure hides a sequence of three transformations of the problem. As we saw in the previous section, the first step is to replace the differential equation with a difference equation to obtain approximate values y_n of the solution at times $t_n = t_{n-1} + h_n$,

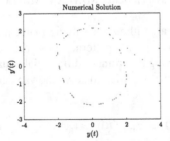

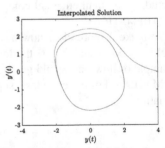

(a) Sequence of phase space values produced by ode45.   (b) Interpolant of the numerical skeleton from (a).

Figure 8.1 Solution of the van der Pol Eq. (8.6) for $\mu = 1$, $y_0 = 4$, $y_0' = 0$ for $t \in [0, 40]$ using MATLAB's ode45 routine

---

[12] Note that this is technically a transformation to a more feasible problem, and is not entirely unproblematic; see Ascher et al. (1995).

[13] Note that `init` is a vector, with the first component being y_0 and the second y_0'.

where the time step h_n can now vary from step to step.[14] But these difference equations are also rarely solvable analytically, and can include implicit equations that require iterative methods to solve. Thus, we take a second problem simplification step, which involves a move to an implementation of the numerical method in the code of a high-level programming language. This transformation has some obvious aspects, such as translating from a mathematical to a programming language, and some subtler aspects, such as replacing the continuum of real numbers with a finite bounded system of floating point numbers, a simulation of the real numbers that a computer can calculate with. Thus, it really is a transformation of the problem involving approximation, not simply a reformulation. But this step does not produce a solution either, since the code needs to be run, which requires a third transformation of the problem into machine code, which can be run on a microprocessor.[15] It is only when the code is run that a solution is computed, when the solution finally is made feasible. The computed solution is a sequence of binary strings, which is then back-interpreted into decimal numbers, which can be interpreted as specifying points in phase space. When we interpolate the skeleton of values with a suitable interpolant, we then obtain a continuous (and differentiable) curve that is our approximate solution to the original equation (as in Figure 8.1b).

We see, therefore, that running six lines of MATLAB code involves a sequence of problem simplification stages that systematically make problem-solving feasible, followed by a back-interpretation of the result through the sequence of transformations to obtain an approximate solution to the original problem, completing the inference. The approximation, or translation, at each step increases the feasibility of the solution. We first replace a solution that varies continuously (differential equations), with one that varies discretely (difference equations), which in principle can be computed step-by-step, an increase in feasibility. Since these equations are rarely solvable in practice, we replace the numerical method with a fully algorithmic procedure written in code, which in principle can be computed by inputting the data for the problem and running the algorithm, a further increase in feasibility. Since such computations are extraordinarily tedious, and error-prone if performed by a human, we translate the algorithm into machine language so that a microprocessor can execute it, which then makes the solution feasible in the conditions we encounter in scientific practice.

---

[14] The routine ode45 actually uses a pair of numerical methods that work together to approximate the error and take a step of the integration, as well as sophisticated step size control to control error and take as few steps as possible.

[15] MATLAB actually uses a combination of compiled kernel code and interpreting of the code the user enters to produce machine code, so a more detailed analysis of problem transformation is possible. The net result, however, is a transformation from high-level code to machine code.

A similar structure appears in any case of the use of numerical methods to solve problems on computing machines, whether we are solving a single linear equation or nonlinear partial differential equations for heterogeneous materials. Thus, the pattern of feasible computation is actually fundamental to scientific computing. It is not restricted to numerical computing, however, appearing also in exact, symbolic computation. Given that pure mathematics uses transformations of problems to equivalent ones to increase feasibility, as pointed out above, it may not be surprising that the feasible computing pattern appears in exact computation. What may be surprising, however, is that exact algorithms can involve forms of error and approximation.

In exact computation, algorithms generally come down to arithmetic of integers, polynomials, and matrices, avoiding the floating point numbers used in numerical algorithms, which use approximate arithmetic. Doing arithmetic in the domains in which calculations are formulated, however, can be very computationally expensive and can lead to the problem of intermediate expression swell, where the problem and solution might be fairly simple but the expressions in between can be large, requiring large amounts of time and space to compute. Thus, the standard approach is to use one of a number of reduction methods to project the problem into a finite domain, computing the result there and then mapping back the result to solve the original problem. This is precisely the feasible computation pattern, which is essential in computer algebra for reducing the time and space complexity of algorithms.

Even the simplest problems, such as arithmetic over the integers $\mathbb{Z}$, can use reduction methods. Intermediate expression swell can be avoided by reducing arithmetic operations to equivalent operations over finite sets of integers modulo m, i.e., $\mathbb{Z}_m = \mathbb{Z}/m\mathbb{Z}$, which can be represented as $\mathbb{Z}_m = \{0, 1, \ldots, m - 1\}$. Since the results stay within this finite set, memory usage is tightly controlled, reducing complexity and avoiding expression swell. For integers of size n, the algebraic complexity of integer arithmetic is $O(n^2)$, but this can also be reduced by moving to modular domains and performing further optimization steps.[16] For sufficiently large integers, fast Fourier transform methods can yield asymptotically fast integer arithmetic ($O(n^{1+\varepsilon})$, for any $\varepsilon > 0$) by a strategy of encoding integers as polynomials with coefficients in $\mathbb{Z}_p$, where p is a well-chosen prime number, computing the product of the polynomials (via a further reduction based on evaluations of the polynomials at primitive roots of unity and an interpolation of the result), and then decoding the resulting polynomial to the integer solution.[17] This process itself is seen to follow the feasible

---

[16] The "size" here is measured in terms of the number of bits in a binary representation, and the algebraic complexity is measured in terms of bit-level operations.

[17] The nth primitive roots of unity in the complex numbers are the numbers $e^{2\pi i/n}$, but in modular fields $\mathbb{Z}_p$ they are distinct elements ω of the field that have the property $\omega^n = 1$. Such distinct elements exist precisely when the condition $n|(p - 1)$ is satisfied.

computing strategy. Further complexity gains can be made by reducing to modular domains $\mathbb{Z}_p$, where p is a machine word prime and arithmetic operations can be performed on the microprocessor effectively in a single clock cycle, reducing operations to $O(1)$.

Methods that reduce to and perform operations in finite domains $\mathbb{Z}_m$ are called *modular methods*. These methods come in two general classes that function in different ways, which can be combined for some problems, such as polynomial factorization. One works by computing multiple modular images of the input, performs the required calculation in each of these images, and then uses a result called the Chinese remainder theorem to combine (back-interpret) the results to obtain the solution to the original problem. The other works by computing a single modular image, performs the required calculation, and then "lifts" (back-interprets) the result by computing a sequence of approximations using a process called the *Hensel construction*, to obtain the solution. Interestingly, the Hensel lifting process is based on Newton's method for numerical root-finding. The single modular image becomes the "initial guess" of the iterative process.[18] This initial guess can be regarded as the "zeroth-order" approximation in a power series expansion (in powers of a prime number p) of the solution. The iteration therefore computes increasingly "higher-order" approximations until the exact solution is reached. Modular methods can be used for many problems beyond those mentioned, including fast polynomial arithmetic, integration, and solving linear difference and differential equations (Gerhard 2004).

Although the sense of "approximation" here is analogical and does not involve error in the manner that numerical or asymptotic approximations do,[19] some uses of the Hensel construction nevertheless introduce error, so that there is actually a loss of information in the use of the feasible computation process, just as in the case of numerical computation. An example of this is the sophisticated algorithm developed by Dahan et al. (2005) for solving nonlinear systems of polynomials. This algorithm makes complexity gains by using a *probabilistic* modular method, which introduces a nonzero probability that a solution will be missed. The algorithm considers the case where the full solution is a set of points, each described as a special set of polynomials called a triangular set. By accepting a small probability that one or more distinct solutions can be mapped to the same modular image, so that when the result is

---

[18] The modular image is the unique, exact first term in a p-adic expansion of the exact solution, so there is no guessing involved. It just functions as an initial guess in that the Newton iteration generates the subsequent terms in the p-adic expansion.

[19] The sense of "approximation" uses an analogy to higher-order terms in power series, where the higher-order terms are generally understood to be smaller in size when $|x| \ll 1$. In the Hensel construction, however, the "higher-order" terms involve higher powers of p and are hence larger in size.

lifted only one of the corresponding solutions is computed, the complexity can be reduced dramatically, rendering the solution of a large number of nonlinear systems feasible. This failure probability can actually be chosen, increasing the complexity as it is reduced, much like an error tolerance in numerical computation, showing that some instances of the feasible computation pattern in exact computation share a surprising degree of structure with the numerical case. This algorithm is implemented for solving nonlinear systems in MAPLE.

In an analogous manner to the implementation of numerical methods, the implementation of symbolic algorithms involves stages of transformation of the problem so that the algorithms can ultimately be reduced to machine language and the feasibly computed results back-interpreted to solve the original problem. Also analogously to numerical methods, which so often reduce to floating point polynomial and linear algebra computations, symbolic methods often reduce to multiprecision integer polynomial and matrix algebra. In recognition of this, there are now efforts to produce optimized low-level tools for symbolic computation. One such project, on which the author is a developer, is the basic polynomial algebra subprograms (BPAS), which aims to be a symbolic analogue to BLAS (Chen et al. 2015).

We now see how the advent of high-performance computing machines has allowed the feasible computing strategy to be employed to great effect to solve mathematical problems with high efficiency and reliability. The feasible computing strategy is used in both the symbolic and numerical branches of scientific computing, involving in both cases the reduction of high-level problems to combinations of optimized low-level algorithms with high-efficiency implementations, which allow solutions feasibly computed by microprocessors to be back-interpreted to high-level solutions. Thus, a strategy that has its roots in the history of physics and pure mathematics now allows computing machines to greatly expand the range of mathematical problems that are feasibly computable, hence greatly expanding the range of feasible scientific inference.

8.4   Consequences for Computability Theory

The contrast between numerical and symbolic computing offers a way of bringing out the implications of feasible computing in computational science for the theory of computation. One consequence is brought out in the common structure between numerical and symbolic computing. In both cases we find the main structure of transformations that increase the feasibility of solution of problems. As such, instances of the feasible computing strategy are "higher-order" algorithms, in the sense that they seek to find optimal transformations *between* problems such that the overall algorithm has minimal computational complexity, rather than simply being concerned with finding optimal algorithms for a particular problem. When we consider that the full feasible computation process involves implementation of mathematical algorithms in a software environment and running of machine code, the process also involves transformation between *kinds* of problems, not simply mappings between problems of a single sort. If we regard problems as points in a space, then feasible computing reduces complexity by strategically moving between points within the space and between spaces for the implementation process. Given the importance of the feasible computing structure in the solution of mathematical problems, the development of a theory of such higher-order computation stands to be of significant value.

A second consequence is one that is brought out by the difference in structure between numerical and symbolic computing. The main difference between the numerical and symbolic computing versions of the feasible computing strategy is that the numerical version is more general. This is because approximations break the exact structure preserved in symbolic computing, but if we restrict the strategy by reducing possible approximations to zero we get the symbolic version. The theory of computation is geared toward the exact case, where we can talk about the space $\mathcal{A}_p$ of all algorithms to solve a particular problem p and the minimal complexity in that space. The symbolic feasible computing strategy provides an effective means of *finding* minimal complexity algorithms, or approximations to them, but the algorithms generated generally stay within $\mathcal{A}_p$. In contrast, the numerical version involves moving to a different problem p' with its own space $\mathcal{A}_{p'}$ of algorithms, which can have a different minimal complexity. Thus, we can make complexity gains by allowing moves to nearby problems, provided the nearby problem provides (approximately) the same information as the original one, at least for the purposes for which the algorithm is being used.[20] It may be that the space $\mathcal{A}_{p'}$ has a strictly lower complexity bound than $\mathcal{A}_p$, in which case the problem p' has a lower complexity, or it may be that the complexity reduction is only epistemic, in the sense that we can *access* lower-complexity algorithms by a shift to p' even though there are equally low-complexity, but feasibly inaccessible, algorithms in $\mathcal{A}_p$. Either way this leads to a different way of thinking about computational complexity in computation.

To bring the issue here into clearer focus, consider the following thought experiment. Suppose that the problem p we wish to solve is only computable in exponential time, but that, generically within some neighborhood of p in a natural space of problems, problems have polynomial time complexity.[21] Letting P denote the (dense) set of nearby problems of polynomial complexity,

---

[20] Note that what counts as a nearby problem depends on the metric (or norm) on the problem space, and the suitability of a given metric (or norm) depends on the context, including what kinds of error are permissible.

[21] Note that "generic" is a technical term that for our purposes can be read as meaning a property holding on a dense set. For a careful definition, see Moir (2010, p. 33).

does it still make sense to consider p to have exponential complexity if the problems in P have no discernible differences in their solutions? Although some of the computational science examples show that this is not a purely hypothetical issue, consider also that heuristic algorithms function precisely by reducing computational complexity at the expense of exact, optimal or complete solutions, viz., they introduce forms of approximation to accelerate running time.[22] Therefore, where heuristic algorithms can reliably provide accurate solutions at substantially reduced computational cost, perhaps this should lead us to change how we think about the complexity of the original problem. Of course the standard definition of complexity of p still applies, but where p models some phenomenon or situation and solutions to nearby problems provide functionally equivalent results, arguably what matters more are the minimum-complexity problems in the neighborhood of p, and not p itself, if solutions to p are not feasible.

A final consideration for computability theory concerns the larger pattern of feasible computation in scientific inference processes. We traced in more detail the roots of the general, approximate feasible computation process to approximation methods in physics. We also saw in a more limited way that the exact version of the process has roots in pure mathematics. The result is that the feasible computation pattern is found extremely broadly, whenever a mathematical problem is not solvable in its original form, whether the solution process sought is purely analytical or involves scientific computation.[23] Since the method stems from the epistemological need to overcome an inferential obstacle to draw feasible consequences, it is likely a very general process throughout scientific inference, possibly involving weaker forms of the strategy of modifying problems to compute feasible solutions in fields that are less mathematized. Given that computability theory and the analysis of algorithms rely heavily on asymptotic methods, they are certainly using a form of the feasible computation pattern. If the feasible computation pattern is found more broadly in the practice of science, then an extension of computability theory to treat higher-order approximate computation presents the possibility of expanding the theory of computation, and possibly even advanced algorithms through developments in computational science, to a theory of the methodology and epistemology of inference in scientific practice.

---

[22] It is interesting to note that there are a variety of heuristic algorithms for a particular problem, such as the Traveling Salesman Problem, each with different trade-offs in terms of complexity, accuracy and stability; see, e.g., Nilsson (2003). As a result, heuristic algorithms are quite strongly analogous to numerical methods in terms of how they provide approximate solutions to problems.

[23] It is shown in Moir (2013, ch. 4) that what we here call the feasible computation pattern obtains also in the process of using data from image plates to solve the orbit determination problem in optical astrometry. Thus, there are also known examples of the pattern in data handling in applied science.

9   Relativistic Computation

Hajnal Andréka, Judit X. Madarász, István Németi, Péter Németi, and Gergely Székely

9.1   Introduction

Two major new paradigms of computing arising from new physics are quantum computing and general relativistic computing. Quantum computing challenges complexity barriers in computability, while general relativistic computing challenges the physical Church-Turing thesis itself. In this chapter, we concentrate on relativistic computers and on their challenge to the physical Church-Turing thesis (PhCT).

The PhCT concerns the belief that whatever physical computing device (in the broader sense) or physical thought experiment is designed by any future civilization, it will always be simulable by a Turing machine, in some sense. In a somewhat more concrete form, the PhCT says that for any function $f : N \to N$, where N is the set of natural numbers, if f is physically realizable then it is also Turing computable. It is worthwhile to make explicit what we mean by "physically realizable." Roughly, f is physically realizable if there is a physical thought experiment in which f is computed according to the present laws of nature. By a "physical thought experiment" we mean the following. We specify a recipe or design for a possibly futuristic experiment, and show that it is consistent with the currently known laws of physics that if our far future descendants were to perform this experiment, the outcome for input $n \in N$ would be $f(n)$. The design we present in the present chapter is an arrangement in the physical world that operates on finite data in accordance with a finite program. Hence it is a challenge to the PhCT as formulated in Copeland (2015): "Whatever can be calculated by a machine (working on finite data in accordance with a finite program of instructions) is Turing machine computable." More detail is given in Etesi and Németi (2002, p. 351). Piccinini (2011) calls this thesis the Modest Physical Church-Turing Thesis.[1]

Intuitively, a Turing machine is a finite-state automaton, TM, which has access to a finite but indefinitely extendible tape. TM can add a cell to the tape,

---

[1] Related work on transcending the PhCT using designs based on physical considerations can be found, e.g., in Tipler (1994, app. G), van Leeuwen and Wiedermann (2001), Cooper (2006), Syropoulos (2008), Stannett (2009), and Beggs et al. (2014).

it can write on a cell of the tape, and it can read a cell of the tape depending on its state. TM can also move along the tape. A typically non-Turing-computable task is one which needs to survey the result of a given algorithm for all natural numbers n before being able to return an answer. Such tasks include deciding recursively enumerable sets. In this chapter, we design a physical computer for a thought experiment operating in general relativistic spacetimes, which can decide any recursively enumerable set. Thus, it can compute a non-Turing-computable task. Our relativistic computer is a team consisting of a computer, a Turing machine, and a programmer all operating in the general relativistic spacetime of a slowly rotating black hole.[2] The computer may work for infinite time, and the programmer can extract the desired result of its infinite work in a finite time. We investigate whether this physical computer could be realized by a future civilization.

A positive contribution of this chapter is that we answer some worries published in Earman (1995), Piccinini (2011), and Manchak and Roberts (2016). Namely, we address the questions whether the probability of error should be one for a machine working for infinite time, whether the transfer of results between the computer and the programmer is possible, and whether spacetimes allowing the operation of such a relativistic computer are physically realistic.

The layout of the present chapter is as follows. In Section 9.2, we outline the idea of a relativistic computer. In Section 9.3, we define what an abstract relativistic computer is, and we investigate what kinds of relativistic spacetimes can harbor such computers. In Section 9.4, we deal briefly with how physically realistic two aspects of this abstract physical computer are. In Section 9.5, we survey results about the complexity of sets of numbers that can be computed by relativistic computers. We close the chapter with a conclusion in Section 9.6.

9.2   Design for a Relativistic Computer

This chapter serves as a motivation for the next, more formal, one. We will use intuitive language, similar to that in Rindler's relativity book (Rindler 2001).[3] In the next section, we will define formally all the notions we use.

9.2.1   *A Fairy Tale*

Imagine a computer C which can compute arbitrarily fast. More precisely, the programmer P can speed up C with an arbitrary (but finite) rate. For example, in the first 10 minutes of its operation C may execute 10 steps. Then in the

---

[2] There are many other general relativistic spacetimes, besides the one of the slowly rotating black hole, in which such an arrangement can be done; see Section 9.3.

[3] The final stage of the intuitive plan for creating an infinite speed-up effect we outline in this section is elaborated in concrete mathematical detail in Etesi and Németi (2002).

next 5 minutes it executes another 10 steps. After this, in the next 2.5 minutes it executes a third 10 steps. And so on. So, in the first $20 = 10+5+2.5+ \ldots$ minutes of its operation, C performs infinitely many steps. Hence, in the first 20 minutes, C performs the same number of computational steps as an ordinary Turing machine would do in the whole of its infinite lifetime. Such a C could be used to decide an uncomputable problem, e.g., it could decide the Halting Problem of Turing machines, or the consistency of ZF set theory, or it could decide any recursively enumerable set of numbers. Clearly, C could be a "super-computer": It could outperform ordinary Turing machines, in particular it could challenge the PhCT.[4]

Unfortunately, according to present-day physics, physically realizing this C is not possible. To alleviate this problem, in the present work we will gradually refine the design of C such that it will approach being physically possible.

9.2.2   Relativity Theory

In the above situation, a hope is offered by relativity theory, as relativity theory makes it possible to manipulate time. It can be arranged, by taking suitable journeys, that for some observers time runs faster, while for others time runs slower. This is usually visualized by saying that wristwatches, or simply clocks, of the first observer run faster while clocks of the other run slower. Of course, this does not mean the speeding up only of clocks, but of all atomic and sub-atomic processes.[5]

Consider first the simpler case of special relativity. The Time Dilation theorem of special relativity states that clocks of a fast-moving observer run slower, as measured by a stationary observer.[6] As the velocity of the observer approaches the speed of light, his clocks gradually freeze. Now, in our design for a relativistic computer, instead of making the clocks of the computer run faster, we make the clocks of the programmer run slower. Let us put the programmer in a spaceship and put the computer on the Earth. If the spaceship moves fast enough, its clocks will run arbitrarily slow, hence the clocks on the Earth will run arbitrarily fast relative to the clocks of the programmer in the spaceship. The programmer can accelerate his spaceship in such a manner that finite time will pass for him while the computer computes for an infinite time

---

[4] Copeland (2002a) calls this computer an Accelerating Turing Machine. Other names used are Zeno machine (Andraus 2016), Zeus machine (Syropoulos 2008), and Plato machine (Earman 1995).

[5] In the next section, wristwatch time will be called proper time along worldlines.

[6] This is connected to the popular theorem known as Twin Paradox of relativity: If one of two twins takes a journey in which he travels fast, he will be younger than his twin at the time of reunion. For an experimental confirmation, see Bailey (1977).

on Earth (according to Earth-clocks). However, to achieve this, the programmer has to endure unbounded acceleration, which according to present-day knowledge will kill him. Because of this and because of other problems we would face following this setting,[7] we abandon special relativity and we refine our design for a relativistic computer by turning to general relativity.

9.2.3   Gravitational Time-Dilation

If we switch from special relativity to general relativity, the problem of unbounded acceleration may be overcome by "letting general relativistic spacetime do the work of distorting time" (instead of using solely the programmer's engine for this purpose).

Let us start out from the so-called Gravitational Time Dilation effect (GTD). The GTD is a theorem of relativity theory: It says, sloppily, that gravity makes time run slow. More sloppily: Gravity slows time down. Clocks that are deep within gravitational "wells" run slower than ones that are farther out. We will have to explain what this means, but before explaining it we mention that GTD is not only a theorem of general relativity. This theorem, GTD, can already be proved in special relativity in such a way that we simulate gravity by acceleration (Madarász et al. 2007). (We transfer the result from special relativity to general relativity by using Einstein's Equivalence Principle, which states the equivalence of inertial and gravitational mass; see, e.g., Rindler [2001, p. 18].) So one advantage of GTD is that why it is actually true can be traced down by using only the simple methods of special relativity. Another advantage of GTD is that it has been tested several times, and these experiments are well known. Actually, the General Positioning System (GPS) of today's technology tests GTD on a daily basis: Each time GPS is used, say, for locating a car, GTD is tested (Pogge 2017).

Roughly, GTD can be interpreted by the following experiment. Choose a high enough tower on the Earth, put precise enough (say, atomic) clocks at the bottom of the tower and the top of the tower, then wait enough time, and compare the readings of the two clocks. Then the clock on the top will run faster (show more elapsed time) than the one in the basement, at each time one carries out this experiment.[8] Gravity "causes" the clock on the top to tick faster

---

[7] Pitowsky (1990, sec. 2) designs a spaceship circling the Earth faster and faster so that finite time passes for him while the computer on the Earth has an infinite time to compute. Assuming that the programmer can endure the unbounded acceleration, he will learn if ZF is inconsistent since the proof of inconsistency will be found at a concrete time and sent to him. However, he will never be in a situation to know that ZF is consistent if it is so, since in his life there is no event "after" the computer's infinite time. In technical terms, special relativistic spacetime is not a Malament-Hogarth one: See Section 9.3.

[8] Actually, this experiment was done at Harvard University's Jefferson Tower in 1960 by Pound and Rebka (Rindler 2001, p. 28).

than the one in the basement. Therefore computers there also compute faster, if only slightly. How could we use GTD for designing computers that compute more than Turing machines can? In the above-outlined situation, by using the gravity of the Earth, it is difficult to make practical use of GTD. However, instead of the Earth, we could choose a huge black hole.

9.2.4   Ordinary Black Hole

A black hole is a region of spacetime with such a strong "gravitational pull" that even light cannot escape from this region. There are several types of black holes.[9] We begin to demonstrate the main ideas of a relativistic computer in the case of the simplest, spherically symmetric black hole, called a Schwarzschild spacetime.

As we move closer and closer to the black hole, the gravitational pull gets stronger and stronger. By this we mean that it is harder and harder to maintain distance from the black hole. Actually, there is a distance where the gravitational pull is so strong that we cannot maintain altitude any more and we are pulled toward the black hole inevitably. These "no-return" places form a sphere around the center of the black hole which is called the "event horizon."

Let us study observers suspended over the event horizon. "Suspended" means that the distance between the observer and the event horizon does not change. This "suspending" of the observers can be done, for example, by letting their spaceships hover over the event horizon, using their rockets for maintaining altitude. (This is rather similar to ourselves being suspended over the Earth, where, instead of rockets, the electromagnetic forces in the crust of the Earth keep us "suspended.") As we approach the event horizon from far away outside the black hole, the gravitational "pull" of the black hole approaches infinity as we get closer and closer to the event horizon. We can measure this gravitational pull as the acceleration of "apples" let go by a suspended observer, or equivalently, by the amount of fuel needed to maintain the height of a suspended observer.

Assume one suspended observer C (for computer) is higher up and another one, P (for programmer), is suspended lower down. So, C sees P below her while P sees C above him. Now the GTD will cause the clocks of P to run slower than those of C, that is, the clocks of C will run faster than those of P. Moreover, they both agree on this if they are watching each other, e.g., via photons (light signals). Let us keep the height of C fixed. If we gently lower P towards the event horizon, the ratio between the speeds of their clocks increases. Moreover, as P approaches the event horizon, this ratio approaches infinity. This means that for any integer n, if we want C's clocks to run n times

---

[9] A good source is Taylor and Wheeler (2000) or Thorne (1994).

as fast as P's clocks, then this can be achieved by lowering P to the right position. To achieve an "infinite speed-up" we could do the following. We could lower P towards the event horizon such that P's clocks slow down (more and more, beyond limit) in such a way that there is a certain finite time-bound, say b, such that, roughly, throughout the whole history of the universe, P's clocks show a time smaller than b. More precisely, by this we mean that whenever C decides to send a photon to P, then P will receive this photon before time b according to P's clocks. We can do this as follows, for example. After 1 hour has passed in C's life, P lowers to a height where, suspended, his time is twice as slow as C's one, so while 1 more hour passes in C's life, in P's life only $1/2$ hour passes. Then P goes even lower such that in his life only $1/4$ hour passes while in C's life 1 hour passes. And so on. Apart from the times needed for lowering, then in P's life $2 = 1 + 1/2 + 1/4 + \ldots$ hours pass while in C's life an infinite time passes.

However, as P gets closer and closer to the event horizon, the gravitational pull tends to infinity and this will kill him. This is analogous to the situation in special relativity where the unbounded acceleration kills our programmer P. If P does not approach the event horizon through stages of being suspended, but instead he falls into the black hole without using rockets to slow his fall, then he does not have to withstand the gravitational pull of the black hole. He would only feel the so-called tidal forces which can be made negligibly small by choosing a large enough black hole. However, his falling through the event horizon would be so fast that some photons sent after him by C would not reach him outside the event horizon. Thus P has to approach the event horizon relatively slowly in order that he be able to receive all possible photons sent to him by C. In theory, he could use rockets for this purpose, i.e. to slow his fall. However, we cannot compromise between the two requirements by choosing a well-balanced route for P: no matter how he chose his route, either P will be crushed by the gravitational pull, or some photons sent by C would not reach him. This is the reason why we cannot base our relativistic computer on a Schwarzschild black hole.[10]

9.2.5   Rotating Black Hole

To solve this problem, we would like to slow down the "fall" of P not by brute force (e.g., by rockets), but by an effect coming from the structure of spacetime itself.

We use a huge, slowly rotating black hole, called slow-Kerr in the physics literature (O'Neill 1995). In our slowly rotating black hole, besides the

---

[10] In technical terms, Schwarzschild black holes are not Malament-Hogarth spacetimes, either; see Section 9.3.

gravitational pull of the black hole there is a counteractive repelling effect coming from the rotation of the black hole. This repelling effect is analogous to "centrifugal force" in Newtonian mechanics and will cause P's descent to slow down at the required rate. So the idea is that instead of using the rockets of P to slow his fall, we use this second effect coming from the rotation of the black hole.

Slowly rotating black holes have two event horizons: these are bubble-like surfaces, one inside the other – see Figure 9.1 and O'Neill (1995, Fig. 2.2).

The outer event horizon is the result of the gravitational pull. It behaves more or less as we described in the previous subsection, and we cannot use it for our purpose. The inner event horizon marks the point where the repelling effect overcomes the gravitational effect. This is like a cushioning effect, and it makes it possible for our P to enter the inner event horizon in a free fall such

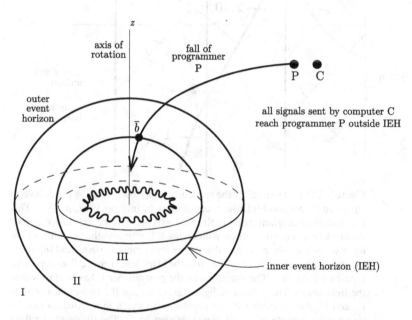

Figure 9.1 Cartoon-like illustration of a slowly rotating black hole. Such a black hole has two event horizons and a ring-shaped singularity. Programmer P can fall towards it such that he receives, before entering the inner event horizon at location $\bar{b}$, any signal the computer C in her infinite lifetime may send after him, and then fall inside and survive forever. The ring singularity can be avoided. The geometry of this spacetime is such that from the inner region P might safely escape to a new infinite universe and repeat the experiment; see Section 9.4 and Figure 9.4

that he can receive, before entering the inner event horizon, all the signals C may send after him in her infinite life.[11]

By this we achieved the infinite speed-up we were aiming for, as represented in Figure 9.2. It can be seen in the figure that whenever C decides to send a photon towards P, that photon will reach P before P meets the inner horizon.

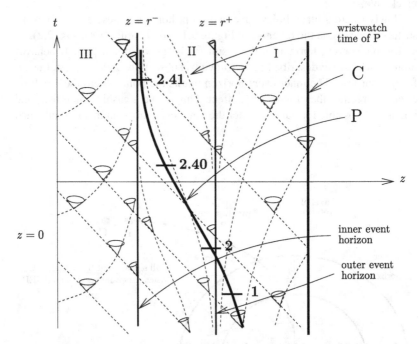

Figure 9.2 The "tz-slice" of the spacetime of a slowly rotating black hole in advanced Eddington-Finkelstein coordinates (d'Inverno 1992, p. 259). This is a spacetime diagram with t as the time-axis and z as the axis of rotation of the black hole. For comparison with Figure 9.1, r^+ marks the worldline of the outer event horizon, r^- marks that of the inner event horizon, and the time-axis marks that of the "center" $z = 0$ of the black hole. Dashed lines represent light-like geodesics. The worldline of the programmer P has to stay within the light cones. The pattern of light cones between the two event horizons r^- and r^+ illustrates that P can decelerate so much in this region that he will receive outside of r^- all messages sent by C. The tilting of the light cones indicates that not even light can escape through the two event horizons. The time measured by P (between the beginning of the experiment and the event when P meets the inner event horizon) is finite while the time measured by C is infinite. A third, different kind of illustration of P's worldline is in Figure 9.4

[11] Details of a possible route are elaborated in Etesi and Németi (2002).

9.2.6   Implementing a Relativistic Computer

We now use the above infinite speed-up scenario to describe a computer that can compute tasks which are beyond the Turing limit. We design a relativistic computer, which, given any recursively enumerable set S and number n as input data, will decide whether n is in S or not.

For input data let us choose the task, for an example, to decide whether ZF set theory is consistent. That is, we want to learn whether from the axioms of set theory one can derive a formula which is false. This formula can be taken to be $x \neq x$. The recursively enumerable set S is the set of theorems of ZF and the computer has to decide whether $x \neq x$ is in S or not. The relativistic computer starts operating as follows: In the beginning, the programmer P and his computer C are together (on Earth), not moving relative to each other, and P uses a finite time-period for transferring the input data, i.e., an enumerating algorithm for the theorems of ZF together with the formula $x \neq x$ to the computer C, as well as for how C should send a signal to P when the formula $x \neq x$ is listed as a theorem of ZF. After this, P boards a huge spaceship, taking all his mathematician friends (relatives, everybody who is important to him, like Noah's Ark), and chooses an appropriate route towards a huge slowly rotating black hole, entering the inner event horizon when his wristwatch shows time b (as in Figures 9.1, 9.2, and 9.4). While he is on his journey towards the black hole, the computer that remained on the Earth checks one by one the theorems of ZF set theory, and as soon as the computer finds a contradiction in set theory, i.e., a proof of the formula $x \neq x$ from the axioms of set theory, the computer sends a signal to the programmer indicating that set theory is inconsistent. If it does not find a proof for $x \neq x$, the computer sends no signal. The programmer falls into the inner event horizon of the black hole, after which he can evaluate the situation. If a signal has arrived from the direction of the computer, of an agreed color and agreed pattern, this means that the computer found an inconsistency in ZF set theory; therefore, the programmer will know that set theory is inconsistent. If the signal has not arrived, and the programmer is already inside the inner event horizon, then he will know that the computer did not find an inconsistency in set theory and did not send the signal, therefore the programmer can conclude that set theory is consistent. So he can build the rest of his mathematics on the secure knowledge of the consistency of set theory. We note that a beneficial aspect of a rotating black hole is that, there, the programmer P does not have to fall into the singularity in a finite time as in an ordinary Schwarzschild black hole. He can avoid the ring singularity, and he can even decide to escape to a new region similar to the one in which he has started out and repeat the whole experiment.[12]

---

[12] This escape requires an extension of Kerr spacetime, as shown in Figure 9.4.

How does the computer C experience the task of this computation? C will see (via photons) that the programmer P approaches the black hole, and as he approaches it, his wristwatch ticks slower and slower, never reaching a wristwatch time b^-.[13] C will see the programmer approaching the black hole in all her infinite time. For C, the programmer shines on the sky for eternity. The only effect of C's time passing is that this image gets dimmer and dimmer, but it will never disappear. Under this sky, C computes away her task consisting of potentially infinitely many steps, i.e., checking the theorems of ZF one by one, in an infinite amount of time.

How does the programmer P experience the task of this computing? He is traveling towards the black hole, and he only has to check whether he has received a special signal from the computer or not. For this task, which consists of finitely many steps, he has a finite amount of time. What would he see if he watched his team-member, the computer? He would see the computer computing faster and faster, speeding up so that when his (P's) wristwatch time reaches b, C would just appear to flare up and disappear. In fact, not just would the computer compute faster and faster, the time in the whole world left behind would tick faster and faster, and would flare up and disappear with the computer. This flare-up would burn P, because it carries the energy of the photons emitted during the whole infinite life of C, which in total is infinite. In fact, we have to design a shield (or mirror) so that only intended signals from C can reach P. This means that we have to ensure that P does not see C! P's task is to watch whether there is one special kind of signal coming through this shield. (We return to protecting P in Section 9.4.1.) All in all, P's task is to do finitely many steps in a finite amount of time.

9.3   Relativistic Computers in a General Setting

Relativistic computers are not tied to rotating black holes, as there are other general relativistic phenomena on which they can be based. An example is anti-de-Sitter spacetime, which attracts more and more attention in explaining recent discoveries in cosmology (Németi and Dávid 2006; Grøn and Hervik 2007). Roughly, in anti-de-Sitter spacetime, time ticks faster and faster at farther away places in such a way that P can achieve infinite speed-up by sending away the computer C and waiting for a signal from her. This scenario is described and is utilized for computing non-Turing computable functions in Hogarth (1996). This example shows that using black holes is not inherent to relativistic computers. It also shows that it is not necessary for the programmer to travel. Wormholes also can be used for constructing relativistic computers;

---

[13] b^- is the time P's wristwatch shows when he falls through the outer event horizon.

see, e.g., Andréka et al. (2012). In this latter scenario, neither the computer nor the programmer has to travel.

From a birds-eye view, the main parts of a relativistic computer as described in the previous section are a computer C (a Turing machine) with an infinite life for computing and a programmer P with two distinguished events in his life, one joint event O with the computer (when P programs C) and another later event E when the programmer can evaluate the results of C's infinite life. This event E is such that C can send a signal arbitrarily late in her life such that it reaches P before E. Intuitively, the event from which C sends a signal which reaches P before E can be thought of as being "earlier" than E, and then one can think that E is "later" than any event in C's infinite life. If we think about it, it is surprising and hard to imagine at all that there can be an event "after" an infinite time has passed (on C's worldline). Such a thing cannot happen in our usual Euclidean space and "normal" time. General relativistic spacetimes with this property help our imagination, since they provide us with a consistent scenario in which this can happen. In this section, we define relativistic computers in abstract terms as consisting of a general relativistic spacetime together with two suitable worldlines. In the next section, we examine some realizability questions this setting raises.

A general relativistic spacetime is a structured set of events. An event is thought of as specified by which entities are present in it. We tend to call these entities (possible) observers, or bodies. The worldline of an observer is the set of all events in which this observer is present. The spacetime structure specifies which subsets of events are possible worldlines of observers and how time passes for these observers on their worldlines. This is done in the following manner. In a spacetime structure $\langle M, g \rangle$, the first component is a differentiable manifold; see, e.g., Wald (1984). It specifies the large-scale topology of the spacetime, and it specifies which subsets of events count as "curves." In more detail, a differentiable manifold is a set M of events together with a coherent system of one-to-one coordinate-functions (or charts) $\lambda_i : M_i \rightarrow R^4$ for $i \in I$, where R^4 is the set of all 4-tuples of real numbers and M is the union of the M_i for $i \in I$. By "coherent" we mean that the worldview transformations $\lambda_j \circ \lambda_i^{-1}$ are smooth functions between open subsets of R^4. A "curve" is a parameterized subset of M that is a smooth curve in all of the charts, or equivalently in some of the charts. The second component in a spacetime $\langle M, g \rangle$ is a metric tensor on M. This means that g at each event specifies which curves passing through this event count locally as "travelable" by an observer, how time flows for this observer, and, further, which curves can be "traveled" by photons. These are called locally timelike and lightlike at the event, respectively. A curve is called timelike if it is locally timelike at each event in it, and similarly for lightlike. The worldlines in the spacetime are the timelike curves. Since the metric tensor g specifies at each event how time locally passes, we can define the "timelike length," called proper time, along worldlines. This is what we called wristwatch time in Section 9.2. A requirement for *g* is that at each event the spacetime locally is like a special relativistic spacetime with a direction of time. A future-infinite worldline is a timelike curve which has infinite length to the future of any of its points. A curve is called causal if it is locally timelike or lightlike. It is thought that information can be sent along causal curves. The causal past of an event E is the set of all events O such that there is a causal curve with O and E on it such that O is earlier than E in the curve.

We are ready to define the notion of an abstract relativistic computer (Definition 9.3.1). For illustration of the definition below, see Figures 9.3(a) and 9.4.

Definition 9.3.1 *A(n abstract) relativistic computer is a tuple $\langle M, g, \gamma_C, \gamma_P, O, E \rangle$, where $\langle M, g \rangle$ is a general relativistic spacetime, γ_C is a future-infinite worldline with O in it, γ_P is a worldline with O and E in it such that E is later than O in γ_P and γ_C lies in the causal past of E.*

A relativistic spacetime is called Malament-Hogarth (MH) if it harbors a relativistic computer. This is equivalent to requiring that it contains an MH event, an event in whose causal past there is a future-infinite worldline. This definition appeared first in Hogarth (1992) and the name was coined in Earman and Norton (1993); see also Earman (1995, ch. 4.3).

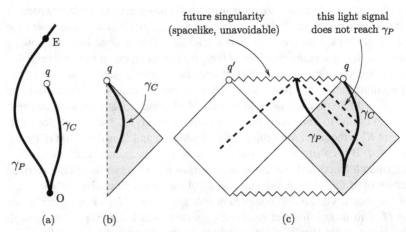

Figure 9.3 A schematic illustration of a relativistic computer as defined in Definition 9.3.1 is shown in (a). Conformal diagrams of Minkowski spacetime and Schwarzschild spacetime are shown in (b), and (c), respectively. It can be seen that these are not MH spacetimes. Lightlike directions are at 45 degrees, and directions which are more vertical than horizontal are timelike. Worldlines of infinite future length are those ending in a hollow circle

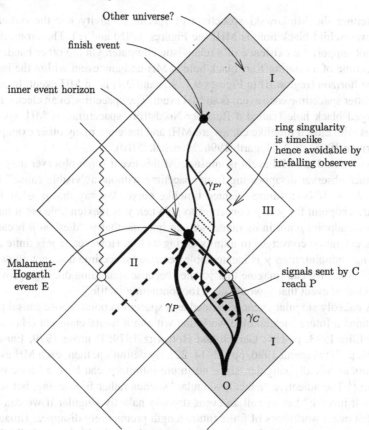

Figure 9.4 Conformal (or Penrose) diagram of a slowly rotating black hole. This is a Malament-Hogarth spacetime. The length of γ_P is finite between O and E, while the length of γ_C is infinite and it lies in the causal past of E. So, $\langle \gamma_C, \gamma_P, O, E \rangle$ is a relativistic computer in this spacetime. We use $\gamma_{P'}$ later, in Section 9.4.1

Conformal, or Penrose, diagrams of spacetimes (see d'Inverno 1992, ch. 17) are especially suitable for illustrating whether a spacetime is MH or not. Conformal diagrams are compactified two-dimensional representations of spacetimes which illustrate causality relations faithfully, but distort metric length. Typically, they represent infinite regions with finite ones. Representing causality relations in conformal diagrams works as follows. Lightlike directions are at 45 degrees, the future direction is upward, and directions which are more vertical than horizontal are timelike. In our diagrams, there are hollow circles, and worldlines ending in these represent future-infinite ones.

Neither the Minkowski spacetime of special relativity nor the ordinary Schwarzschild black hole is MH; see Figures 9.3(b) and (c). Therefore, they do not support the existence of a relativistic computer. On the other hand, the spacetime of a rotating Kerr black hole is MH and any event within the inner event horizon (region III in Figures 9.1, 9.2, and 9.4) is an MH event. In anti-de-Sitter spacetime each event is an MH event. The spacetime of an electrically charged black hole (called a Reissner-Nordström spacetime) is MH, space-times with closed timelike curves are MH, and there are many other examples (Earman 1995, ch. 4; Hogarth 1996; Manchak 2010).

As we said, an MH event is unusual: At this event some observer may see another observer disappearing from spacetime without a "visible cause." Let $\gamma : I \to M$ be a future-oriented timelike curve. We say that $p \in M$ is a future endpoint for γ if γ converges to p. When γ is inextendible but it has a future endpoint p not in its range, we can imagine that γ "dies" at p because its acceleration converges to infinity during its life. However, if γ is finite and has no endpoint, then γ is inextendible because "spacetime does not allow its extension." We can imagine that γ "dies because spacetime does not provide the kind of event that γ would need for continuing its life."

A nakedly singular event is defined as a spacetime point whose causal past contains a future-inextendible worldline without a future endpoint (Hawking and Ellis 1973, p. 184; Geroch and Horowitz 1979; Penrose 1979; Earman 1995, p. 74; Hogarth 1996, pp. 13–14, 22). By definition, then, each MH event counts as nakedly singular, since no future-infinite γ can have a future end-point.[14] The adjective "nakedly singular" sounds rather frightening, but what does it involve? Let us call an event strongly nakedly singular if we can see at that event worldlines of finite future length prematurely disappear (imagin-ing that some horrible disruption of the spacetime called "singularity" killed the observers at that point of their lives). Our MH events are only "weakly" nakedly singular in that the worldlines that we see disappear are of infinite future length. We may see nothing unusual about an observer having this worldline. Why do we call MH events singular, then?

Singularities, or singular behavior, in the general relativistic spacetime lit-erature are connected with break-down of predictability and determinism (Earman 1995, p. 66). A function assigning real numbers to the events in a spacetime is called a global time. We say that a global time supports Lapla-cian determinism, roughly, when there is a time-instance such that (a) events

---

[14] Here is a simple proof that arose in a correspondence with J. B. Manchak: Let $\gamma : I \to M$ be a future-oriented future-infinite timelike curve and let $p \in M$ be arbitrary. There is a neighborhood O of p which does not contain future-infinite timelike curves. This follows from g being locally special relativistic and from the continuity of g. Let $t \in I$ be arbitrary. Then γ restricted to $\{s \in I : s \geq t\}$ is still a timelike curve of infinite length, thus O cannot contain it. This shows that γ does not converge to p.

with this time-tag form a reasonable "now," and (b) the arrangement of certain quantities computable in the spacetime at the events belonging to this "now" determines the whole spacetime. A requirement in (a) is that this "now" forms a spacelike hypersurface and the quantity in (b) is the Einstein-tensor. For a concrete definition, see e.g., Earman (1995, ch. 6.3). This is a rather strong, global kind of determinism and spacetimes satisfying it are called globally hyperbolic. Hogarth (1992) proved that no MH spacetime is globally hyperbolic. This is the theorem that justifies calling MH events "singular."[15]

Thus, in order to have MH events, we have to give up the existence of a global time supporting this strong determinism. However, assuming a weaker type of global time is consistent with MH spacetimes: see the definition of stable causality in Section 9.4.1.

How strange are MH spacetimes from a physical perspective? First, there are many MH spacetimes without strongly nakedly singular events, e.g., anti-de-Sitter spacetime. Here, as well as in Gödel's rotating universe, each event is weakly nakedly singular and there are no strongly nakedly singular events. Further, global hyperbolicity has not been generally accepted as a necessary condition for a spacetime to be physically realistic (Wald 1984; Earman 1995, pp. 97–99; Manchak 2011; Etesi 2016). Hogarth (1996, p. 83) emphasizes and explains that the presence of weakly nakedly singular events in and lack of global hyperbolicity of MH spacetimes does not render these spacetimes physically unrealistic. E.g., MH spacetimes do not have to be singular, which, following Hawking and Ellis (1973), means being timelike or lightlike geodesically incomplete. Hogarth emphasizes that there are nonsingular MH spacetimes in which carrying out the computation seems to be unproblematic. For example, anti-de-Sitter spacetime is such. Moreover, Manchak (2017) shows that there is a clear sense in which general relativity allows for MH spacetimes to be brought about.

The above clears, we hope, several worries in the literature. E.g., Piccinini (2011) writes that MH spacetimes contain singularities and this might be an obstacle for relativistic computing. This worry is also quoted in Manchak and Roberts (2016). As we outlined in the discussion above, it is mainly due to a matter of definition that MH events count as (weakly) nakedly singular: the breakdown of the kind of determinism they bring about (i.e., lack of global hyperbolicity) is not generally accepted as necessary for being physically realistic.

---

[15] Penrose's strong Cosmic Censorship Hypothesis (CCH) prohibits the existence of nakedly singular events. Etesi (2013) describes many other interesting connections between variants of the CCH and relativistic computing.

Is our own universe MH? Many of the MH spacetimes, like rotating black holes, may be built by a future, advanced civilization inside our usual "standard" universe according to high-precision cosmology (Grøn and Hervik 2007, ch. 12). Further, astronomers have observed objects that they think with great confidence are large rotating black holes (Niedzwiecki and Miyakawa 2010; Abbott et al. 2016). A typical example is the rotating 100-million-solar mass black hole at the center of galaxy MCG-6-30-15.

9.4   Physical Realizability

In this section, we briefly discuss some realizability/engineering questions in connection with relativistic computation. We concentrate on the plausibility of having reasonable communication between C and P, and on the assumption of C having an infinite life.

9.4.1   The Blue-Shift Problem

The blue-shift problem concerns communication between the computer and the programmer. Copeland (2002a) emphasizes the importance of this communication between C and P, pointing out that an accelerating Turing machine as in Section 9.2.1 cannot compute anything more than an unaccelerating one without a means to communicate with the "external world."

The blue-shift problem is a side-effect of our effort to make an infinite speed-up, but this side-effect can be cured with sufficient care. The essence of the problem is that if C sends a message of, say, 100 successive signals in a time period of 1 second (C's time), the programmer P may receive these 100 signals in an arbitrarily small time interval (P's time). By the same token, a light signal (photon) sent by C can get arbitrarily blue-shifted when received by P. Thus the energy of a photon sent by C can increase arbitrarily. This in turn might kill the programmer and also the programmer might not be able to decode the message.

For formulations of the blue-shift problem, see Earman (1995, p. 111) and Manchak (2010, p. 279). In these formulations, one assumes that the programmer and the computer communicate via light signals: signals whose worldlines are lightlike geodesics. We say that we have a bounded blue-shift for an MH event together with the infinite worldline of C if there is an arrangement for P's worldline together with a series of possible light signals from a cofinal set of events of C to P such that the rate of frequency of a light signal in this arrangement as observed by C and P, respectively, is bounded. Otherwise, we say that we have a divergent blue-shift (Earman 1995; Manchak 2010).

A spacetime $\langle M, g \rangle$ is called stably causal if it possesses a global time function $t : M \to R$, where R is the set of real numbers, which respects the flow of time on worldlines, i.e., whenever two events p and q are connected by a future-directed worldline such that p is earlier than q in this worldline, then $t(p) < t(q)$.[16] Earman (1995, ch. 4.5) proves that, among the stably causal spacetimes, the MH ones are exactly those with a divergent blue-shift, if we assume the following hypothesis: The family of light-signals that C can send to P forms a two-dimensional integral submanifold in which the order of omission from C matches the order of reception at P. However, Manchak (2010) constructs a stably causal MH spacetime with bounded blue-shift, and his construction exhibits many other desirable properties for a relativistic computer. He thus shows that in order to avoid the blue-shift problem we do not have to give up even the existence of a "good" global time-function, we only have to give up the above-cited hypothesis about a strong concert between the light-signals.

However, it is not necessary for C and P to communicate via light-signals.[17] Moreover, we do not need to stick with "geodesic messages." Several alternative means are offered in Németi and Dávid (2006, sec. 5.3).[18] Here we recall one of those.

First of all, C does not send any photons towards P. When she wants to send a message to P, the computer C sends a postman P' after P and asks P' to carry her message to P. This P' is in a second spaceship chasing P and meeting P behind the inner event horizon sometime after the MH event but before an agreed-upon time limit, e.g., before P re-enters region I at the "finish event" as shown in Figure 9.4, where the worldline of postman P' is also indicated. This P' can be engineered in such a way that P' catches up with P and hands the message to P, e.g., in writing on paper (or some other material coding).[19] So, after the finish event, P comfortably reads the message – say, the proof of the inconsistency of ZF – and can decide whether or not it is correct. There are helpful "roadsigns" along this spacetime helping P to navigate. E.g., there are observable effects indicating that P has reached the finish event, namely, the light of the "new universe" (the second appearance of region I) becomes visible for P when he reaches it. P does not have to wait for any information after the "finish event." Summing up: P can decide whether ZF is consistent in

---

[16] This is not the original definition; we use Earman (1995, p. 166).

[17] Hogarth (1996, p. 87) already suggests communicating by other means than photons. He suggests to consider "switching the lights off" and to "operate and communicate as purely mechanical devices (in a kind of billiard ball model)." Manchak (2010, p. 279n) also hints at this possibility.

[18] A rather ingenious alternative way of communication is explored by Wüthrich (2015). He proposes that C keeps one of an entangled pair of particles while P takes the other with him.

[19] The details depend on whether there are limits to compressing information coded in a given amount of physical material. If there are no such limits then P' carries the whole proof to P. If there is such a limit, P' carries only selected information about the existence and nature of the proof.

bounded time (as we wanted). Namely, if ZF is inconsistent, a postman P′ will appear before the finish event, hence P learns about the inconsistency. (Even before this event, P will receive a light signal soon after crossing the MH event from P′ indicating "you have mail, a message is coming to you.") If, on the other hand, ZF is consistent, P′ does not show up, and after hitting the finish event, P will conclude that ZF is consistent.

An advantage of using a postman P′ for carrying the message to the programmer over using a photon or inertial body for carrying the message is the following. As opposed to the photon, P′ can navigate intelligently and can adapt to situations arising during his carrying of the message. Further, P′ can match velocities with P before sending the "you have mail" warning and before transferring the message itself.

A future task is to check whether C can send postman P′ after P with the message arbitrarily late in her life so that there is a common bound for the acceleration for these possible postmen. We believe that in the slow-Kerr spacetime this can be arranged.

9.4.2   Infinity

A worry about the present project might be that it requires unbounded time and space for the computer C to carry out her task. Luckily, present-day cosmological data (results of measurements) point in the direction that probably our universe is infinite both in time and space (Grøn and Hervik 2007, chs. 11.10, 12.8). For a similar conclusion, we also refer to Earman (1995, p. 119) and Németi and Dávid (2006, p. 137). Operating C also seems to require an unbounded amount of energy or matter. This worry is dealt with in detail in Németi and Dávid (2006, sec. 5.4.3). Here we concentrate on one specific problem connected to infinity.

Piccinini (2011, 2017) finds it unlikely that any physical machinery can flawlessly operate for an infinitely long time. He says that the probability of error for such a machine is 1. We think that the example of the infinite jar with more and more balls in Manchak and Roberts (2016, sec. 1.3) shows that one has to be careful when generalizing from cases of unbounded finite time to infinite time. We briefly recall the example from Manchak and Roberts (2016). Suppose we have a jar with the capacity to hold infinitely many balls. We also have a countably infinite pile of balls, numbered $1, 2, 3, 4, \ldots$ First we drop balls 1–10 into the jar, then remove ball 1. Then we drop balls 11–20 in the jar, and remove ball 2. Suppose that we continue in this way ad infinitum, and that we do so with ever-increasing speed, so that we will have used up our entire infinite pile of balls in finite time. How many balls will be in the jar when this task is over? One may answer: none. Namely, ball 1 was removed at the first stage, ball 2 was removed at the second stage, ball n was removed at the *n*th stage, etc. Since each ball has a label *n*, each ball was removed at some stage, thus there will be no balls left in the jar at the end. This is so in spite of the fact that the number of balls in the jar tends to infinity as the number of stages approaches infinity. The case with probability 1 in Piccinini's argument is analogous to this example. The MH event is analogous to the end of the drawing process when there are no balls in the jar, and the error probability is analogous to the number of balls tending to infinity at every particular stage. Below we briefly describe a kind of management of time and space which seems to allow for infinite operation with a small error probability.

Analogously to the way life works on the Earth, instead of implementing the computer C by a single Turing machine, the programmer can use a fleet of self-reproducing robots which carry out the task of C in a massively parallel self-correcting fashion. Each robot lives for a finite time, and in each time instant there are finitely many robots. However, both numbers are unbounded, as our fleet of robots follow the strategy that sometime after each time, the computation is independently started from the beginning by more and more new robots with more and more life-expectancy. This kind of organization seems to ensure that if P did not receive a signal, then he would have confidence that there does not exist a proof of $x \neq x$ which the "society" of robots overlooked, e.g., because an accident killed all the robots. If one of the robots found a proof of the inconsistency of ZF, this would not be sent to the programmer P automatically, but instead would be double-checked by the other robots and only in the case of a consensus would it be sent to P.[20]

9.5   Computational Power

In computability theory, there is a hierarchy for how hard it is to compute a task. We recall the basic notions from Cooper (2004). First, one codes the task to be about a set of numbers. For example, one gives an algorithm for coding the set-theoretic formulas as numbers, and then one considers the task of deciding whether the code of the formula $x \neq x$ is in the set of codes of theorems of ZF. After this, we deal only with sets of numbers.

A set *S* of natural numbers is called arithmetical if it can be defined by a formula of the first-order language having function symbols for zero, successor, addition, and multiplication, and a relation symbol for the ordering of natural numbers. This means that *S* is called arithmetical when there is a formula $\varphi(x)$ of this language with one free variable *x* such that $\varphi(n)$ is true in the structure of natural numbers if and only if $n \in S$. A similar definition holds for an *n*-place

[20] This kind of management of time and space with massive parallelism, redundancy, and self-correcting processes like the way evolution of life on Earth works is presented in Németi and Dávid (2006, sec. 5.4).

relation S of natural numbers. An n-place relation is called Σ_1, or recursively enumerable, if it is definable by a formula of the form $\exists \bar{y}\varphi(\bar{y}, \bar{x})$, where φ is a quantifier-free formula, $\bar{y}$ and $\bar{x}$ are sequences of variables, and the length of $\bar{x}$ is n. Continuing this way, assume that we have defined Σ_k relations. A relation is called Π_k if its complement relation is Σ_k, and a relation is called Δ_k if it is both Σ_k and Π_k. We define a Σ_{k+1} relation as one that can be defined by a formula of the form $\exists \bar{y}\varphi(\bar{y}, \bar{x})$, where $\varphi(\bar{y}, \bar{x})$ defines a Π_k relation. It can be proved that a relation is arithmetical if and only if it is Σ_k or Π_k for some k. A Δ_1 relation is called decidable. Thus, a relation is decidable if and only if both it and its complement are recursively enumerable. It can be proved that a relation is decidable in the just-defined sense if and only if there is a Turing machine that can decide about a sequence of numbers whether it is a member of the set, and a problem is recursively enumerable if there is a Turing machine that can enumerate the members of the set. There are recursively enumerable but not decidable sets, e.g., the set of all valid first-order formulas. A decidable task is called solvable or computable while all others are called non-computable or non-solvable.

The train of thought outlined in Section 9.2 can be used to show that any recursively enumerable set can be decided by a relativistic computer (Etesi and Németi 2002). Thus, relativistic computers can compute more than Turing machines can. The PhCT is formulated in Copeland (2015) as "whatever can be calculated by a machine (working on finite data in accordance with a finite program of instructions) is Turing-machine-computable." Since our relativistic computer operates on finite data and according to a finite program, it challenges the PhCT. Actually, relativistic computers can do more than decide the recursively enumerable sets. Computability limits connected with relativistic computers are addressed by several authors (Wiedermann and van Leeuwen 2002). For example, Welch (2008, thm. B) proves that the relations computable in Kerr black holes as presented in Section 9.2.6 form a subclass of the Δ_2 relations, and this is a proper subclass if and only if there is a fixed finite bound on the number of signals sent to the programmer on the finite length path. On the other hand, Hogarth (1996, 2004) proves that there are other MH spacetimes in which the members of the arithmetic hierarchy can be decided: for all k he constructs a spacetime in which all Σ_k relations can be decided by a relativistic computer.

9.6   Conclusion

We have used results of general relativity and cosmology to design a relativistic computer that can compute a non-Turing computable function. We described, in the spacetime of a slowly rotating black hole, a scenario of a team consisting of an ordinary Turing machine computing for an infinite time and a programmer who can survey in finite time the result of this computation. We investigated how close this scenario can be brought to being physically realizable. A contribution of the present chapter is that we answer some concerns voiced in the literature.

Piccinini (2011) closes his discussion of relativistic computing with two worries, namely that the error probability of an infinite computation is 1, and that spacetimes supporting relativistic computers contain singularities. We answered the first concern in Section 9.4.2, and we answered the second one in Section 9.3. Another worry in the literature concerns communication between the computer and the programmer. The new feature we propose, in Section 9.4.1 – relative to Earman (1995, p. 118), Manchak (2010), or Etesi and Németi (2002) – is that we send the message not via light signals but via a postman (not inertial and slower than the speed of light). It is a future task to check whether in the Kerr spacetime there is a sequence of worldlines for postmen starting from a cofinal set of events in the worldline of the computer such that there is a common bound for the acceleration in these worldlines.

Part IV

Computational Perspectives on Physical Theory

10   Intension in the Physics of Computation: Lessons from the Debate about Landauer's Principle

James Ladyman

Thermodynamics is a funny subject. The first time you go through it, you don't understand it at all. The second time you go through it, you think you understand it, except for one or two small points. The third time you go through it, you know you don't understand it, but by that time you are so used to it, it doesn't bother you any more.

–Arnold Sommerfeld

10.1   Introduction

The physics of computation is a broad subject that involves quantum physics, the Church-Turing thesis, and other deep issues. There are very divergent views about its philosophical implications. At one extreme is the putative reduction of physics to computation that reifies notions such as data and information, and takes them to be fundamental and not reducible to standard physical properties (John Wheeler's [1990] "it from bit" exemplifies such a view). At the other end of the spectrum information is regarded as entirely anthropocentric and physical computation is taken to depend on the interpretation of physical processes by intelligent agents.

Computers are defined in terms of data, information and logical operations. There are no physical constants or magnitudes that directly correspond to such concepts. Actual computers are physical systems of one kind or another. As such they are subject to the laws of thermodynamics. Prima facie there is no reason to think that there should be systematic connections between the thermodynamic properties of processes and the abstract features of the computations they implement.

Landauer's principle (LP) posits a direct connection between thermodynamic entropy and *irreversible* computations.[1] LP is hotly contested, and even its formulation is contentious. Most crudely put, LP says that destroying information produces entropy. Rolf Landauer is often said to have had an important insight that "[i]nformation is physical" (Landauer 1991). Philip Ball takes this

---

[1] What is meant by "irreversible" is defined in Section 10.4.

219

literally and says LP "in effect implies an equivalence between information and heat: In other words, information itself can be converted into heat" (Ball 2013, p. 39). These are striking claims and their assessment involves both the foundations of computation and the foundations of physics.

In this paper the key issues in the debate about LP are identified, and important lessons are drawn about how our view of computation and physics relates to and is informed by the interpretation of thermal physics. It is argued that there are several respects in which modality and how processes and systems are represented are crucial to both LP and the application of thermodynamics. Exactly what LP says, and why we might believe it, is explained in further detail in Section 10.3 and thereafter. The formulations considered do not entail strong claims about the ontological status of information, but it is true that on any reading of LP, it is at the center of what John Norton calls "the new science of the thermodynamics of computation," which is, he continues, "no science at all" (2013a, p. 4433). The orthodoxy among physicists seems to be that LP is correct and that thermodynamics and computation are linked (see, for example, Blundell and Blundell 2010), but Norton marshals powerful counter-arguments against the arguments for LP. He argues that the putative experimental confirmation of LP (Berut et al. 2012) does not confirm it, just standard thermodynamics.

What does it mean to say a physical system implements a computation?[2] Without an answer to this question, one cannot even precisely formulate LP, let alone assess its truth. Landauer generalizes from one example of an idealized device that is naturally taken to implement the resetting of one bit of data. In order to formulate a general proof of LP, Ladyman, Presnell, Short, and Groisman (2007) (hereafter "LPSG") analyzed implementation. Their model of the latter in terms of the notion of an L-machine is described in Section 10.4 as an exemplification of a general account of computation and its physical realization. Section 10.5 reviews the debate that followed the LPSG argument for LP, and shows how the way putative protocols and processes are modeled is pivotal, particularly in respect of the modal properties of states, the processes that connect them, and the role of representation.

Norton's response to LPSG involved a much broader critique of idealizations that are standard in the literature. Section 10.6 argues that Norton's no-go theorem for the thermodynamics of computation presupposes a particular view of how the physical processes in question are supposed to proceed and argues against it. Section 10.7 considers LP in the light of work by David Wallace (2014) on the nature of the Second Law of thermodynamics, and the relationship between thermodynamics and statistical mechanics. Section 10.8 considers the relationship between thermodynamic entropy and information

---

[2] This is the title of Ladyman (2009).

theoretic entropy, and argues that thermodynamic states are intensional, in the sense that what we know about them and how they are represented matters to the physics of what can be done with them. In the conclusion, it is argued that both computational and thermodynamic states are intensional, but that the modal structure of physical states that represent them can nonetheless be taken to be an objective component of implementation. In the next section, it is explained why representation and modality are fundamental for our understanding of what is required for realism about computation.

10.2   Computation

Realism about computation is the view that whether or not a particular physical system is performing a particular computation is at least sometimes an objective matter that does not depend at all on our beliefs, desires, and intentions. Realism about computation is challenged by Hilary Putnam (1988, pp. 120–125) and John Searle (1992), among others. According to Putnam and Searle, physical systems do not implement computations just in virtue of the patterns of physical activity that actually occur within them. If these are the only features of physical systems about which we should be realists, it follows that we should not be realists about computation. On their view, systems only ever implement particular computations in virtue of being interpreted as doing so. They both argue this by arguing that physical systems of sufficient complexity can be construed as implementing any computation whatsoever. For example, Putnam argues that every ordinary open system realizes every finite automaton, because any sequence of the successive temporal states of such a system can be mapped onto the successive computational states of the automaton.

As Chalmers (1996) discusses, one response to the above form of argument is to claim that for a particular computation to be implemented by some system depends not only on what it does, but on its modal features. It is not sufficient that the states that actually form the particular time evolution of the system that takes place on that occasion have enough complexity to be mapped to the sequence of computational states of the computation in question. This is because for the sequence of physical states to instantiate the computation in question requires both that there is a corresponding set of physical states for the other branches of the computation resulting from a different initial computational state, and that they would have occurred had the system been in the corresponding initial physical state. For example, consider a handheld calculator. Suppose the appropriate buttons are pressed to calculate the sum of 7 and 5 and the screen outputs 12. This only counts as a calculation of the sum because, had different buttons been pressed, it would have given a different output. Similarly, a logic gate that gives the correct output "1" when the inputs represent "1" and "1" is only computing AND if, *had* the inputs represented

"1" and "0," it would have given the output "0." Computation is modal (cf. Maudlin 2011, p. 149).

It is possible to consider a computation that takes a single state to a single state (it can be thought of as the identity map on the empty set), but obviously, in general, functions are maps from many different values in the domain of the function to values in its range. Any particular physical process can only directly represent a particular instance of the function, because it is an evolution from a state that represents one value in the domain to a state that represents the appropriate value in the range. Hence, in any interesting case of computation, strictly speaking, a physical process cannot be said to implement a logical transformation at all, because all it can ever do is implement the part of the map that takes one of the logical input states to another logical input state. In terms of the truth table that represents the logical transformation AND, for example, clearly a particular physical process could only be said to implement a single row of it. (These points are recapitulated in Section 10.4.)

It follows that, for there to be facts of the matter about what computation a physical system implements on some occasion, there must be facts about what it would have done were its initial state different. The upshot would appear to be that realism about modality is a necessary component of realism about computation, where realism about modality is the claim that there are mind-independent facts about counterfactual as well as factual matters.

However, even this is not sufficient for realism about computation, because we must also suppose that there is a fact of the matter about what the physical states of the system represent. In the above examples, it is only because the interpretation of the states of the physical systems in question is held fixed that they can be said to implement a particular computation. Hence, realism about computation is closely related to realism about representation. Putnam and Searle are right that there is at least something arbitrary about the map that takes bits, or other units of information, to physical states. For example, an AND gate can be used as an OR gate, if the interpretation of the input and output states is inverted.

Standardly, signifiers seem to be arbitrary. For example, the words on this page only mean what they do because of contingencies in the evolution of written English. Such representation seems not to exist independently of human beliefs, desires, intentions, and the history of the system, and so to undermine realism about computation. On the other hand, maps and pictures have structural similarities with what they represent. Hilary Putnam (1981) considers the case of an ant that crawls across sand leaving a trace that happens to look like a passable caricature of Winston Churchill. He argues that it is nonetheless not a representation of the man because the ant did not intend it as such, and in general rejects the possibility of naturalizing representation. On different grounds, van Fraassen (2008) argues that scientific representation, while illuminatingly considered in terms of isomorphisms and other kinds of mapping relations, is pragmatic in the sense of depending on agents and their perspectives and purposes.

However, many philosophers and scientists, such as Ruth Millikan (1984), have advocated theories according to which at least some representations can be considered natural. Animal signaling systems and human natural language are understood using "teleosemantics," according to which a property or state of an organism represents some feature of the world in virtue of the adaptive value of carrying the relevant information about the world. On such views, representative content is tied to biological function.[3] This would imply realism about the computations that occur in the nervous systems of organisms, and about natural computation in general. The states of systems that have not evolved according to natural selection could not then be assigned objective representational roles on the basis of their natural functions. However, some extension of teleosemantics to those used by biological agents might be viable.

Realism about computation, as defined at the start of this section, seems to require both realism about modality and realism about representation. If realism about representation is dropped, it can still be maintained that both the aptness of systems to be taken to represent computations of various kinds, and what follows once representations are fixed, are nonetheless objective. It may also be argued that once it is required that the labeling system should be less complex than the computation imputed to the physical system, it is far from trivial to regard a system as computing. For a physical system to be used in practical computation it is required to have a modal structure of considerable complexity, with physical states that can be reliably prepared and observed, and an emergent effective dynamics between them that can be reliably initiated. The above analysis is incorporated into the definition of L-machines in Section 10.4. Section 10.8 argues that LP becomes less contentious if realism about computation is weakened. The next section explains LP and how it is usually motivated.

10.3   Landauer's Principle

LP posits a surprising direct connection between logic and thermodynamics. The qualitative version of LP says roughly that "irreversible" computations are dissipative/increase entropy, and the quantitative version says that "erasure" of a single bit increases the entropy of the system by $k \ln 2$ (where k is Boltzmann's constant). It is widely assumed to connect thermodynamics and information-theoretic/computational reasoning in diverse areas of physics,

---

[3] See also MacDonald and Papineau (2006). Sprevak (2005) defends realism about computation using a teleosemantic account of representation.

including cosmology. It is also often appealed to in discussions about the status of the second law of thermodynamics (SL). This is because some models of Maxwell's demon posit an agent that finds out about the state of a system and controls a device that makes random changes in the microstate add up to a macroscopic violation of the SL. LP is taken to be relevant because the agent must wipe its memory for such a system to perform a cycle.

Both versions of LP are apt to mislead. The qualitative form does not make clear what it means for there to be an irreversible computation in the physical world, and the ideas of heat dissipation and increasing entropy are different, with the entropy definable in many different ways – the most important differences being between the thermodynamic entropy originally introduced to model changes in the properties of bulk matter in cycles run by heat engines, and extensions and analogues of thermodynamic entropy in statistical mechanics and information theory.

The quantitative form of LP is ambiguous, because there is more than one way of erasing information. The more general notion of erasure requires only that the final state is probabilistically independent of the initial state. On the other hand, "RESET" requires that the final state is a fixed particular state. Clearly RESET is a form of erasure, but the converse is not true. Randomizing data is a form of erasure, but not a form of RESET. In practical computation it is usually RESET that is meant by deleting information not randomizing. LP was originally formulated by Landauer (1961) in terms of RESET, but the literature is often not clear about this or the difference between different forms of erasure (in particular, Norton does not take account of this distinction).

10.3.1   Arguing for LP

There are two main kinds of arguments for LP:

1. **The compression of phase space argument**

   This argument is based on the Liouville theorem, which is fundamental to both equilibrium and non-equilibrium statistical mechanics and has an analog in quantum statistical mechanics. It states that the distribution function, which determines the probability of the system being found in some phase space volume, is constant along any trajectory in phase space. This is glossed as the claim that the volume of phase space *accessible* to a closed system is constant.

   In the implementation of RESET, the trajectories that represent different initial states of the degrees of freedom taken to bear the information have to merge since both have to arrive at the same final state. This means that the phase space accessible to the "information-bearing degrees of freedom" must decrease. If the Liouville theorem is applied to the system alone, it would appear to be violated, but if it is applied to the system plus the environment, then it can be maintained if there is an increase in the phase space accessible to these "non-information-bearing degrees of freedom" (Landauer 1991, p. 24).[4] In statistical mechanics, entropy is associated with phase space volume, so if the phase space volume of the environment has to increase on RESET, then this amounts to the production of entropy.

2. **Consideration of particular systems**

For example, a logical bit can be represented by a one-molecule gas in a box with a partition. If the state is erased by the partition being pulled out and a piston being used to compress the gas before the partition is reinserted, then work must be done and there is a heat flow to the environment and hence an entropy increase.

10.3.2   Norton's Critique

Earman and Norton (1999) point out that there are two ways to connect LP and SL. What they call the "profound" approach, which is alluded to above, seeks to show that SL holds despite the apparent possibility of a Maxwell demon by arguing that an intelligent demon must end an SL-violating cycle in the same state in which it began. This requires that it wipe its memory and so, it is argued, LP shows that such a cycle must increase entropy after all. Earman and Norton and Norton in further work (2005; 2013c) point out lots of confusion in the literature about such use of LP in putative exorcisms of Maxwell's demon. Norton makes many acute criticisms of the latter project, which he interprets in terms of the perceived need to reconcile SL with the fact that, if sufficient microscopic "violations" of it, in the sense of trajectories that take the system away from equilibrium, are accumulated, it would be false. This leads him to consider demons that are supposed to accumulate fluctuations from equilibrium by some mechanical device such as the famous ratchet and pawl device considered by Feynman et al. (1966, ch. 46). However, it is tendentious to describe such devices as being intended to work by the accumulation of "molecular-scale violations of the second law" (Norton 2013a, p. 4433) to give a macroscopic violation, since it is not at all clear that fluctuations should be treated as violations of SL; see Section 10.7.

In any case, all this is irrelevant to the merits of accounts of LP that take it to be a consequence of the SL, these following what Earman and Norton call the "sound" approach. Sound approaches deny that LP can be used to demonstrate

---

[4] As discussed in the previous section, physical degrees of freedom are "information-bearing" only if they are taken as representing computational or logical states. The significance of this is discussed further in Sections 10.8 and 10.9.

that there are no Maxwell demons, and allow that at most LP can be used to illustrate why certain Maxwell demons don't work. Bennett (2003) concedes that only the sound approach is viable.

Proponents of the sound approach still have to demonstrate LP somehow and Norton argues against the two forms of argument in the previous subsection as follows.

1. The compression of phase space argument is invalid, he argues, because the volume of accessible phase space is the same before and after RESET, because both before and after RESET the system is in one of two states, each of which has the same phase space volume associated with it. The fact that before RESET we do not know which of the two states it is in, whereas after RESET we do, is not taken to be relevant (but see Section 10.7).
2. Consideration of specific examples of implementation of erasure that produce entropy does not show that LP holds generally.

Both of these arguments deserve further attention. For example, in respect of the second, consideration of specific examples might form the basis of a reasonable inductive inference, especially if nobody is ever able to exhibit a schema for performing erasure that is thermodynamically reversible.[5] In any case, improving on the above arguments for LP and advancing the discussion requires an analysis of the notion of the implementation of a computation by a physical system so that LP can be precisely formulated. This is the subject of the next Section.

10.4   Computation and Physics: Implementation and *L*-machines

Norton's critique of LP correctly diagnoses confusion in the literature discussing it. This is in part because there is confusion about what is meant by a computation, and certainly much discussion of LP fails to make use of an explicit account of implementation, and so makes even a precise statement of LP impossible.

Computation is often represented in terms of Turing machines which are abstract structures that can be identified with sets. Turing introduced his machines to capture the idea of what could be computed by an effective procedure that could in principle be followed by a human being. The class of Turing-computable functions is equivalent to the class of functions computable by the Lambda calculus, Post machines, Markov algorithms, and Kleene's Formal systems. However, there are notions of computation with respect to which

---

[5] Norton's putative counter-examples are shown not to satisfy the conditions for an implementation developed below and in Ladyman and Robertson (2013).

Turing non-computable functions are computable. Whether such hypercomputation is physically realizable is left as an open question here. In any case, the idea of Turing machines does not help us at all with the physics of computation as it includes no account of implementation.

The approach taken here is to isolate the minimal conceptual structure of computation and implementation, and to avoid features not compelled by the analysis. The model developed below, while designed to address LP, is completely general. The key features of computations and their implementation in physical systems are as follows.

1. Paradigmatic computations include arithmetical sums, finding prime factors, sorting lists, compressing data, and simple logical operations such as AND and COPY.
2. In all such cases a computation can be thought of as a single-valued map L from a finite set X of input states, into a finite set Y of output states. The states and the map are abstract entities.[6]
3. Often the maps of interest are logical transformations. For example, consider the case of binary-valued logic, in which the input and output states are bit strings (with 0 and 1 usually representing "false" and "true," respectively); the mapping L could be a truth table, or a digital circuit constructed from some set of universal gates (e.g., NAND and COPY).

   LP says there is something physically different about logical operations depending on whether they are reversible, in which case knowledge of the operation and the output is sufficient for knowledge of the input, or irreversible, in which case more than one input state is mapped to a single output state.
4. A logical transformation is *logically reversible* if and only if $L : X \to Y$ is a one-to-one (injective) mapping. If L is a many-to-one mapping, then it is *logically irreversible*.[7]

   Computations are actually implemented and realized only if abstract computational states are represented by physical systems. For example, numbers are encoded by bit strings, which are then directly represented by electromagnetic structures within chips that evolve in time in such a way as to represent the computation of, say, addition.
5. Implementation means that a physical system is taken to represent abstract mathematical or logical states. The representation relation is between physical degrees of freedom and the input and output states of the computation.

---

[6] Landauer (1991, p. 23) is explicit about this.

[7] Note that whether or not L is surjective is irrelevant because if there are output states that do not get mapped to from any input states they are irrelevant to thermodynamic considerations about the implementation of the computation because there is no need to take the physical system to have states that represent them. One-to-many maps are ruled out by feature 2 above.

There is also a derivative representation relation between logical operations and physical processes.

6. The same physical degrees of freedom may or may not represent both input and output logical states.

The analysis of implementation should be neutral about whether representation is a matter of stipulation and thus completely anthropocentric and conventional, or is natural and independent of our choices.[8]

In general, computations have multiple inputs and outputs. Hence, any logical transformation is a map from a *set* of logical states to a *set* of logical states, but, as pointed out above, any actual physical process is a change in a physical system whereby it goes from a *particular* physical state to a *particular* physical state. Actual physical processes are unique time evolutions of token physical states.[9]

For example, RESET is the map that takes 0 to 0 and 1 to 0. Any actual physical process instantiates a particular initial-to-final state transformation. If the initial state is taken to represent 1 and the final state 0 then that physical process, that "computation," is just one of the two transformations that constitute ⸱RESET. Hence to say that this is an example of the implementation of RESET requires also that, had the initial state been that corresponding to the other input state of the computation, the resultant physical final state would have been the right representative output state. It is important to bear in mind this model or some alternative conceptual structure satisfying the above four features when discussing physical implementations of RESET in the context of LP. It is not sufficient just to imagine some particular physical process that could represent a particular part of a computation such as $1 \rightarrow 0$.

7. For a physical system to implement (a nontrivial) logical transformation there must be a *family* of processes and each of the physical states that represent the logical input states must be taken by one member of the family to the physical state that represents the appropriate logical output state.

A process in some physical system can only be said to implement a computation if the former is one of a family such that, had the input been different, the relevant other member of the family would have evolved into the appropriate output. The notion of implementation

---

[8] Note that it is possible to hold an intermediate position according to which there is natural, in the sense of non-anthropocentric, computation only in some physical systems such as DNA and associated molecules in cells, or other biological systems, because only for them does teleosemantics allow for realism about representation as suggested in Section 10.2.

[9] Respecting the above distinction between the abstract states of computations and the physical states that implement them, to avoid confusion it is best not to refer to logical "processes," but rather to logical transformations and their implementation by physical processes.

of a logical transformation by a physical device presented in the next section incorporates this essential feature of implementation and further explains it.

8. The dynamics of a system is essential to implementing a computation. The family of processes described in feature 7 above reliably occur when the device is on and the system is put in the initial state. This dynamics is often, if not always, an effective dynamics between states that are coarse-grained with respect to some underlying microstructure. The significance of this is made apparent in Section 10.7.

The next subsection summarizes a model of implementation that has the above features.

10.4.1 L-machines

LPSG introduced the idea of an *L-machine* to model implementation and prove LP. This is a hybrid physical-logical entity that combines a physical device, a specification of which physical states of that device correspond to various logical states, and an evolution of that device which corresponds to the logical transformation *L*.

The *L*-machine analysis incorporates the essential features of implementation argued for in Section 10.2 and above. Formally, an *L*-machine is:

- A physical device D, situated in a heat bath at temperature T.
- A set of input states of the device, which are distinct thermodynamic states of the system (i.e., no microstate is a component of more than one thermodynamic state). $D_{in}(x)$ is the representative physical state of the logical input state *x*.
- A set of distinct thermodynamic output states of the device. $D_{out}(y)$ is the representative physical state of the logical output state *y*. Note that the set of input states and output states may overlap.
- This device is assumed to have a dynamics that drives it deterministically, or at least very nearly fully predictably, between initial and final states.[10] The model includes a time-evolution operator Λ_L for the device, such that $\forall x \in X$, $\Lambda_L(D_{in}(x)) = D_{out}(L(x))$. (See Figure 10.1.)

The most general form of LP can be precisely stated in terms of *L*-machines, namely, the logical irreversibility of *L* implies the thermodynamic irreversibility of every corresponding *L*-machine. The next subsection further clarifies the statement of LP.

---

[10] See Anderson (2010) for a probabilistic version of *L*-machines.

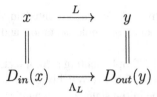

Figure 10.1 The relationship between the logical states x and y and their representative physical states $D_{in}(x)$ and $D_{out}(y)$, and the logical transformation L and the physical time evolution operator Λ_L

10.4.2   Known Versus Unknown Data

In any practical computation, the physical system must be such that it can be prepared in any of the physical states that represent the logical states, and there must be a time evolution that can be set in motion and which will bring about the physical process that results in the appropriate final physical state whatever the given initial state. As Landauer says, "The designer need not anticipate all the possible computations carried out by the machinery: that is what makes a computer more than the mechanical equivalent of looking things up in a table" (1991, p. 24). Rather, everything involved in the computation must be included in the operation of the device, because, as LPSG show, it is trivial that LP is made false if an external agent is allowed who has knowledge of the system, and who chooses what to do with the device in a way that depends on what the initial state is.

To see this, consider a bit represented by a particle on one side or other of a partition. One can perform RESET by doing nothing if the device is in the state corresponding to 0, and by rotating the device frictionlessly if it is in the other state. Norton gives the example of shifting a one-molecule gas from one side to the other of a box adiabatically (granting the defender of LP that such operations are admissible, though he argues against this, as explained in Section 10.6). Similarly, it is no coincidence that Norton's schemes to violate LP all involve such an external agent.[11]

For example, a physical system implementing RESET must not have another copy of the input state and use it to perform a different choice of operation, because if this is done, the system is not implementing a logically irreversible computation; rather it is effectively implementing UNCOPY (which might be loosely called the erasure or resetting of "known" data).

---

[11] Norton continues to regard the distinction between known and unknown data as at best confused and at worst irrelevant, and even claims that it is something mysterious to do with the past history of the data cell supposedly making a difference to the physics (2013a, p. 4446).

Data is "known" when the system has a copy of the data which can be used as part of a controlled operation, or when an external agent is able to measure the system and use the information gained to chose which operations to perform. LP only applies, if at all, to the RESET of data not known in either sense. As Feynman puts it, "only if we *do not know* which side of the compartment the atom is in do we expend free energy" (Feynman et al. 1996, p. 144).

In the next section another proof of LP and the debate about it with Norton is explained.

10.5   Ladyman et al. Versus Norton on the Proof of LP

LPSG offered a new proof of LP based on the construction of a thermodynamic cycle, arguing that, if there was an *L*-machine that violated LP, it would be possible to harness it to violate SL. Hence, LPSG argued as follows: If not LP then not SL; SL therefore LP. (Thus they follow the sound horn of the dilemma mentioned in Section 10.3.2 because they assume SL in arguing for LP, rather than claiming that LP supports SL by exorcising Maxwell's demon.) LPSG's cycle requires a repertoire of processes in a model involving a one-molecule gas, and includes controlled operations to and from it and the putative LP-violating *L*-machine that depend on the states of each respectively. Hence, LPSG's argument is really: If the processes and controlled operations in the cycle are admissible, and not LP, then not SL; SL, therefore, LP or at least one of the processes or controlled operations in the cycle is not admissible.[12]

Norton (2011) argues that LPSG's proof is not sound for (at least) three reasons:[13]

1. The processes LPSG use can be used to construct a counterexample to LP, which he refers to as "dissipationless erasure."
2. It is possible to use the processes that LPSG allow to violate the SL, so LPSG use inconsistent assumptions.
3. One of the controlled operations requires a thermodynamically reversible measurement of the state of the gas, but this is not admissible because no process can determine the position of a particle without dissipation.

Norton agrees that LPSG's processes are "the standard repertoire of idealized processes in the literature about the thermodynamics of computation" (2011, p. 185). He argues that these and other processes are problematic from the point of view of statistical mechanics; see Section 10.6. Ladyman and Robertson (2013) argue that in fact his counterexample to LP uses a process that is not used in LPSG's proof, and furthermore that the process in

---

[12] This section summarizes some of Ladyman and Robertson (2013).
[13] He also objects that LPSG use a statistical form of SL with no attempt to ground it in the underlying physical properties of the system; see Sections 10.7 and 10.8.

question is not admissible. They also argue that the processes LPSG allow do not violate SL. In the following subsections, the processes assumed by LPSG are enumerated and it is argued that Norton mischaracterizes them and their use.

10.5.1   Processes

1. *Isothermal expansion.* This is a reversible process that takes place at constant temperature; a piston is inserted and the gas does work on it, increasing the volume of the gas. The amount of work done by the gas is equal to the heat flow from the reservoir to the gas. In an isothermal expansion to twice its initial volume, the work done by the gas is $-kT \ln 2$.
2. *Isothermal compression.* A reversible process at a constant temperature. The piston reduces the volume of the gas. The work done on the gas by the piston is equal to the heat delivered to the heat bath. The work done by the piston is $-kT \ln 2$.
3. *Removal of the partition.* The partition (which traps the gas on one side of a box) is removed.
4. *Insertion of the partition.* The partition is inserted, trapping the gas in a smaller volume. The partition is inserted halfway along the box. (There is no heat flow.)
5. *Detection.* The location of the molecule can be determined without any thermodynamic cost.
6. *Detect and Trigger.* "According to whether the outcome of the detection is L or R, $Process_L$ or $Process_R$, respectively, may be initiated, without the initiation passing heat to the heat bath, where these are any two admissible processes" (Norton 2011, p. 189).

    This is a controlled operation. This process is used in step 2 and 3 of LPSG's cycle.
7. *Shift.* "If a system has states M_1 or M_2 of equal thermodynamic entropy, then a shift process moves the system from one state to the other without passing heat to the heat bath" (Norton 2011, p. 189).

10.5.2   Controlled Operations

A controlled operation is a logical operation that maps (at least) two bits in such a way that how (at least) one (target) bit is transformed depends on the value of the (control) bit(s). Ladyman and Robertson (2013) argue that the target bit and control bit must be represented by distinct physical degrees of freedom, because otherwise it would not be possible to make a process that transforms a bit according to some rule in the same way regardless of the value of other bits. For example, in the case of CNOT, it is required that the physical process that transforms the target bit depend only on the value of the control bit and not on the value of the target bit itself.

At a crucial step in Norton's (2011) putative LP-violating cycle there are two different processes enacted depending on the state of device. In one case, Shift is performed, while in the other, nothing is done. This is only possible if there can be a controlled operation from a degree of freedom onto itself (or if an external agent whose choice of action depends on the state of the bit is tacitly assumed). Norton's (2011) putative cycle showing that the standard inventory of processes used by LPSG violates SL also uses a controlled operation of a bit onto itself. Norton's (2013b) response to Ladyman and Robertson does not defend the claim that LPSG's processes can be used in the way he argues in his (2011). Rather, he appeals to a general critique of all such processes that is introduced in his (2013a) and (2013c). This is the no-go result discussed in the next section.

10.6   Norton's No-Go Theorem

Norton charges LPSG with inconsistency in their treatment of fluctuations at the microscopic scale, claiming that they and others selectively ignore them. He generalizes this, arguing that the whole field of the thermodynamics of computation is revealed to be bogus once such fluctuations are taken into account. Norton's no-go theorem extends his critique of LP significantly because he claims it shows that "the new science of the thermodynamics of computation is no science at all" (2013a, p. 4433). The conclusion of the theorem is that it is not possible to implement the kind of processes envisaged in Section 10.5.1, because there can be no thermodynamical reversibility at very small scales, due to the processes being "disrupted by thermal fluctuations" (2013a, p. 4434); hence, according to Norton, "fluctuations will fatally disrupt all efforts to complete thermodynamically reversible processes at molecular scales" (2013a, p. 4453).

Norton argues that the no-go theorem renders LP not false but irrelevant, because it shows that computation "can be achieved only by irreversible processes that create thermodynamic entropy in excess of the Landauer limit" (2013c, p. 1182). This implies that microscopic implementation of both reversible and irreversible computations cannot be done in a thermodynamically reversible way so LP in its qualitative form is obviated, and the quantitative LP lower bound is significantly exceeded by all actual implementations of computation. Norton concludes: "Alas, this image of a well-developed science is an illusion. The thermodynamics of computation is an underdeveloped muddle of vague plausibility arguments and misapplications of statistical physics" (2013c, p. 1183).

Recall that Norton regards fluctuations away from equilibrium as microscopic violations of SL; see Section 11.3 of this volume. As with the Smoluchowski trapdoor mechanism for a Maxwell demon which he analyzes in detail, he argues that any process at the molecular level that was based on the accumulation of such fluctuations to bring about an aggregative effect, whether to violate the SL in a macroscopically appreciable way, or to perform a step in the implementation of a computation, would be fatally disrupted by other fluctuations.

Ladyman and Robertson (2014) argue that Norton's analysis of fluctuations is problematic in various ways, discuss in detail his use of the Einstein-Tolman method for calculating the probability of fluctuations away from equilibrium, and criticize what they regard as inappropriate application of the concepts of equilibrium and fluctuation away from it. The first of the following three points is also made by them.

1. Processes in thermodynamics do not proceed only spontaneously by fluctuations, but also by some driving potential. Norton assumes that the processes implementing computations should be modeled as a series of fluctuations from equilibrium. Hence, he argues it is illicitly selective to ignore the fluctuation in the position of the piston used with the one-molecule gas, because he supposes fluctuations in the density of the gas (the molecule moving from one side of the box into the other) cause the process of isothermal expansion to proceed. Norton interprets the processes of Section 10.5.1 as proceeding from the result of the aggregation of fluctuations. Norton uses statistical mechanics to analyze the dynamics of the system. In his model, the transition probability for a process is determined solely by the difference in the free energies of the initial and final states. But, of course, if one state were intrinsically more probable than the other, then the time reverse of the process in question could not also be a viable process. However, both isothermal compression and isothermal expansion are among the processes of Section 10.5.1. Prima facie, isothermal expansion is the time-reversal of isothermal compression. The sequence of states is identical but the same states are passed through in opposite temporal directions. If the processes were driven solely by properties of each state, there could be only isothermal expansion or isothermal compression. Furthermore, there would be no way to control the rate at which processes happened, which undermines the idea of reversible/quasi-static processes. (For more on the importance of control see below and the next section.)

2. Thermodynamics is based on idealizations that involve the operations of agents using systems to build cycles reliably to perform operations. The concepts of equilibrium, adiabatic and diathermal walls, and of quasi-static

processes that are supposed to be virtually at equilibrium at every stage are clearly highly idealized – the one-molecule gas even more so. It is not precisely specified how long systems connected by diathermal walls should take to reach equilibrium, nor how long exactly states must stay the same to count as equilibrium states. Norton argues that they are "perfectly at equilibrium" and that they must therefore be impossible. Spontaneous heat flow between thermal equilibrium states requires a disequilibrium between them that is maintained somehow, and Norton argues that at the microscopic level this cannot be idealized away. He would be right that the processes that are supposed to occur in thermodynamics are impossible idealizations if it were true that they must proceed spontaneously by fluctuations, but it is part of the idealization that they can be driven by the dynamics of the computational device so that it can be deliberately and reliably used for tasks (see the next section).

3. Norton denies that Berut et al. (2012) experimentally verified LP because they "make no accounting of the quantities of thermodynamic entropy created by the many ancillary devices, such as the piezoelectric motor used to drive the compression" (2013a, p. 4449). However, if these devices are not at the same temperature as the system performing the computation then it is reasonable to neglect the heat and the entropy associated with them.[14] The processes of Section 10.5.1 may be run by big machines that work away in the corner of the laboratory generating lots of heat, but if they can be kept effectively thermally isolated from the system, then they are not relevant to the physics of the implementation in the device.

Norton may yet be right that fluctuations prevent thermodynamically reversible processes occurring at the microscopic scale, and his "no-go theorem" raises many important issues about the interpretation of thermodynamics and its relationship with statistical mechanics. In the next section, LP and Norton's views are discussed in the light of (Wallace 2014).

10.7   Wallace on Thermodynamics as Control Theory

David Wallace (2014) points out that thermodynamics is not a "dynamics" in the sense of classical mechanics or electromagnetism, but rather the theory of heat and work. As such it is a theory about various processes that we can put together in cycles in engines, and does not have its own time evolution laws. Rather, contrary to Norton's models discussed above in which the states in

---

[14] Thanks to John Preskill, who in discussion drew my attention to the fact that the driving devices need not be at the same temperature as the system.

thermodynamic cycles are supposed to proceed spontaneously by fluctuations, as pointed out by Tatjana Ehrenfest-Afanassjewa, the states of thermodynamic systems change because of our actions (1925, p. 942; 1956, p. 6). Wallace echoes Ehrenfest-Afanassjewa when he argues that "[w]hen the states of thermodynamical systems change, it is because we do things *to* them" (2014, pp. 699–700). Hence, in accordance with the points made in Section 10.6, Wallace rejects Norton's view that thermodynamic systems are supposed to evolve spontaneously to bring about some process in virtue of their internal states. Rather, there is some external driver of the processes that occur represented by some Hamiltonian (Wallace 2014, p. 704). Hence, the statistical mechanics of the internal dynamics is not sufficient to show that the process occurs as Norton requires.

In the light of this Wallace considers thermodynamics to be a theory about "tasks" that can be performed and combined, and crucially also about tasks that cannot be performed, by adiabatic manipulation of devices, like a gas in the box complete with pistons and a partition, and by bringing systems into thermal contact with each other. Similarly, Myrvold (2011) argues that Maxwell took thermodynamics to be "means-relative" in the sense that it concerns what can be done by agents with systems given their knowledge of them.[15] Note that information theory and computation theory are also about tasks and protocols. For example, cloning and compression are tasks in information theory, and RESET is a computational task. (See also Skrzypczyk et al. [2014], who define thermodynamics in terms of "resources" for performing various tasks.) Note also that the idea of a task involves temporal asymmetry. For example, the time reverse of RESET is randomization of known data, which is a completely different task. Formulated in such terms, SL says that the task of turning heat wholly into work cannot be performed, although the task of wholly turning work into heat can. Whatever the ultimate source of this asymmetry, this thermodynamic form of SL is empirically adequate.

Statistical mechanics considers what happens to the time evolution of vast systems of microstates by defining coarse-grained states and describing their probabilistic time evolution. Even though the underlying dynamics is deterministic, it is possible rigorously to derive an indeterministic dynamics of macrostates, and a version of SL in terms of them. This derivation depends on a time-asymmetric boundary condition concerning the lack of correlations in the system. It is possible to see how the laws of thermodynamics relate to statistical mechanics given that entropy is non-decreasing in the latter due to time-asymmetric assumptions in the way coarse-grained probabilities over

---

[15] It is interesting to consider how this relates to the idea that thermodynamics is a "principle theory."

microstates are defined. However, it is a theme of Wallace's work on probabilistic and thermal physics that there is much more to statistical mechanics that the beloved project of philosophers to provide a foundation for thermodynamics (2014, p. 702). For example, the time taken for a particular kind of system to relax to equilibrium may be calculated in statistical mechanics but not in thermodynamics.

In the light of this way of thinking about the relationship between thermodynamics and statistical mechanics, SL should not be seen as a single law. There is a version of SL in statistical mechanics that is statistical and, as such, microscopic fluctuations away from equilibrium are not violations of it, contra Norton. On the other hand, construed as a law about what we can do with thermodynamic processes, SL is also not violated by microscopic fluctuations. SL and the associated prescription against perpetual motion machines of the second kind were first proposed for "bulk matter," and no violations of SL in the behavior of bulk matter can be obtained by microscopic fluctuations for the reasons Norton gives.

LPSG analyzed LP and the thermodynamic entropy in terms of tasks (though without using the word). Thinking of thermodynamics as a control theory, as Wallace does, makes it clear that LP is concerned with what can reliably be done with certain resources. As argued in Section 10.4, the resources in question are physical systems with both their states and their time evolutions. The states are construed as representing logical states. The known-versus-unknown-data distinction shows that what we can do with such systems depends on what we know about them. This does not mean that it only depends on what we know about them, because, as argued in Sections 10.2, 10.4, and 10.8, we cannot perform computations with systems unless they have an appropriate array of states connected by appropriate dynamics, and these are objective constraints. In the next section it is argued that the fact that thermodynamics is concerned with what we can do with physical systems given what we know about them has implications for how the thermodynamic entropy should be assigned to systems.

10.8   The Relationship between Knowledge and Thermodynamics

Ladyman, Presnell, and Short (2008) argue that the information we have about a system is relevant to the thermodynamic entropy we should assign to it, because what we know about a system affects how we can use it.

They consider a generalization of this simple example. We are told that a system C has been prepared in either state Ω_A or state Ω_B with equal probability, with known distinct entropies S_A and S_B. We need to carry out a thermodynamic calculation involving C. What should we say about its entropy S_C?

Consider the following three possible answers to this question:

1. We cannot proceed since S_C is either S_A or S_B, and definitely not anything else.
2. S_C is the weighted average of S_A and S_B, namely $\frac{1}{2}(S_A + S_B)$.
3. S_C is the weighted average of S_A and S_B plus an additional term that represents a contribution to the thermodynamic entropy due to the probability distribution itself. In this particular case, $\frac{1}{2}(S_A + S_B) + k \ln 2$.

They argue that the correct entropy to assign to a probabilistic mixture is the generalization of the third one (and indeed LPSG use this formula in an alternative proof of LP). This result is derived using a probabilistic version of SL that says that there is no process that *on average* has the sole effect of turning heat into work. This version is of broader scope than standard formulations since it also applies to probabilistic theories. Norton (2013a, p. 4447) claims that the Ladyman, Presnell, and Short version of SL is a weakening but, as Wallace's analysis makes clear, it follows from how SL should be understood in thermodynamics construed as control theory. In that context, the point about a task is that, if someone offers a protocol for enacting it, they must be able to do it on demand, repeatedly and reliably, in a way that is counterfactually robust in accordance with the discussion of Section 10.4. This version of SL is arguably what most physicists have taken it to be since Maxwell argued that SL is statistical. (See Myrvold [2011] for discussion of Maxwell on SL, and see especially p. 238 for a statement of SL similar to the one above.)

10.9   Conclusion

Thermodynamics differs from other parts of physics in being at least partly about what agents can reliably do with systems given their knowledge of them. In general, a bit can be reset and heat can be turned into work, if an external agent is assumed to know the state of the system and can choose what operation to perform accordingly, or if the agent guesses correctly what the state is. However, in any practical context, both the thermodynamical form of SL and LP concern what can be done reliably when the state of the system is not known.

LP is about the thermodynamics of computation, which concerns physical states being taken as representing logical states. A physical state can be taken to represent different logical states, so reasoning about the latter always involves a physical state under some mode of presentation or other. What counts as "information-bearing degrees of freedom" depends on how systems are taken to represent abstract states.[16]

Hence, both computational and thermodynamic states are intensional in the sense that they are sometimes individuated at least in part by what is known about them, how they are represented and/or by to what use they are being put. (This is not to say that statistical mechanical states are too, though that also seems plausible because of the time-asymmetric assumptions that are built into the theory.) In thermodynamics the free energy can be defined as the energy available to do work. The total energy is then the heat plus the free energy, and heat is the energy unavailable to do work. But what counts as work depends on how the system is delimited (Coopersmith 2010, p. 326).

Hence, in thermodynamics physical states are thought of differently depending on how the boundary of the system is defined. Furthermore, particular microstates are thought of as instances of macrostates. There is a clear presumption that operations can be performed adiabatically and the theory is formulated in terms of idealizations such as diathermal walls and equilibrium states.

The way that we think about a physical state may affect what we can do with it. The relevant notion of "accessible phase space" may be taken to be what is epistemically possible given what we know about the system. It is less surprising that there be a connection between logic and thermodynamics via LP once it is recognized that it is concerned with physical states understood intensionally. If realism about computation is taken to apply only up to the choice of a representation relation between physical and logical states, then the objective content of LP is confined to the modal structure of the physical states and their respective time evolutions.

Acknowledgments

Thanks very much to Berry Groisman, James Lambert, Stuart Presnell, Katie Robertson, and Tony Short, with whom I have worked on the issues addressed in this paper, and to Michael Cuffaro, Sam Fletcher, and an anonymous referee for detailed comments on a draft. Many thanks also to John Norton.

---

[16] Furthermore, the implications for practical computation only concern the parts of the system that are in thermal equilibrium with the "information-bearing" degrees of freedom as in Section 10.6, point 3.

11 Maxwell's Demon Does not Compute

John D. Norton

11.1 Introduction

Is it instructive to model some physical process as a computational process or, more generally, as one that processes information? That it would be so is a hypothesis that needs to be tested case by case. Sometimes it will be very instructive. Shannon's information theory applied to communication channels is a striking success. There can be failures, however. This chapter will describe a lingering and striking failure.

A Maxwell's demon is a device that can reduce the thermodynamic entropy of a closed system, in violation of the Second Law of Thermodynamics, by means of molecular-scale manipulations. The received view since the mid-twentieth century is that such a device must fail for reasons most instructively captured by theories of information and computation. This received view of the demon's exorcism, I will argue here, is misdirected and mistaken.

First, there are many proposals for Maxwell's demons in which there is no obvious computation or information processing. As a result, the exorcism of the received view cannot be applied to them. It is no general exorcism.

Second, the received view depends variously on dubious principles, Szilard's principle and Landauer's principle. They are at best interesting speculations in need of precise grounding, or, at worst, mistakes propped up by repeated misapplications of thermal physics.

Third, prior to the emergence of the received view, we already had a serviceable and generally applicable exorcism that made no use of notions of information or computation. In 1912, Smoluchowski had argued cogently that efforts to reverse the second law by manipulations at molecular scales will fail since they will be disturbed fatally by the very thermal fluctuations they seek to exploit.

Finally, I shall show here that the long-entrenched focus on information and computation-theoretic notions has distracted both supporters and opponents of the received view from a simpler exorcism, even stronger than Smoluchowski's arguments of 1912. A simple specification of what a Maxwell's demon must do turns out to be incompatible with the classical Liouville theorem of statistical

240

physics or its quantum counterpart. Hence the demon must fail, and its failure is established without any recourse to notions of information or computation. The exorcism does not even require serious engagement with the notion of thermodynamic entropy.

The early Sections 11.2, 11.3, and 11.4 will review Maxwell's invention of his demon, its naturalization with the discovery of fluctuation phenomena, and Smoluchowski's argument that these same fluctuations defeat the demon. In Section 11.5, I will report on the appearance of the idea that an intelligent demon may need special accommodations. Sections 11.6 and 11.7 trace briefly how the ensuing idea of a naturalized, intelligent demon came to dominate the Maxwell's demon literature, with exorcisms focusing first on a supposed entropy cost in acquiring information and then in erasing it. This is, I will argue, a failing literature.

In Sections 11.8 and 11.9, I will report a new, stronger, and simpler exorcism based on the contradiction between what the demon must do and Liouville's theorem of statistical physics. The exorcism reported is limited to classical physics. Sections 11.10, 11.11, and 11.12 will show a closely analogous exorcism using the quantum analog of Liouville's theorem.[1]

11.2   Maxwell's Fictional Demon

Maxwell (1871, pp. 308–309) unveiled his demon in print in 1871. He used it to make a point about the character of the second law of thermodynamics. We cannot reverse the second law, Maxwell sought to establish, merely because we have no access to individual molecules. Instead we must treat molecular systems en masse. To make his point, he imagined a quite fictitious being who could access molecules individually. By carefully opening and closing a door in a dividing wall as the molecules of a gas approached it, this demonic being could accumulate slow molecules on one side and fast molecules on the other. The first side cools while the second warms, yet no work is done. The normal course of thermal processes is reversed, in contradiction with the second law.[2]

11.3   Fluctuations Bring Naturalized Demons

A major change in the demon's role came with the recognition in the early twentieth century that thermal fluctuations are microscopically observable. They could no longer be dismissed as an artifact of molecular theory of no practical import. They realize, it was concluded, a microscopic violation

---

[1] The content of Sections 11.11 and 11.12 can also be found in Norton (2014). I thank Joshua Rosaler and Leah Henderson for helpful discussion of the quantum material.

[2] For an account of Maxwell's original proposal and conception, see Myrvold (2011).

of the second law of thermodynamics, which could at best hold only for time-averaged quantities. The celebrated example is Einstein's (1956 [1905]) analysis of Brownian motion. The larger movements of the Brownian particle arise through a transfer of the heat energy of the surrounding water into the particle's kinetic energy. It might then be converted to gravitational potential energy, a form of work energy, if the motion lifts the particle vertically. This is a momentary, microscopic violation of the second law of thermodynamics: ambient heat energy has been fully converted to work.

Maxwell had given no account of just how his demon might be constituted. Since the point was that his demon was fictional and intended to display vividly what we cannot do, there was no need for it. With the new recognition about thermal fluctuations, Maxwell's demon was moved from the realm of impossible fiction to a candidate physical possibility. If momentary, microscopic violations of the second law are possible, might we devise a real machine that can accumulate them and eventually lead to macroscopic violations of the second law? Such a machine would be a naturalized Maxwell's demon. That is, it would be one whose workings conform with the known natural laws of microscopic systems.

What followed were numerous proposals for naturalized Maxwell's demons of simple design. Some were intended to be realized in the laboratory. Such was Svedberg's (1907) colloid demon. In it, the Brownian motion of electrically charged colloid particles would lead them to radiate their thermal energy, which would be trapped in a carefully designed system of casings. The colloid would spontaneously cool, while the casing heated. Smoluchowski's (1912) paper contained a range of more schematic proposals. One was a one-way valve that would allow gas molecules to pass in one direction but not the other. This one-way transport was effected by a hole with a ring of hairs, or by a valve with a flapper.

This last proposal entered later literature in modified form as the Smoluchowski trapdoor. In his original thought experiment, Maxwell employed a fictional demon to open and close the door in the dividing wall of the chamber. Smoluchowski's trapdoor was an automatic device. It was lightly spring-loaded and configured so that molecules moving in one direction would flip it open and pass, whereas molecules moving in the opposite direction would slam it shut and be obstructed. For more discussion of these proposals, including what would later become Feynman's "ratchet and pawl" demon, see Norton (2013a, sec. 2).

11.4   Fluctuations Defeat Maxwell's Demon

The main point of Smoluchowski's analysis was that all these proposals for Maxwell's demons fail. For they are machines operating at molecular scales where fluctuation phenomena dominate. In each case, some fluctuation-driven process would reverse the normal course of thermal processes. The individual molecular collisions that flip open the valve flapper or the Smoluchowski trapdoor are pressure fluctuations in the gas. Smoluchowski then showed that, for each case, there was a second fluctuation process that undid the anti-entropic gains of the first. In the case of the Smoluchowski trapdoor, if the device is to operate as intended, the flapper must be so light that collisions with individual molecules can open it. But such a light flapper will have its own fluctuating thermal energy, which will lead it to flap about randomly, allowing molecules to pass in both directions. On average there is no accumulation of violations of the second law.

Smoluchowski made his case by examining many examples of candidate mechanisms and showing that they all failed in the same way. The analysis provided no principled proof of the generalization that all demon proposals must fail this way. However, once one sees the one mode of failure repeated again and again, in the range of examples treated by Smoluchowski, the generalization is hard to resist.

There is another way to see that fluctuations are a formidable obstacle to efforts to realize a Maxwell's demon. Such a demonic device will operate at molecular scales and will be composed of a series of steps, each of which must be brought to completion before the next can start. In recent work (Norton 2011, sec. 7; Norton 2013a, part II), I have shown that the completion of *any* single process at molecular scales, no matter how simple or complicated, intelligently directed or otherwise, involves dissipation. For any such process must overcome the thermal fluctuations that disrupt its orderly execution. They can only be overcome by the dissipative creation of entropy, if completion is to be assured, even just probabilistically. The quantities of entropy involved are great enough to swamp the entropy reduction envisaged in the operation of a Maxwell demon.

These considerations of fluctuations are not a deductive proof from first principles of the impossibility of a Maxwell's demon. However, they make it quite plausible that a molecular-scale demon cannot overcome the disrupting effects of thermal fluctuations. They give us a simple and proven recipe for demonstrating the failure of any new proposal for a Maxwell's demon: Look for the neglected effects of fluctuations.

11.5   The Distraction of Intelligent Intervention

Smoluchowski's 1912 verdict on the possibility of a naturalized Maxwell's demon provides a resolution that is still illuminating today. Naturalized demons will likely fail because thermal fluctuations will disrupt their intended operations. Smoluchowski's paper was delivered as a lecture at the 84th

Naturforscherversammlung (Meeting of Natural Scientists) in Münster. The discussion that followed is reported at the end of the journal printing of Smoluchowski's lecture. In it, Kaufmann directed a quite awkward question to Smoluchowski:

Kaufmann: The lecturer has indicated why presumably also no mathematical selection [among molecules of different speed] that contradicts the second law can be brought about by means of an automatic valve. The relations are otherwise for a valve with something like a sliding bar, whose motion requires no work in theory. Then there is an intelligence operating the valve and ensuring that the opening and closing is in the right moment; I believe that, for Brownian molecular motion, something like this is practically achievable. Then the second law would be violated by the participation of an intelligent creature. [This is] a conclusion that one possibly could regard as proof, in the sense of the neo-vitalistic conception, that the physico-chemical laws alone are not sufficient for the explanation of biological and psychic occurrences.

This is the sort of question any speaker dreads. Smoluchowski had just based his lecture on the presumption that a Maxwell's demon is naturalized, that is, it is subject to the normal physico-chemical laws. Then the demon will fail. Now he is asked to contemplate the case of a neo-vitalist demon; that is, an intelligence whose actions are not governed by those laws but is animated by some kind of vital force. It is even suggested that this might lead to an experiment that vindicates vitalism. The suggestion is far-fetched. If an intelligent organism – a human, for example – accumulates microscopic violations of the second law in Brownian motion in a real laboratory experiment, one must also account for the entropy created in the organism's metabolism. To ignore it through some vitalist commitment would make the vitalist interpretation of the experimental result circular.

Smoluchowski gives the best reply he can muster:

Lecturer: What was said in the lecture certainly pertains only to automatic devices, and there is certainly no doubt that an intelligent being, for whom physical phenomena are transparent, could bring about processes that contradict the second law. Indeed Maxwell has already proven this with his demon.

This grants the tacit presumption of the question: that a vitalistic demon, were there such a thing, could succeed. However, Smoluchowski then awkwardly reminds the questioner of the background assumption of Smoluchowski's entire analysis. He continued:

However, intelligence extends beyond the boundaries of physics. On the other hand, it is not to be ruled out that the activity of intelligence, the mechanical operation of the latter, is connected with the expenditure of work and the dissipation of energy and that perhaps after all a compensation still takes place.

Intelligence, presumably in the abstract, disembodied sense, is something that lies outside physics. But intelligence that can act in the world will do it through a physical system and this is still a system that will be governed by the familiar laws. The wording is hesitant – it should not be ruled out. However, I attribute the hesitancy merely to the politeness required to respond to a question clearly outside the scope of the speaker's talk.

11.6   Szilard's Principle

What happens if the intervening demon is an intelligence unconstrained by normal physico-chemical laws? This was a question best left to die quietly. If one allows such an intelligence, then no physical law is secure. If, however, the intelligence is embodied in a physical system, then Smoluchowski has already provided a quite serviceable answer: whether the system is intelligent or not, thermal fluctuations will likely preclude its operation. The question of an *intelligent* intervening demon is a distraction, since *all* demonic intervention will fail.

Unfortunately Leo Szilard was unable to resist the temptation of pursuing the distracting question. His 1929 "On the decrease of entropy in a thermodynamic system by the intervention of intelligent beings" responded directly to Smoluchowski's work and quoted liberally from it. It initiated a decline in the literature on Maxwell's demon from which we have still to recover.

The details of Szilard's analysis are quite complicated and even obscure. See Earman and Norton (1998, sec. 7) for a review. What survived into the ensuing literature were a few ideas in a form somewhat simpler than Szilard's formulation. The most important idea was that one need not provide physical details of the mechanism that animates the intelligent demon. All one needs to know is that its operation requires the gaining of information. The mere fact of gaining information, however it is done, creates enough entropy to defeat the demon.

To illustrate the point, Szilard introduced an ingeniously simplified arrangement in which the demon cyclically manipulates a one-molecule gas. Each cycle requires the demon to discern whether the molecule is trapped on the left or the right side of a partition. This discerning – in later literature the gaining of one bit of information – was, Szilard asserted, necessarily a dissipative process that creates entropy and protects the second law from violation.

How much entropy does this gaining of information create? If the second law is to be protected, then the process must create at least $k \log 2$ of thermodynamic entropy for each bit of information gained, where k is Boltzmann's constant. This principle was later called "Szilard's principle" (Earman and Norton 1999). That this amount suffices to protect the second law was assured by the expedient of working backwards. Assume that the second law is preserved and compute from that assumption how much entropy must be created.

Szilard's principle ensues. While Szilard and others after him did try to justify the principle by examining particular detection processes, working backwards remained the simplest and most general justification.

The principle in this form supported a flourishing literature in the 1950s. It proclaimed a deep truth in the connection between information and thermodynamic entropy. This insight, it assured us, explains why a Maxwell demon must fail, even though its core claim of Szilard's principle was commonly derived by circular reasoning from the very presumption that a Maxwell's demon must fail.

For a synoptic discussion of this new literature and the ensuing literature in the thermodynamics of computation, and for reproductions of key papers, see Leff and Rex (2003).

11.7   Landauer's Principle

The success of this last exorcism was short-lived. It was replaced within a few decades by a modified version that drew on computational notions. The modified version retained the idea that one should abstract away all of the details of the demon's constitution excepting its treatment of information. But now the unavoidable dissipative step was not the acquiring of information. It was the erasure of information. To function, a demon must remember what it has learned. In the case of Szilard's example, the demon must remember that the molecule was trapped on the left or the right side of the partition, and that memory must be captured in some physical change in the demon. To complete the thermodynamic cycle, the demon's memory must be returned to its initial state. That return is the moment of dissipation. The erasure of this one bit of information is associated with $k \log 2$ of thermodynamic entropy, which is just the amount needed to protect the second law. The statement of this erasure cost is "Landauer's principle," drawn from the work of Rolf Landauer (1961). It is the central result of what soon came to be known as the "thermodynamics of computation."

The new computation-theoretic exorcism was laid out in Bennett (1982, sec. 5). In order to secure its primacy, the new exorcism needed to overturn the old exorcism. Its proponents, we were now told, had simply erred in attaching the necessity of dissipation to information acquisition. All the clever arguments and manipulations of the old exorcism were deceptive mirages. Bennett (1982, sec. 5; 1987) sketched new thought experiments in which information about the states of target systems could be gained by processes claimed to be thermodynamically reversible.

This computation-theoretic exorcism has now settled in as the standard in the literature. Although there have been amendments offered that draw on notions of complexity and quantum theory,[3] the basic ideas of the exorcism have survived with some stability. One might be excused for taking this stability as a sign of cogency. Alas, the computation-theoretic exorcism of the 1980s was no improvement on the fragile information-theoretic exorcism of the 1950s. It had merely rearranged some of its parts.

To begin, the essential problem remains. There are many proposals for Maxwell's demon in which there is no overt collection of information and no overt computation that employs a memory that must be erased. These processes, for example, are simply not present in the canonical Smoluchowski trapdoor or Feynman's ratchet and pawl demon. Therefore, neither information-theoretic nor computation-theoretic exorcism can touch them. However, Smoluchowski's original, thermal fluctuation-based exorcism applies to them and all the rest.

Second, the information-theoretic exorcism had been supported by ingenious thought experiments that illustrated how gaining information is thermodynamically costly. In a thought experiment reminiscent of the celebrated Heisenberg microscope of the quantum uncertainty principle, Brillouin (1951) had computed that dissipation compatible with Szilard's principle must occur, if a photon with energy above the thermal background is used to locate a particle. In spite of the luminaries of physics like Brillouin who had supported them, these thought experiments were all misleading and mistaken, we were now told. The trouble was that the thought experiments that replaced them were no better. Bennett's (1982, sec. 5; 1987) illustrations of devices that could gain information dissipationlessly all required devices of delicate sensitivity. It takes only the most cursory of inspections to see that their operations would be fatally disrupted by thermal fluctuations, just as Smoluchowski envisaged. (See Norton [2011, sec. 7.3]). One defective set of thought experiments had merely been replaced by another.

Finally, the computation-theoretic exorcisms draw on Landauer's principle. When Landauer (1961) introduced the principle, it was little more than a promising speculation, supported by a sketchy plausibility argument. Over half a century later, one might imagine that this would be sufficient time to place the principle on a more secure foundation. This has not happened. It is not for want of trying. However, as I have documented in detail elsewhere (Norton 2005, 2011) and summarized in Norton (2013a, sec. 3.5), the now burgeoning literature on Landauer's principle persists in committing repeatedly a small set of interconnected errors in thermal analysis.

---

[3] See Earman and Norton (1999).

11.8   Asking the Right Question

These failed traditions are driven by the belief that a successful exorcism of Maxwell's demon abstracts away all details of the demon's operation, other than its processing of information. As the discussion of the previous sections illustrates, this belief has presided over a descent into a feckless, convoluted, and confused literature. As long as the attention of authors in the field, proponents and critics alike, remains focused on information processing, this descent is likely to continue. Here, ruefully and regretfully, I include much of my own writing over more than a decade on the topic. At best I have been able to show what does not work in exorcising the demon. What I should have asked is what does work.

Let us start again. Let us set aside information and computation-theoretic notions and take stock of what we know. We have known since Smoluchowski's work of 1912 that disruptions by fluctuations present a formidable barrier to all efforts to realize a Maxwell's demon. We now also have strong empirical indications of the impossibility of such a demon. Nanotechnology has given us abilities to manipulate individual atoms far beyond anything Maxwell or Smoluchowski could have imagined. In 2013, scientists at IBM made a stop-motion video of a stick figure boy playing with a ball.[4] The figures were drawn by lining up individual carbon monoxide molecules on a copper surface in a scanning tunneling microscope. Even with such prodigious capacities to manipulate individual molecules, no fully successful Maxwell's demon has been made. Rather, all work at nanoscales struggles to overcome thermal fluctuations. They are the nemesis of nanoscience, just as Smoluchowski argued. The molecules of the IBM stop-motion video were cooled to $-268°C$ to suppress fluctuations.

There have been other empirical clues. The biochemistry of a cell involves molecular processes of comparable refinement. The operation of a ribosome in a cell is a marvel of miniaturized molecular machinery. It was brought into being by the creative powers of evolution. Yet these same prodigious powers have failed to construct a demonic device in the cell, in spite of the obvious advantage to the cell of a process that converts ambient heat energy to useful work.

With some reasonable expectation that a Maxwell's demon is impossible, let us ask the question that has been neglected: is there a simpler way to demonstrate the impossibility of a Maxwell's demon that avoids the convolutions of the present literature?

---

[4] See www-03.ibm.com/press/us/en/pressrelease/40970.wss.

11.9   A Better Exorcism

It came as a sobering surprise when I found recently (Norton 2013a, sec. 4) that there is a very simple exorcism of Maxwell's demon that requires only elementary notions from statistical physics. There is no need for notions of information or computation or erasure, or tendentious principles like Szilard's or Landauer's. One need not even mention the ever-troublesome notion of entropy. The exorcism shows that a description of what a Maxwell's demon must do is incompatible with Liouville's theorem of statistical physics.

Here, in brief, is how it works. When presented with a target thermal system such as a gas in a vessel, it is presumed that a Maxwell's demon is able to drive the system away from its normal state of thermal equilibrium into what would otherwise be judged a disequilibrated state were there no interaction with the demon, and that the system remains in that state. For example, Maxwell's original demon or the Smoluchowski trapdoor takes a gas at uniform temperature and separates the hotter, faster molecules from the slower, colder ones. Once its work is done, we have the disequilibrated gas, with the hotter part on one side of a partition and the colder part on the other side. To ensure that there is no compensating hidden thermal dissipation or degradation in the demon itself or any supporting systems it uses, we require that the demon and these supporting systems are returned to their original states at the end of the process. Such a process reverses the second law of thermodynamics.

If we redescribe this process in the context of standard statistical physics, we quickly see that it is impossible. In that context, systems are presumed to be governed by Hamilton's equations, versions of which cover virtually all physical theories considered. The state of a system is fixed by determining a large number of generalized position and momentum variables. These variables are the coordinates of a space, known as a phase space. The state of a Hamiltonian system at one moment corresponds to a single point in the phase space. As the state changes, it traces a trajectory in the phase space.

A closed system will revert spontaneously to equilibrium states. For example, a gas confined in an isolated vessel will evolve to a state of uniform pressure, temperature, and density. These equilibrium states occupy virtually all of the system's phase space. The non-equilibrium states with non-uniformities occupy only a tiny fraction of the volume of the phase space. This difference of volumes is the rough and ready explanation for why closed thermal systems revert to their equilibrium states. As the phase point of the system migrates in time through the phase space, it almost always ends up in the much larger part of phase space where equilibrium systems are found. The non-equilibrium states are mere temporary intermediates on the way to equilibrium.

When we couple a Maxwell's demon and its support systems to some target system in thermal equilibrium, we form a larger system with its own, larger phase space. If the demon operates as intended, the target system will evolve from an equilibrated to a disequilibrated, intermediate state, while the demon and its support systems revert to their original states. (Since the supposition of successful action of the demon upsets the normal notions of equilibrium and disequilibrium, henceforth these disequilibrated states will be labeled more neutrally as "intermediate states.") This evolution is required to happen no matter which is the equilibrium microstate of the target system; or at least for most of the equilibrium microstates of the target system. That is, the operation of the demon must compress the phase space volume of the target system down to a very much smaller volume, while leaving the phase space volume of the demon and supporting systems unchanged. The overall effect is that the successful operation of the demon must compress the phase space of the combined system.

The combined system is governed by Hamilton's equations. An early and easily gained property of such systems is Liouville's theorem. It states that time evolution leaves phase space volumes unchanged. That is, if we select some set of states forming a volume in the phase space, over time, as the systems evolve, the set of states occupied will move around the phase space. However, the volume that they occupy in phase space remains unchanged.

In sum, the successful operation of a Maxwell's demon must compress phase space. Liouville's theorem of statistical physics asserts that this is impossible. Therefore a Maxwell's demon is impossible.

11.10   Classical or Quantum?

The exorcism just sketched informally was developed formally in Norton (2013a, sec. 4) and the main derivations will be reproduced again below. There is a weakness in this exorcism. The processes involved occur at molecular scales, where the quantum mechanical properties of systems can be important. Yet the exorcism employs classical physics.

The remaining analysis below rectifies this weakness. The bulk of the original analysis remains the same and an analogous result of comparable simplicity is recovered. All that is needed is to substitute quantum analogs for those parts of the argument that depend essentially on classical physics. The main substitution is to replace the conservation of phase volume of classical physics by its analog in quantum theory, the conservation of dimension of a subspace in a many-dimensional Hilbert space. This substitution will be described in Section 11.11. The following section will then list the premises of the classical exorcism along with their quantum counterparts.

11.11   Conservation of Volumes

The statistical treatment of thermal systems in classical and quantum contexts is sufficiently close for it to be possible to develop the relevant results in parallel, as in the two columns below. Corresponding results are matched roughly horizontally.

Classical Hamiltonian Dynamics

The state of a system is specified by $2n$ coordinates, the canonical momenta $p_1, \ldots, p_n$ and the canonical configuration space coordinates $q_1, \ldots, q_n$ of the classical phase space Γ. The time evolution of the system is governed by Hamilton's equations:

$$\dot{p}_i = \frac{dp_i}{dt} = -\frac{\partial H}{\partial q_i}, \quad \dot{q}_i = \frac{dq_i}{dt} = \frac{\partial H}{\partial p_i},$$
$$(11.1)$$

where $i = 1, \ldots, n$ and $H(q_1, \ldots, q_n, p_1, \ldots, p_n)$ is the system's Hamiltonian.

Classical Liouville Equation

If $f(q_i, p_i, t)$ is a time-dependent function defined on the phase space, then the total time derivative of f, taken along a trajectory $(q_i(t), p_i(t))$ that satisfies Hamilton's equations, is:

$$\frac{df}{dt} = \frac{\partial f}{\partial t} + \sum_{i=1}^{n} \left(\frac{\partial f}{\partial q_i} \frac{dq_i(t)}{dt} + \frac{\partial f}{\partial p_i} \frac{dp_i(t)}{dt} \right)$$

$$= \frac{\partial f}{\partial t} + \sum_{i=1}^{n} \left(\frac{\partial f}{\partial q_i} \frac{\partial H}{\partial p_i} - \frac{\partial f}{\partial p_i} \frac{\partial H}{\partial q_i} \right)$$

$$= \frac{\partial f}{\partial t} + \{f, H\}.$$

Quantum Statistical Mechanics

The system state $|\psi(t)\rangle$ is a vector in an n-dimensional Hilbert space, with orthonormal basis vectors $|e_1\rangle, \ldots, |e_n\rangle$. The time evolution of the system is governed by Schrödinger's equation:

$$i\hbar\frac{d}{dt}|\psi(t)\rangle = H|\psi(t)\rangle,$$
$$(11.1')$$
$$-i\hbar\frac{d}{dt}\langle\psi(t)| = \langle\psi(t)|H,$$

where H is the system's Hamiltonian.

Quantum Liouville Equation

In place of the classical probability density ρ, we have the density operator ρ, which is a positive, linear operator on the Hilbert space of unit trace. It may be written in general as:[5]

$$\rho(t) = \sum_{\alpha} p_\alpha |\psi_\alpha(t)\rangle \langle\psi_\alpha(t)|,$$

where $\sum_\alpha p_\alpha = 1$ for some set $\{|\psi_\alpha\rangle\}$ of state vectors, which need not be orthogonal. This operator represents a "mixed state," that is a situation in which just one of the states in the set $\{|\psi_\alpha\rangle\}$ is present, but we do not

[5] For a proof, see Nielsen and Chuang (2000, sec. 2.4.2).

Set f equal to a probability density $\rho(q_i, p_i, t)$ that flows as a conserved fluid with the Hamiltonian trajectories. Now ρ satisfies the equation of continuity:[6]

$$0 = \frac{\partial \rho}{\partial t}$$
$$+ \sum_{i=1}^{n} \left(\frac{\partial}{\partial q_i}(\rho \dot{q}_i) + \frac{\partial}{\partial p_i}(\rho \dot{p}_i) \right)$$
$$= \frac{\partial \rho}{\partial t} + \{\rho, H\}.$$

Combining with the expression for the total derivative $d\rho/dt$, we recover the classical Liouville equation

$$\frac{d\rho}{dt} = 0. \qquad (11.2)$$

It asserts that the probability density in phase space evolves in time so that it remains constant as we move with a phase point along the trajectory determined by Hamilton's equations.

know which, and our uncertainty is expressed as the ignorance probability p_α.

If the state vectors $|\psi_\alpha(t)\rangle$ evolve in time according to the Schrödinger Eq. (11.1'), the quantum Liouville equation follows:[7]

$$i\hbar \frac{d\rho(t)}{dt} = H\rho(t) - \rho(t)H = [H, \rho(t)].$$
$$(11.2')$$

Alternatively, we can write the integral form of the Schrödinger equation with the unitary operator $U(t)$ as

$$|\psi(t)\rangle = \exp(-iHt/\hbar)|\psi(0)\rangle$$
$$= U(t)|\psi(0)\rangle,$$
$$\langle\psi(t)| = \langle\psi(0)| \exp(iHt/\hbar)$$
$$= \langle\psi(0)|U^{-1}(t). \qquad (11.1'')$$

From it, we recover the integral form of the quantum Liouville equation:[8]

$$\rho(t) = U(t)\rho(0)U^{-1}(t). \qquad (11.2'')$$

---

[6] This follows, since:

$$\sum_{i=1}^{n} \left(\frac{\partial}{\partial q_i}(\rho \dot{q}_i) + \frac{\partial}{\partial p_i}(\rho \dot{p}_i) \right) = \rho \sum_{i=1}^{n} \left(\frac{\partial \dot{q}_i}{\partial q_i} + \frac{\partial \dot{p}_i}{\partial p_i} \right) + \sum_{i=1}^{n} \left(\frac{\partial \rho}{\partial q_i} \dot{q}_i + \frac{\partial \rho}{\partial p_i} \dot{p}_i \right).$$

Using Hamilton's Eqs (11.1), the first term on the right vanishes since:

$$\sum_{i=1}^{n} \left(\frac{\partial \dot{q}_i}{\partial q_i} + \frac{\partial \dot{p}_i}{\partial p_i} \right) = \sum_{i=1}^{n} \left(\frac{\partial^2 H}{\partial q_i \partial p_i} - \frac{\partial^2 H}{\partial p_i \partial q_i} \right) = 0$$

and the second term is

$$\sum_{i=1}^{n} \left(\frac{\partial \rho}{\partial q_i} \dot{q}_i + \frac{\partial \rho}{\partial p_i} \dot{p}_i \right) = \sum_{i=1}^{n} \left(\frac{\partial \rho}{\partial q_i} \frac{\partial H}{\partial p_i} - \frac{\partial \rho}{\partial p_i} \frac{\partial H}{\partial q_i} \right) = \{\rho, H\}.$$

[7] Applying the Schrödinger equation to each $|\psi_\alpha\rangle\langle\psi_\alpha|$ in the expression for ρ yields

$$i\hbar \frac{d}{dt} \sum_{\alpha} (|\psi_\alpha(t)\rangle\langle\psi_\alpha(t)|) = \sum_{\alpha} (H|\psi_\alpha(t)\rangle) \langle\psi_\alpha(t)| - |\psi_\alpha(t)\rangle (\langle\psi_\alpha(t)|H) = H\rho - \rho H.$$

[8] $\rho(t) = \sum_{\alpha} p_\alpha |\psi_\alpha(t)\rangle\langle\psi_\alpha(t)| = \sum_{\alpha} p_\alpha U(t)|\psi_\alpha(0)\rangle\langle\psi_\alpha(0)|U^{-1}(t) = U(t)\rho(0)U^{-1}(t)$

A quantum analog of classical phase space volume is the dimension of a subspace of the Hilbert space. It is measured by a trace operation. That is, the projection operator

$$P = |e_1\rangle\langle e_1| + \ldots + |e_m\rangle\langle e_m|$$

projects onto an m-dimensional subspace of the n-dimensional Hilbert space, spanned by the orthonormal basis vectors $|e_1\rangle \ldots |e_m\rangle$, where $m < n$. We can recover the dimension of the subspace as

$$\text{Tr}(P) = \sum_{i=1}^{n}\langle e_i|P|e_i\rangle = ((\langle e_1|e_1\rangle)^2 + \ldots + ((\langle e_m|e_m\rangle)^2 = m.$$

Since the numbering of the basis vectors is arbitrary, the result holds for any subspace, which is closed under vector addition and scalar multiplication.

If the total dimension n of the Hilbert space is small, the dimension of a subspace is a coarse measure of size in comparison with the finer measurements provided by volume in a classical phase space. However, in the present application, the dimension of the Hilbert space is immense, with n at least the size of Avogadro's number, that is, at least 10^{24}. We need to assess the relative size of the thermal equilibrium states in the Hilbert space, in comparison with the non-equilibrium states. The equilibrium states are *vastly* more numerous than the non-equilibrium states. Our measure need only be able to capture this difference for the exorcism to proceed. While the dimension of the subspaces in which the equilibrium and non-equilibrium states are found is a coarse measure, it is fully able to express the great difference in the size of the two.

We convert the forms (11.2), (11.2'), and (11.2'') of the classical and quantum Liouville equation into expressions concerning conservation of volume by introducing analogous special cases of the probability density and density operator:

Classical Hamiltonian Dynamics	*Quantum Statistical Mechanics*
Consider a set of states that forms an integrable set $S(0)$ in the phase space at time 0 of phase volume $V(0)$. Under Hamiltonian evolution, it will evolve into a new set $S(t)$. Define a probability density that is uniform over $S(0)$ and zero elsewhere. That is,	The projection operator $P_{S(0)}$ projects onto a closed subspace $S(0)$ of the Hilbert space. Since $P_{S(0)}$ is a projection operator, it is idempotent:
	$$P_{S(0)} = P_{S(0)}P_{S(0)}.$$
	The dimension of the subspace onto which it projects is
$$\rho_{S(0)}(q_i, p_i) = (1/V(0))I_{S(0)}(q_i, p_i),$$	$$V(0) = \text{Tr}(P_{S(0)}).$$

where $I_S(q_i, p_i)$ is the indicator function that is unity for phase points in the set S and zero otherwise.

The classical Liouville Eq. (11.2) tells us that the probability density remains constant in time along the trajectories of the time evolution. Hence if the initial probability density is a constant $1/V(0)$ everywhere inside the set $S(0)$ and zero outside, the same will be true for the evolved set $S(t)$. That is, the probability density will evolve to

$$\rho_{S(t)}(q_i, p_i) = (1/V(0))I_{S(t)}(q_i, p_i).$$

Since the new probability distribution must normalize to unity, we have[9]

$$1 = \int_\Gamma \rho_{S(t)}(q_i, p_i)\, d\gamma$$
$$= \frac{1}{V(0)} \int_{S(t)} 1\, d\gamma = \frac{V(t)}{V(0)},$$

which entails that

$$V(t) = V(0). \qquad (11.3)$$

Hence the phase volume of a set of points remains constant under Hamiltonian time evolution.

The uniform density operator corresponding to $P_{S(0)}$ is

$$\rho_{S(0)} = (1/V(0))P_{S(0)}.$$

Over time, using the quantum Liouville Eq. (11.2″), this density operator will evolve to a new density operator

$$\rho(t) = (1/V(0))U(t)P_{S(0)}U^{-1}(t)$$
$$= (1/V(0))P_{S(t)},$$

where $P_{S(t)} = U(t)P_{S(0)}U^{-1}(t)$ is the projection operator to which $P_{S(0)}$ evolves[10] after t. We confirm that $P_{S(t)}$ is idempotent since

$$P_{S(t)}P_{S(t)} = U(t)P_{S(0)}U^{-1}(t)U(t)P_{S(0)}U^{-1}(t)$$
$$= U(t)P_{S(0)}P_{S(0)}U^{-1}(t)$$
$$= U(t)P_{S(0)}U^{-1}(t) = P_{S(t)}$$

and define $S(t)$ as the subspace onto which it projects. Hence we can write

$$\rho(t) = \rho_{S(t)}.$$

Finally, density operators have unit trace, so that

$$1 = \mathrm{Tr}(\rho_{S(t)}) = (1/V(0))\mathrm{Tr}(P_{S(t)})$$
$$= V(t)/V(0),$$

where $V(t)$ is the dimension of $S(t)$. It follows that

$$V(t) = V(0). \qquad (11.3')$$

---

[9] $d\gamma$ is the canonical phase space volume element $dq_1 \ldots dq_n dp_1 \ldots dp_n$.

[10] The derivation of this rule of time evolution closely parallels that of the density operator in Eq. 11.2″.

Hence the dimension of a sub-space remains constant as the states in it evolve over time under the Schrödinger equation.

The derivation of the quantum result, Eq. 11.3', was carried out in a way that emphasizes the analogy with the classical case. The same result can be attained more compactly merely by noting that the trace of a projection operator is invariant under Schrödinger time evolution:[11]

$$V(t) = \text{Tr}(P_{S(T)}) = \text{Tr}(U(t)P_{S(0)}U^{-1}(t)) = \text{Tr}(U^{-1}(t)U(t)P_{S(0)})$$
$$= \text{Tr}(P_{S(0)}) = V(0).$$

11.12   Two Versions of the Exorcism

With the parallel results for the classical and quantum cases in hand, we can now restate the original assumptions of the classical exorcism, listed as (a)–(f) below. Quantum surrogates are needed only for (d)–(f) and are indicated on the right.

(a) A Maxwell's demon is a device that, when coupled with a thermal system in its equilibrium state, will, over time, assuredly or very likely lead the system to evolve to one of the intermediate states; and, when its operation is complete, the thermal system remains in the intermediate state.

(b) The device returns to its initial state at the completion of the process, and it operates successfully for every microstate in that initial state.

(c) The device and thermal system do not interact with any other systems.

(classical)   *(quantum)*

(d) The system evolves according to Hamilton's equations (11.1) with a time-reversible, time-independent Hamiltonian.

(d') The system evolves according to the Schrödinger equation, (11.1') and (11.1''), with a time-reversible, time-independent Hamiltonian.

(e) The equilibrium state upon which the demon will act occupies all but a tiny portion α of the thermal system's phase space, V, where α is very close to zero.

(e') The equilibrium state upon which the demon will act occupies all but a tiny subspace of dimension α' of the thermal system's Hilbert space, where

---

[11] The third equality uses the invariance of trace under cyclic permutation: $\text{Tr}(ABC) = \text{Tr}(CAB)$. The fourth uses unitarity: $U^{-1}U = I$.

the dimension α' is much smaller than the dimension of the thermal system's Hilbert space.

(f) The intermediate states to which the demon drives the thermal system are all within the small remaining volume of phase space, αV.

(f′) The intermediate states to which the demon drives the thermal system are all within the small remaining subspace of Hilbert space of dimension α'.

It is assumed in (e′) that the Hilbert space of the thermal system and, tacitly, of the demon, have a finite, discrete basis. This is the generic behavior of systems such as these that are energetically bound, such as a gas completely confined to a chamber.

The analysis now proceeds as in Norton (2013a, sec. 4). In brief, according to the behavior specified in (a)–(c), a demon is expected to take a thermal system that we would, under non-demonic conditions, consider to be in thermal equilibrium and evolve it to an intermediate state, that is, one which we would under non-demonic conditions consider to be a non-equilibrium state.

When coupled with the physical assumptions of (d)–(f)/(d′)–(f′) that behavior requires a massive compression of phase space volume or Hilbert space volume that contradicts the classical result of the conservation of phase space or the quantum analog for Hilbert subspace dimensions.

The key assumption is expressed in (e)/(e′). A thermal system that has attained equilibrium under non-demonic conditions occupies one of many states that all but completely fill the phase space or Hilbert space. The demon must operate successfully on all of these states, or nearly all of them. The intermediate states to which the demon should drive them must occupy the tiny, remaining part of the phase space or Hilbert space. Changes in the demon phase space or Hilbert space can be neglected, since the demon is assumed to return to its initial state.

12 Quantum Theory as a Principle Theory: Insights from an Information-Theoretic Reconstruction

Adam Koberinski and Markus P. Müller

12.1 Introduction

Quantum theory has a long history of axiomatization. Von Neumann (1932) began by placing Schrödinger's wave mechanics and Heisenberg's matrix mechanics within a more general Hilbert space framework. Hilbert space quantum theory has since become the dominant formalism, due in large part to its abstract and general characterization. Von Neumann's "axioms" – and later modifications – focus on the representation of quantum systems within that formalism. As such, the axioms tend to serve as a recipe for constructing a mathematical model of a given quantum situation. These recipes leave much open in terms of physical interpretation and explanation. For example, a quantum state is represented by a ray in a Hilbert space, and undergoes unitary time evolution within the space; there is no explanation regarding *why* quantum systems evolve the way they do, or *what* a quantum state is.

This explanatory gap represents one of the main motivations for the proliferation of different *interpretations* of quantum theory, including the historically dominant Copenhagen interpretation (Faye 2014), Bohm's pilot wave theory (Dürr and Teufel 2009), Everettian many-worlds interpretations (Saunders et al. 2010), dynamical collapse theories (Ghirardi 2014) and several others.[1] A substantial part of this research is characterized by a common goal: Given the accepted mathematical structure of quantum theory, provide a comprehensible ontology with only minimal modification to its empirically confirmed predictions. A hoped-for consequence of this project is that the interpretation will solve certain conceptual problems, like the measurement problem. A major difference among interpretations is the role that the wave function plays in their ontology. Following Leifer (2014) and Cabello (2015), we can distinguish "ψ-ontic" interpretations, which regard the quantum state as an element of reality, from "ψ-epistemic" interpretations, which see the quantum state as

---

[1] The citations listed above are not to the originators of the interpretations; rather, they provide general overviews of each interpretation as each currently stands.

a "representation of knowledge, information, or belief" (Leifer 2014).[2] Except for the Copenhagen interpretation, all of the traditional interpretations listed above fall into the ψ-ontic camp.[3]

The appearance of further ψ-epistemic interpretations, including the views of Wheeler (1983), Brukner and Zeilinger (Zeilinger 1999; Brukner and Zeilinger 2009; Brukner 2017), and "Quantum Bayesianism" (QBism) (Fuchs and Schack 2013), went hand in hand with a broader and more pragmatic development in the foundations of quantum mechanics. Physicists such as David Deutsch (1985, 1989) and Richard Feynman (1982) became aware of the potential computing power possessed by quantum systems. They realized that entanglement was not just a peculiarly quantum phenomenon, but had interesting uses as a computational resource. Over the late 1980s and throughout the 1990s the field of quantum computation saw major progress, guided by the idea of transmitting and harnessing *quantum information* rather than classical information. The promise of improved computing power with quantum computers was a driving force behind the development of quantum information theory, which in turn shed new light on the field, leading to greater insight into the structure of quantum theory.

A key figure in merging quantum information theory with foundational work in quantum mechanics was John A. Wheeler (1983), whose "It from Bit" doctrine – that the fundamental *stuff* of the universe is information – led to attempts to provide a coherent ontological story for quantum theory in information-theoretic terms. Clifton, Bub, and Halvorson (2003) (CBH) appeared to have gone a long way towards this goal in their 2003 paper. They assume a background theory space characterized by C^*-algebras, which are general enough to include both classical and quantum theories, and derive the algebraic structure of standard quantum theory from information-theoretic postulates. In response to Wheeler, CBH *"are suggesting that quantum theory be viewed... as a theory about the possibilities and impossibilities of information transfer"* (2003, p. 1563, emph. added).

One of the major drawbacks of the CBH approach (later acknowledged by one of the authors [Bub 2016]) is that it relies on a C^*-algebraic framework for theories.[4] Though the framework is general enough to include both classical

---

[2] See also Leifer's talk "Is the wavefunction real?" at the *12th Biennial IQSA Meeting*, Olomouc, Czech Republic.

[3] Feintzeig (2014) has questioned the applicability of the ontological models framework to Bohmian mechanics. For the purpose of our argumentation, however, it is sufficient to consider a qualitative distinction between ψ-epistemic and ψ-ontic theories, since none of our arguments rely on the formal mathematical definition of these notions in terms of the ontological models framework.

[4] An early criticism of CBH's C^*-algebraic framework is due to Duwell (2007b). While we broadly agree with most of his conclusions, we also disagree with some of his arguments. For

and quantum theories,[5] its structure is rather restrictive and builds in many assumptions that are crucial to quantum theories. Given the CBH reconstruction of quantum theory, the natural question to ask is, "Why C^*-algebras for physical theories, rather than something else?"

Independently from the philosophers' attempts, physicists and mathematicians had already developed an operational framework that generalizes quantum theory in a much more universal way than the CBH approach. This framework dates back to at least the 1960s, and has been known in the mathematics and physics communities under several different names (order unit/base norm spaces, convex-operational framework, or *generalized probabilistic theories* (GPTs), the name that we will use in this paper), and in slightly different versions. The goal of this framework is to capture all conceivable situations in physics that can be cast in operational language, in particular typical laboratory situations where there is a *preparation procedure*, followed by a *measurement* (and possibly *transformation procedures* in between both). Ideally, the framework is so general that there are no constraints on its applicability,[6] other than the stipulation that it ought to describe the *statistical or probabilistic* properties of a physical theory, and nothing else.

We will describe this framework[7] in more detail in Section 12.3; for now, it is important to note that the GPT framework is comprehensive enough to allow for a variety of conceivable physical phenomena that are more general than those predicted by quantum theory. For example, some of the theories in this framework admit stronger forms of nonlocality than allowed by any version of quantum theory (Popescu and Rohrlich 1994; Barrett 2007), a notion

example, we think that it is not necessary for the information-theoretic constraints to be generally true; for the explanatory power of a reconstruction, it is sufficient that they are true in the regime of applicability of standard quantum theory (excluding any conjectured hidden-variables regime). The same reasoning applies to expectation value additivity.

[5] This includes more general quantum physics, such as quantum field theory and quantum statistical mechanics. The former is somewhat controversial, given that only idealized, nonphysical models of quantum field theory have been constructed in the C^*-algebra framework. The majority view in the field is that the framework *can* accommodate more realistic models, though their construction is elusive. Fraser and Wallace have debated the relationship between algebraic and conventional approaches to quantum field theory (Fraser 2009, 2011; Wallace 2011).

[6] It is a misunderstanding to say that this framework would not apply in the relativistic regime, or to say that the reconstructions of quantum theory in the GPT framework (as described in Section 12.4, for example) would be inherently "non-relativistic." Instead, the GPT framework itself is ignorant about the structure of spacetime: it can apply both to relativistic or non-relativistic scenarios, or to even more general situations (exactly as the Hilbert space and C^*-algebraic frameworks, which are special cases). Applying a certain GPT in a specific relativistic situation will in general involve *additional effort*, for example by modeling Lorentz transformations in the corresponding theory, but there is no a priori obstruction to doing so. The only practical restriction in most current work on GPTs is the limitation to finite-dimensional state spaces for technical reasons, which may have to be overcome for some applications.

[7] Philosophically inclined readers with an interest in some more details may find the exposition by Myrvold (2010) helpful, which includes a comparison of the GPT and C^*-algebraic approaches.

of "higher-order interference" (Sorkin 1994) that has already been tested experimentally (Sinha et al. 2010), and beyond-quantum computational and information-theoretic properties like trivial communication complexity (van Dam 2013) or a violation of information causality (Pawlowski et al. 2009). This framework provides a much more general starting point than the CBH reconstruction, from which to impose physically motivated postulates to single out quantum theory. In particular, instead of asking *"Why quantum mechanics, rather than classical mechanics?"* the focus has shifted towards asking a different question: *"Why quantum mechanics, and not something possibly even more general?"*

Broadly speaking, there have been two different approaches to deriving quantum theory within the GPT framework. The first approach, pioneered by Lucien Hardy (2001),[8] starts with the (large) set of all probabilistic theories with arbitrary sets of states, measurements, and transformations (i.e., dynamics), and then imposes a small set of physically reasonable postulates on top of that framework, followed by a proof that quantum theory is the unique theory in the framework that satisfies these postulates. This approach has led to several fully successful reconstructions of quantum theory, e.g., by Hardy (2001), Chiribella et al. (2011), Dakić and Brukner (2011), Hardy (2011), Masanes and Müller (2011), Masanes et al. (2013), Höhn (2014), and Höhn and Wever (2017); one major goal of this paper is to describe one of these reconstructions (Masanes et al. 2013) in an accessible way, and to discuss its implications (Section 12.4). Another (sometimes called "device-independent") approach disregards dynamics, and tries to single out the set of (static) *quantum correlation tables* via some set of principles from the larger set of all non-signaling correlations; cf. Section 12.3.2. Despite important successes, this approach has not been able to *exactly* single out quantum correlations so far (Navascués et al. 2015). In Section 12.4, we will speculate that this may tell us something important about physics.

Following the former approach, one has a framework that admits the *full* encoding of the essence of quantum theory into a set of simple postulates, singling out quantum theory from the "landscape" of all probabilistic theories. What is remarkable is that these postulates are to a large extent expressed in *computational* terminology. As explained in detail in Section 12.4, quantum theory is singled out by postulating that the state and time evolution of every physical system can be reversibly encoded into a number of interacting "universal bits"; time evolution can be reversible and continuous; global states are uniquely determined by their local properties and their correlations; and one universal bit can carry one binary unit of information and not more.

---

[8] However, there has been a long tradition of reconstruction attempts before Hardy's work; for a historical overview, see, e.g., Hardy (2011, sec. 1.2) and the references therein.

These are properties which are directly linked to the possibility of having a *universal computing machine*, like the quantum Turing machine, which is constructed in a modular way by composing a large number of universal bits. This universal machine is in principle able to simulate the time evolution of any physical system whatsoever, thereby rendering the empirically accessible content of any physical system – that is, its (quantum) state – ultimately "substrate-independent." It is interesting to see that this notion of "universal computation," as formalized in our postulates, is powerful enough to uniquely determine the state space, time evolution, and possible measurements (and thus also other properties like the maximal amount of non-locality) of quantum theory. In this sense, the hypothesis of "physics as computation," interpreted in a suitable way, demonstrates remarkable explanatory power.

The remainder of this paper is organized as follows. Section 12.2 outlines our view that information-theoretic reconstructions of quantum theory provide a fruitful, albeit only partial, interpretation of quantum theory. We describe several alternatives for arriving at a full-fledged interpretation of quantum theory, using the information-theoretic reconstructions as a starting point. Furthermore, we describe why we think that the reconstructions pose a challenge to existing ψ-ontic interpretations. In Section 12.3 we introduce the GPT framework and highlight some of the generalizations beyond quantum theory by examining the gbit and its possible correlations. This is, in some sense, the most general fundamental unit of information within the GPT framework. In Section 12.4 we present the postulates that allow us to single out quantum theory uniquely from the space of GPTs, as first shown by one of the authors (Masanes et al. 2013). In doing so, we highlight the physical significance of the postulates, and some of their potential conceptual consequences. We conclude that information theory – specifically, constraints arising from universal computation – teaches us a lot about the structure of quantum theory. This goes a long way toward answering the question, "Why the quantum?"

12.2   Partially Interpreting Quantum Theory as a Principle Theory

We begin with a distinction first made by Einstein between *principle* and *constructive* theories. Reflecting on his methods for developing the special and general theories of relativity (SR and GR, respectively), Einstein contrasted his principle approach with the usual constructive methods in physics:

We can distinguish various kinds of theories in physics. Most of them are constructive. They attempt to build up a picture of the more complex phenomena out of the material of a relatively simple formal scheme from which they start out. Thus the kinetic theory of gases seeks to reduce mechanical, thermal, and diffusional processes to movements of molecules – i.e., to build them up out of the hypothesis of molecular motion. When we say that we have succeeded in understanding a group of natural processes, we invariably mean that a constructive theory has been found which covers the processes in question.

Along with this most important class of theories there exists a second, which I will call "principle theories." These employ the analytic, not the synthetic, method. The elements which form their basis and starting-point are not hypothetically constructed but empirically discovered ones, general characteristics of natural processes, principles that give rise to mathematically formulated criteria which the separate processes or the theoretical representations of them have to satisfy. Thus the science of thermodynamics seeks by analytical means to deduce necessary conditions, which separate events have to satisfy, from the universally experienced fact that perpetual motion is impossible. (Einstein 1954, p. 228)

Others have noticed the similarity between information-theoretic reconstructions of quantum theory and SR (Zeilinger 1999; Brown and Timpson 2006; Harrigan and Spekkens 2010; Felline 2016), and Bub has been vocal about the utility of information theory for making quantum mechanics a principle theory (Clifton et al. 2003; Bub 2005, 2016). Unlike constructive theories, principle theories explain by showing that features of the world are deductive consequences of a small set of general principles. This leads us to formulate the following thesis, which we will elaborate in more detail in the remainder of this section:

Thesis: *Information-theoretic reconstructions provide a partial interpretation of quantum theory as a* principle theory of information,[9] *by identifying a small set of information-theoretic principles that render our world a quantum one. This leaves several alternatives for extending to a full-fledged interpretation, some of which we describe below. Furthermore, the reconstructions represent a challenge for existing "ψ-ontic" interpretations of quantum theory by highlighting a relative deficiency of those interpretations in terms of their explanatory power.*

Standard textbook formulations often phrase quantum theory as a set of axioms (e.g., that states are given by rays on a Hilbert space, the Born rule, etc.), but these axioms are purely mathematical and abstract, and thus do not correspond to the kind of physical principles that one would expect from a principle theory in Einstein's sense. Furthermore, the standard formulation also does not give an account in the sense of a constructive theory, and as such is "doubly unsatisfactory." On the other hand, if the quantum formalism is derived from simple physical postulates, and cast as a principle theory based on information, then one gains an explanation in the sense that quantum theory is found to be a unique consequence of easily understandable constraints.

---

[9] This is very closely along the lines of Jeffrey Bub's suggestion (Clifton et al. 2003; Bub 2005).

The mathematical ingredients therefore become less arbitrary and provide a partial interpretation in terms of the role they play in satisfying the more comprehensible principles.

Nonetheless, the information-theoretic reconstructions do *not* provide explanatory power in an even stronger sense: They do not typically tell us *what quantum states are*, or *what is really going on in the world* when we perform a Bell experiment, for example. In particular, information theory provides a foundation for understanding quantum theory as a principle theory, but *not* as a constructive theory. However, we highlight three alternatives for a potential generalization to a full interpretation of quantum theory, using information-theoretic reconstructions as a starting point.[10]

First, quantum information theory coupled with ontic structural realism (OSR) (Ladyman and Ross 2007) could provide a full-fledged interpretation of quantum theory by rejecting the ultimate need for an entity-based constructive account of a fundamental theory of physics. This would consist in an ontology of structural relations in some sense – simply of the relational structure uniquely picked out by the information-theoretic postulates from the space of GPTs. This option denies the need for a constructive account of quantum theory, though it does not rule out the possibility of discovering a constructive successor to quantum theory, in particular since ontological stability across theory change is a characteristic of OSR.[11]

Second, one could adopt a subjective view of quantum theory, what Cabello (2015) calls participatory realist interpretations of quantum theory. Though these fall into various forms, subjective interpretations of quantum theory all share the common point of view that quantum theory is not directly about properties of our world. Most subjective interpretations, including the Copenhagen interpretation,[12] the view of Brukner and Zeilinger (Zeilinger 1999; Brukner and Zeilinger 2009, 2003), and Wheeler's interpretation (Wheeler 1983), claim that quantum theory is a theory about our *knowledge* of the world but not

---

[10] Laura Felline (2016) has given a detailed analysis of the notion of explanation that applies to information-theoretic reconstructions of quantum theory, and her conclusions are to a large extent compatible with what we outline in this paper. Felline has also described two of our options (for a full-fledged interpretation), namely the possibility of a "structural" explanation (in conjunction with the rejection of the necessity of causal explanations) and the option to find a constructive successor of quantum theory.

[11] Lucas Dunlap's current work is concerned with a detailed analysis of this approach; see, e.g., his talk "The Information-Theoretic Interpretation of Quantum Mechanics and Ontic Structural Realism" at *Foundations 2016: The 18th UK and European Conference on Foundations of Physics*, London School of Economics. Note that quantum theory has been an important motivation in the development of OSR (Ladyman 1998), but it has also been argued that OSR in itself is not sufficient to constitute an interpretation of quantum mechanics (Esfeld 2013).

[12] At least Heisenberg's view can be characterized in this way (Margenau and Park 1967). "*The* Copenhagen interpretation" is a bit misleading, as there is only a loose set of common elements amongst thinkers like Bohr and Heisenberg. See Faye (2014) for a more nuanced discussion.

the world itself. Another possibility is given by the QBist interpretation that quantum theory is an extension of Bayesian rules for subjective, rational *belief* updating (Fuchs and Schack 2013; Fuchs et al. 2014). Like the first option, this route "explains away" the need for a constructive account of quantum theory, although some of its proponents (in particular, QBists) express the hope for insights into the actual objective world as a major motivation for their approach.

The final option is inspired by the quote above highlighting Einstein's thoughts on principle and constructive theories. Some might agree with Einstein's view that successful constructive theories carry more explanatory weight, and would prefer a constructive version of quantum theory. Consequently, the third option is to come up with a constructive successor of quantum theory, describing a regime of our world that is currently empirically inaccessible, but which gives rise to the information-theoretic principles (and thus indirectly to quantum theory) after some approximation or coarse-graining. A paradigmatic historical example, as explained by Einstein, is the kinetic theory of gases as a constructive successor of thermodynamics. It is encouraging to see how, at least in this instance, the development of a constructive theory was guided and constrained by the presence of a principle theory. Thus, one might hope that the information-theoretic reconstructions could serve as a similar guideline in the case of quantum theory. In fact, thermodynamics and kinetic theory ended up working *together* to form a fuller, more explanatorily powerful framework for the properties of bulk materials. We consider this additional power – in terms of prediction, or unification – a crucial signature of success of the constructive theory, and a main reason to adopt it in addition to the bare principle theory.

This third option seems to be particularly attractive for proponents of "non-subjective" ψ-epistemic interpretations of quantum theory (or "ψ-epistemic type I" in Cabello's [2015] nomenclature, i.e., those that "view the quantum state as representing knowledge of an underlying objective reality"), insofar as they regard their interpretations as an open research program. Proponents of this view would agree that a quantum state describes our (limited) *information* about the world, and could thus be happy with a principled formulation of quantum theory in information-theoretic terms as a starting point. They could see the information-theoretic reconstructions as a constraining guideline for the development of the sought-for constructive underpinning of quantum theory, in particular since they would consider these information-theoretic principles as more directly related to properties of the real world than people in the subjective camp would.

We note that these options are neither completely exhaustive nor mutually exclusive, but they all go beyond what is directly provided by any reconstruction. However, even if one takes *only* the partial interpretation that a principle theory provides, we think that its explanatory power is sufficient to pose a challenge to proponents of more traditional ψ-ontic interpretations. None of Bohmian mechanics, Everettian quantum theory, or collapse theories fill the explanatory role of a principle theory. For example, Everettian quantum theory does *not* start with a broad general framework of "theories of many worlds," put simple principles on top of that, and prove that quantum theory is the unique theory of many worlds that satisfies these principles.

One can certainly object that ψ-ontic theories are not meant to be principle theories in the first place – their ambition is to exhaustively describe what is going on in a quantum world, and as such they would more accurately be described as candidate constructive theories (possibly with Bohmian mechanics coming closest to Einstein's vision of a constructive theory). Nevertheless, given that they do not provide the specific explanatory power of the information-theoretic derivations that we have just described, one should demand that they give us something else in replacement in order to be considered successful: some additional element of empirical or theoretical success that goes *beyond* the standard formulation of quantum theory. Clearly, the aforementioned example of kinetic theory (underpinning thermodynamics) satisfies this demand in a quite spectacular way, but we do not think that any of the existing ψ-ontic interpretations of quantum theory currently does.

Now the question of *what exactly* would or should constitute an empirical or theoretical success which would cause physicists to accept a given ψ-ontic theory is clearly a difficult and controversial question. Some supporters of ψ-ontic theories will see the *realist* worldview that these theories provide as enough of a success to accept them. However, in contrast to kinetic theory, this acceptance currently comes at the price of *not also* being provided with additional elements of explanation or prediction which, for example, could help to empirically validate the specific claims of one ψ-ontic theory (say, Bohmian mechanics) over another one (say, the many-worlds interpretation). What the information-theoretic derivations do is to emphasize, and extend, this explanatory deficiency: They demonstrate how a (partial) interpretation of quantum theory as a principle theory of information can provide novel explanations of the quantum formalism that are not supplied by the current ψ-ontic theories.[13]

Is this argument also a challenge for ψ-epistemic interpretations? We do not think so. Interpretations that treat the quantum state as a state of knowledge or belief are conceptually more closely related to the view of quantum theory as a principle theory of information, which has led some physicists (e.g.,

---

[13] Since the information-theoretic reconstructions are largely agnostic about the physical substrate underlying quantum theory, they are in principle entirely compatible with existing ψ-ontic interpretations. Our objection is that current ψ-ontic interpretations do not empirically supplement the principle theory in a similar way as the kinetic theory of gases empirically supplements thermodynamics.

Brukner[14]) to argue that the success of the latter is evidence for the validity of the former. Another major difference lies in the fact that the ψ-ontic interpretations make much more specific claims about the real world. For example, Bohmian mechanics stipulates that the universe is in some (unknown) configuration, which (together with a pilot wave) evolves according to specific differential equations. On the other hand, while QBists view quantum states as subjective states of belief, they are rather agnostic about the underlying real world that gives rise to these beliefs, and see the question of ontology as much more of an open research program. Thus, different demands for predictive power in the empirical regime follow from a simple doctrine: that concrete, specific claims about the real world should ultimately be backed up by successful empirical predictions which do not also arise identically from rival theories.

12.3   The GPT Framework: General No-Signaling Theory (Boxworld)

The empirical content of quantum theory lies in the statistical prediction of measurement outcomes: If the preparation of a quantum system in state ρ (a density matrix) is followed by an n-outcome measurement with projection operators $P_1, \ldots, P_n$, then the probability of the ith outcome is given by $\mathrm{tr}(\rho P_i)$. This is the minimal content on which all interpretations of quantum mechanics agree. The framework of generalized probabilistic theories (GPTs) allows one to describe basically *any* theory that fits within this basic operational prescription, namely that a preparation procedure is followed by a measurement (and possibly a transformation procedure before the measurement takes place), yielding one of several possible outcomes probabilistically. Classical probability theory[15] and quantum theory are special cases of GPTs, but there are many others which are neither classical nor quantum.

Here, we will *not* give a complete formal definition of the GPT framework; interested readers can find such a definition, for example, in Hardy (2001) or Barrett (2007). Instead, we focus on a simple illustrative example of one specific non-classical and non-quantum theory within this framework, sometimes called "general no-signaling theory" (Barrett 2007), or, more colloquially, "boxworld."

---

[14] For example, at the conference "Quantum Theory: from Problems to Advances" in Växjö, Sweden (June 9–12, 2014), Brukner argued as follows: "The *very idea of quantum states as representatives of information* – information that is sufficient for computing probabilities of outcomes following specific preparations – *has the power* to explain why the theory has the very mathematical structure it does. This in itself is the message of the reconstructions."

[15] Here, "classical probability theory" is not meant to describe the scientific discipline, but concretely stands for the state space of discrete probability distributions, in contrast to quantum theory where the states are density matrices. That is, it is the technical notion of a specific GPT where the set of states Ω is the set of probability vectors $(p_1, \ldots, p_n)$ (where $n \in \mathbb{N}$ is fixed), $p_i \geq 0$, $\sum_i p_i = 1$, and the reversible transformations are the permutations of entries. Geometrically, these ("classical") state spaces are simplices.

First we introduce a single system in boxworld, sometimes called the "gbit" (generalized bit), and we will see in what way it differs from quantum theory's "qubit." Then we describe a composite state space of two gbits, which turns out to be the infamous "no-signaling polytope." Finally, we discuss some of its properties and point out crucial differences with the state space of two quantum bits. This example will also serve to illustrate the use and the meaning of the information-theoretic postulates which allow one to reconstruct quantum theory, as we will discuss in Section 12.4.

12.3.1   The "Gbit": A Single Generalized Bit in Boxworld

Consider a single spin-1/2 particle, for example, an electron. The quantum physics allows us to prepare the spin degree of freedom of that particle in any quantum state (described by a 2×2 density matrix ρ), and later on to measure the spin in any spatial direction $\vec{n}$ that we choose – for example, by setting up a Stern-Gerlach device with a magnetic field gradient in direction $\vec{n}$. The measurement outcome will always be either spin-up or spin-down. Since we always have two possible outcomes, this represents an elementary binary alternative within quantum theory – a "qubit."

Now imagine a world where there are particles that behave in a somewhat similar way: We can prepare them in some state ω (for now remaining silent about what mathematical object this is), and later on decide to perform a two-outcome measurement (which we will denote by x). However, let us assume that there are only *two* possible choices of two-outcome measurement, denoted $x = 0$ and $x = 1$. This is in contrast to the quantum particle, where we have infinitely many possible two-outcome measurements, corresponding to the different choices of quantization axis $x = \vec{n}$. Let us denote the possible outcomes by $\uparrow$ and $\downarrow$; in the quantum case, these correspond to "spin-up" and "spin-down," respectively.

If we decide to perform measurement x on one of these hypothetical particles, we get outcome a with some probability $p(a|x)$. Clearly $p(\uparrow |x) + p(\downarrow |x) = 1$ both for $x = 0$ and for $x = 1$. This means that the two real numbers $p(\uparrow |0)$ and $p(\uparrow |1)$ tell us everything that we need to know to predict the outcome of any measurement that we could possibly perform on a particle. We can thus write down the state ω simply as

$$\omega = \left(p(\uparrow |0), p(\uparrow |1)\right).$$

Now we make one further assumption: namely, that the hypothetical physics that describes our hypothetical particle does not restrict these two probabilities in any way. In other words, we imagine that for *any* given choice of these two numbers (in the unit interval $[0, 1]$), we can in principle prepare a parti-cle in the corresponding state ω. Then the set of all possible states, the *state space*, can be visualized as all the points in two dimensions with coordinates between 0 and 1, that is, the unit square. Note that the state itself does not tell us anything about *what that particle is*, or *what happens during a measurement*. All it does is to allow us to compute the outcome probabilities of all possible future measurements – it formalizes the operational content of the physics of the hypothetical particle, and not more.

Figure 12.1 illustrates this and compares it to the quantum bit. For a qubit, the state can be described by a density matrix

$$\rho = \frac{1}{2} \begin{pmatrix} 1 + a_1 & a_2 - ia_3 \\ a_2 + ia_3 & 1 - a_1 \end{pmatrix}.$$

Density matrices must have non-negative eigenvalues. Since the two eigenval-ues are $\frac{1}{2}\left(1 \pm \sqrt{a_1^2 + a_2^2 + a_3^2}\right)$, we obtain the condition $|\vec{a}| \leq 1$ for the vector $\vec{a} := (a_1, a_2, a_3)$. We can thus visualize the quantum bit as the unit ball in three dimensions, which is known as the *Bloch ball*. As we see in Figure 12.1, the gbit and the qubit state spaces thus look very different, but both have the important property of *convexity*. This formalizes the idea of a "mixed state." Imagine a procedure where one tosses a fair coin, and, depending on the out-come, prepares either state ω or another state φ. Repetition of this procedure will statistically correspond to the preparation of the state $\frac{1}{2}\omega + \frac{1}{2}\varphi$ – a *con-vex combination* of ω and φ. Therefore, convex combinations of states are also states, which forces state spaces to be convex sets. In principle, *any* convex,

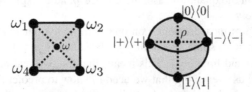

Figure 12.1 The state spaces of a gbit (left) and a qubit (right). The gbit state space has four pure states (that is, extremal points of the convex state space), namely $\omega_1 = (0, 1)$, $\omega_2 = (1, 1)$, $\omega_3 = (1, 0)$, and $\omega_4 = (0, 0)$. For any given state ω (for example $\omega = (\frac{1}{2}, \frac{1}{2})$), the decomposition into pure states is not unique (and for this example, $\omega = \frac{1}{2}\omega_1 + \frac{1}{2}\omega_3 = \frac{1}{2}\omega_2 + \frac{1}{2}\omega_4$). This is a signature of nonclassicality, and is also present in the quantum bit. For example, for the maximally mixed state $\rho = \frac{1}{2}\mathbf{1}$, we have $\rho = \frac{1}{2}|0\rangle\langle 0| + \frac{1}{2}|1\rangle\langle 1| = \frac{1}{2}|+\rangle\langle +| + \frac{1}{2}|-\rangle\langle -|$, where $|\pm\rangle = \frac{1}{\sqrt{2}}(|0\rangle \pm |1\rangle)$. The qubit state space has infinitely many pure states (all states on its boundary)

closed, bounded set in a vector space over the real numbers[16] corresponds to a valid state space in the framework of GPTs.

Many important physical properties of a conceivable physical system can be inferred from the shape of its state space. First, one can see that both the gbit and the qubit are *nonclassical* in the sense that states have in general more than one convex decomposition into pure states. Classically, a state corresponds to a probability distribution over mutually distinguishable alternatives (such as probability p for "heads" and $1 - p$ for "tails" for a coin), and there is always a unique way to decompose that state into its alternatives. However, this is not true for the qubit or for the gbit, as Figure 12.1 illustrates.

Another important property of a state space is to ask what kinds of *reversible transformations* are possible. We can think of a "transformation" as any operation that we apply to the particle, e.g., switching on a magnetic field for some time, or performing a "gate" in a computation. In particular, *time evolution* from some initial state ω_i to some final state ω_f is an example of a transformation T, i.e., $\omega_f = T(\omega_i)$. To preserve the statistical interpretation of mixtures, transformations T must necessarily be affine-linear (Holevo 1982; Barrett 2007). A transformation is *reversible* if it can be inverted, and if its inverse T^{-1} is a transformation too. In the quantum case, the reversible transformations are exactly the *unitary transformations* U, mapping an initial state ρ_i to a final state $\rho_f = U\rho_i U^\dagger$. In the Bloch ball picture, unitary transformations correspond to *rotations* of the Bloch ball. In other words, for every unitary U there is a 3×3 rotation matrix R such that the final Bloch vector $\vec{a}_f$ is related to the initial Bloch vector $\vec{a}_i$ via $\vec{a}_f = R\vec{a}_i$. This illustrates the general fact that reversible transformations in a GPT are *affine-linear symmetries* of the state space.

Clearly, a three-dimensional ball can be continuously rotated, which means that we can have *continuous reversible time evolution* (for example, according to the Schrödinger equation) of the form $\vec{a}(t) = R(t)\vec{a}_i$, where $t \in \mathbb{R}$ denotes time. On the other hand, the only possible reversible transformations on a gbit are the symmetries of the square – rotations by multiples of 90° and reflections. The time evolution of the state would have to pass in discrete steps; continuous reversible evolution is impossible. This is related to the fact that the qubit is "round" and the gbit is not; or, more formally, that the qubit has *infinitely many pure states*, while the gbit has only four pure states (the corners).

---

[16] In the examples above, this vector space will be $\mathbb{R}^2$ (for the gbit) and the real vector space of Hermitian complex 2×2 matrices (for the qubit). See Holevo (1982, ch. 1) for a detailed explanation of why one can always without loss of generality represent state spaces as subsets of a real vector space.

12.3.2   Two Gbits: The No-Signaling Polytope

If we have two distinguishable quantum systems, described by Hilbert spaces $\mathcal{H}_A$ and $\mathcal{H}_B$, then there is always a unique corresponding composite quantum system, namely the one that is described by the tensor product Hilbert space $\mathcal{H}_A \otimes \mathcal{H}_B$. General state spaces in the GPT framework can be composed analogously, with the crucial difference that *there are in general infinitely many possibilities on what the composite state space can be*. In more detail, if Ω_A and Ω_B are two state spaces (for example, gbits), then any state space Ω_{AB} is considered as a valid composite, as long as it satisfies the following postulates:

- **Local preparations:** For every two states $\omega_A \in \Omega_A$ and $\omega_B \in \Omega_B$, there is a state in Ω_{AB} that corresponds to the independent local preparation of ω_A and ω_B. Here we denote this state by $\omega_A \omega_B$.
- **Local measurements:** For every measurement on Ω_A that has possible outcomes a_i, and measurement on Ω_B that has possible outcomes b_j, there is a joint measurement on Ω_{AB} with pairs of outcomes (a_i, b_j) that describes a situation where the two measurements are performed independently on Ω_A and Ω_B. Denoting the probability of an outcome a_i if one measures on state ω_A by $p_{\omega_A}(a_i)$, the corresponding measurement satisfies $p_{\omega_A \omega_B}(a_i, b_j) = p_{\omega_A}(a_i) p_{\omega_B}(b_j)$.
- **No-signaling:** If local measurements are performed on an arbitrary state in Ω_{AB}, then the outcome probabilities on Ω_A do not depend on the choice of measurement on Ω_B, and vice versa.

While these postulates are always taken as background assumptions in any GPT, the following postulate is sometimes introduced in addition:

- **Tomographic locality:** States on Ω_{AB} are uniquely determined by the statistics of local measurements.

All these postulates are satisfied in quantum theory: If ρ_A, ρ_B are density matrices on Hilbert spaces $\mathcal{H}_A$ and $\mathcal{H}_B$, then the corresponding local preparation is given by the density matrix $\rho_A \otimes \rho_B$ on $\mathcal{H}_A \otimes \mathcal{H}_B$. Similarly, if we have projectors[17] $P_i^{(A)}$ and $P_j^{(B)}$ that define measurements on $\mathcal{H}_A$ and $\mathcal{H}_B$, then the projectors $P_i^{(A)} \otimes P_j^{(B)}$ define the corresponding local measurement on the composite quantum system. Since projectors of this product form span the full linear space of Hermitian matrices, their outcome probabilities $\mathrm{tr}(P_i^{(A)} \otimes P_j^{(B)} \rho_{AB})$ determine the state ρ_{AB} uniquely, which implies tomographic

---

[17] We can more generally consider *positive operator-valued measures* where the P_i are not necessarily projectors, but arbitrary positive semidefinite operators.

locality. These postulates are also satisfied by classical probability theory, i.e., the state spaces of classical discrete probability distributions.

Now consider two gbit state spaces Ω_A and Ω_B (i.e., the same square state space twice, but with different labels). We define a composite state space Ω_{AB} in the following way:

The composite state space Ω_{AB} shall contain all non-signaling probability tables of the form $p(a, b|x, y)$, where $x, y \in \{0, 1\}$ denote the local choices of measurements, and $a, b \in \{\uparrow, \downarrow\}$ denote the two local outcomes.

Intuitively, no-signaling means that the choice of experiment for one system has no effect on the statistics obtained for the second system. Mathematically, this is equivalent to the marginals $p(a|x)$ and $p(b|y)$ being well-defined and independent of the choice of experiment y and x, respectively. This leads to

$$p(a|x) = \sum_b p(a, b|x, y) = \sum_b p(a, b|x, y') \text{ for all } a, x, y, y', \quad (12.1)$$

$$p(b|y) = \sum_a p(a, b|x, y) = \sum_a p(a, b|x', y) \text{ for all } b, x, x', y. \quad (12.2)$$

These conditions are supplemented by the conditions of non-negativity, $p(a, b|x, y) \geq 0$, and normalization: $\sum_{a,b} p(a, b|x, y) = 1$ for all x, y.

A state ω_{AB} of two gbits corresponds to a list of sixteen probabilities $p(a, b|x, y)$. However, the conditions above reduce the number of affinely independent probabilities to eight. This is similar to the single-gbit case, where we have started with four probabilities $p(a|x)$, but only two of them were affinely independent. Since we have a set of linear equalities and inequalities, the vectors (probability tables) that satisfy them – that is, the state space Ω_{AB} – will form a polytope.

The vertices of this polytope fall into two types. The first type are local deterministic states, which correspond to local preparations where all probabilities are either zero or one:

$$p(a, b|x, y) = p(a|x)p(b|y), \text{ and } p(a|x), p(b|y) \in \{0, 1\}.$$

Since there are four possible choices for deterministic $p(a|x)$ (the four pure states of the gbit), and similarly for $p(b|y)$, we obtain sixteen vertices of this kind.

The second type of vertex is sometimes called a "Popescu-Rohrlich box" (PR-box) (Popescu and Rohrlich 1994). There are eight vertices of this type. Let us only describe one of them; the other seven can be obtained from it by relabeling the measurement choices and outcomes. A PR-box state is

$$p(a, b | x, y) = \begin{cases} 1/2 & \text{if } (x, y) = (0, 0) \text{ and } (a, b) \in \{(\uparrow, \uparrow), (\downarrow, \downarrow)\}, \\ 1/2 & \text{if } (x, y) \in \{(0, 1), (1, 0), (1, 1)\} \text{ and } (a, b) \\ & \in \{(\uparrow, \downarrow), (\downarrow, \uparrow)\}, \\ 0 & \text{otherwise.} \end{cases}$$

That is, if both parties decide to perform the 0-measurement, then the outcomes are correlated; in all other cases, they are anticorrelated. The PR-box exhibits correlations that are impossible to obtain within quantum theory: A PR-box would violate the CHSH Bell inequality by more than any quantum state (Popescu and Rohrlich 1994). This illustrates one way in which general theories in the GPT framework can describe physics that differs from quantum physics.

The state space Ω_{AB} is the polytope that is spanned by these twenty-four pure states (i.e., their convex hull, cf. Figure 12.2). It is eight-dimensional, and it has been extensively studied in quantum information theory (Barrett 2007). In our context, it will be particularly illuminating to study the possible *reversible transformations* on Ω_{AB}.

As we have argued above, every transformation of this kind must be an affine-linear symmetry of the state space. Thus, in order to find the reversible transformations, we need to mathematically classify these symmetries. This has first been done in Gross et al. (2010), where it was shown that the *only* symmetries in boxworld (for any number of measurements, outcomes, and parties) correspond to relabelings of measurement outcomes on individual subsystems, and permutations of subsystems. In particular, *there is no reversible transformation that maps a local deterministic state to a PR-box state*. Figure 12.2 shows a caricature: If the state space were in fact identical to that in the figure, then it would be geometrically obvious that no linear symmetry can map one

Figure 12.2 A caricature of the no-signaling polytope Ω_{AB} (which, in contrast to this schematic figure, is eight-dimensional and has 24 vertices). There are two classes of pure states: local states (labeled "L"), and states of PR-box type. Since no linear symmetry of this polytope can map a local vertex to a PR-box vertex, there cannot be any reversible transformations in boxworld that generate a PR-box state from a local state

of the PR-box vertices to any of the local vertices. This is because the local and PR-box vertices as depicted in the caricature are geometrically inequivalent: Every local vertex connects, via edges, to four other vertices, but every PR-box vertex is connected to three other vertices only.

The actual no-signaling polytope Ω_{AB} is more complicated than the caricature, but one can analyze its structure by using the PolyMake software (Gawrilow and Joswig 2000). We can construct a description of Ω_{AB} within PolyMake, and then use it to count the number of vertices that are connected to a given vertex via edges. What we find is that the local vertices connect to 17 other vertices: 13 other local vertices and 4 PR-box vertices. The PR-box vertices, in contrast, connect to 8 other vertices, all of which are local vertices. In summary, this implies that *reversible time evolution cannot map a deterministic state of two gbits to a PR-box state of two gbits, or vice versa.*

We can contrast this to the state space of two *quantum* bits, Ω_{AB}^{QM}. This is the state space of 4×4 density matrices, which is not a ball (even though the qubit was), but a rather complicated 15-dimensional convex set. In contrast to Ω_{AB}, it has infinitely many pure states: some of them corresponding to local preparations (like $\rho_A \otimes \rho_B$ with ρ_A, ρ_B pure states) and some of them not (like $|\psi_+\rangle\langle\psi_+|$, with $|\psi_+\rangle = \frac{1}{\sqrt{2}}(|\uparrow\downarrow\rangle + |\downarrow\uparrow\rangle)$) an entangled state). However, all these pure states are geometrically equivalent: If ρ_{AB} and σ_{AB} are any two pure states, then there is always a reversible transformation (namely, a unitary U) such that $\rho_{AB} = U\sigma_{AB}U^\dagger$.[18] In particular, pure product states can be mapped to entangled states – time evolution can create entanglement, and can do so continuously in time. This means that Ω_{AB}^{QM} is much more symmetric than Ω_{AB}: Every pure state looks like every other one.

Thus, the two-bit state spaces of quantum theory and boxworld have many features in common: they are non-classical (i.e., there is non-uniqueness of convex decompositions), they contain entangled states, and they satisfy the principle of tomographic locality. However, two-qubit quantum theory satisfies the following principle, while two-gbit boxworld does not:

- **Reversibility:** Given any two pure states of the system, there is a reversible transformation that maps the first to the second.

12.4   Deriving Quantum Theory

Given the GPT framework, how do we arrive at quantum theory? One of us has shown (with further coauthors) in previous works that quantum theory is the unique consequence of a set of four principles, at least for finite-dimensional

---

[18] For a more thorough introduction to the geometry of quantum state spaces, see Bengtsson and Życzkowski (2006). For a more general overview of quantum information theory, see Nielsen and Chuang (2000).

systems (Masanes and Müller 2011; Masanes et al. 2013). We outline the assumptions and principles behind the derivation (for which see the list below), and offer an analysis of the insight they provide into the structure of quantum theory. We start by discussing the principle of the *Existence of an Information Unit*. Qualitatively, this means that there exists some universal information quantity (the "universal bit"); given a sufficient number of these units, we can encode, model and decode the state of any physical system.

More specifically, if we have any state ω of a physical system with a finite-dimensional state space Ω, and we would like to apply a transformation T to arrive at a state $\varphi = T\omega$, then we can also achieve this by doing it in the following way:

- reversibly encode $\omega \in \Omega$ into a state $\omega^{(n)}$ on the state space of n universal bits (where n is large enough),
- perform a reversible transformation $T^{(n)}$ on n universal bits to arrive at $\varphi^{(n)} = T^{(n)}\omega^{(n)}$, and
- reversibly decode the resulting state to obtain $\varphi \in \Omega$.

In quantum theory, we have such a universal bit, the *qubit*. Furthermore, we can in principle simulate reversible evolutions (that is, unitary transformations) of *any* physical system by encoding its state on a quantum computer, performing a unitary transformation on its quantum memory, and decoding the state back into the physical system in question. Postulate 3 below formulates this property in the GPT framework, without specifically assuming quantum theory. Consequently, *classical probability theory*, where the states are discrete probability distributions, satisfies the principle too: The universal bit is simply the bit.

This can be interpreted as providing a necessary condition for nature being computable in a certain sense: Whatever model of computation we may come up with (like, for example, the quantum Turing machine), it is necessarily a crucial aspect of that model to be "modular," i.e., to consist of a (large) collection of a small set of primitive building blocks. These are the universal bits. The scheme above says that *all* physical systems and their dynamical evolutions can in principle be reliably simulated using a sufficient number of universal bits and their dynamical interaction, which admits simulation of physics on a computer. Note that this encoding-decoding scheme, and its formulation in terms of the postulates below, does not require *all* ingredients that are important for universal (quantum) computation; for example, it does not postulate that all dynamical evolutions can be approximated by a fixed finite "universal gate set" (which is the case in quantum theory [Nielsen and Chuang 2000, app. 3]) – instead, this property follows as a consequence of the postulates. As a further consequence, since all systems can be encoded in a sufficient set of these information units, the task of characterizing the theory's state spaces is reduced to that of characterizing the state space of n universal bits, for all

$n \in \mathbb{N}$. In this sense, information and computability are central to deriving quantum theory.

The full set of principles are listed and described below. Together, they suffice to uniquely single out quantum theory:

Theorem *(Masanes et al. 2013): The unique GPT that satisfies the postulates below is quantum theory.*

That is, the state space of n universal bits is exactly the state space of n qubits: the states are the $2^n \times 2^n$ density matrices, the reversible transformations are the conjugations by unitary matrices, $\rho \mapsto U\rho U^\dagger$, and the possible measurements are the positive operator-valued measures (POVMs). The postulates are:

1. **Continuous Reversibility**. Given two pure states ω and φ, there is a family of reversible transformations $G(t)_{0 \leq t \leq 1}$ such that $G(t)$ is continuous in t, $G(0) = 1$ and $G(1)\omega = \varphi$.

In other words, any pure state ω can be reversibly and continuously ("in time," if t is interpreted as a time parameter) transformed into any other pure state φ. This is a stronger version of the "Reversibility" postulate introduced in Section 12.3.2. It is well-motivated by experience with physical systems: States may be adjusted to any of a continuum of settings, and physical evolution is often continuous and reversible in time.

2. **Tomographic Locality**.

This postulate is explained in Section 12.3.2. It is assumed in conjunction with the background assumptions of the GPT framework mentioned there (i.e., existence of local preparations and local measurements, and the no-signaling principle).

3. **Existence of an Information Unit**. There exists a fundamental information unit such that the state of any system can be reversibly encoded in a sufficiently large number of these information units, as explained above. It must satisfy these properties:[19]

- State tomography is possible: the state space of the information unit is finite-dimensional.
- Units can interact: On the state space of *two* information units A and B, there exists at least one reversible transformation that does not act independently on A and B (or equivalently, at least one locally prepared state $\omega_A \omega_B$ is mapped to a state that cannot be locally prepared).

---

[19] We skip one of the properties, "all effects are observable," since this postulate only applies in a slightly more general situation (where some mathematically possible measurements in the GPT framework may be declared physically forbidden) that we have not introduced in this paper for reasons of brevity.

If information units could not interact, then no useful computation would ever be possible: Initially uncorrelated universal bits in a state $\omega_A \omega_B$ (in quantum theory, we could write the state as $\rho_A \otimes \rho_B$) would always be transformed to an uncorrelated state $\omega'_A \omega'_B$ (resp. $\rho'_A \otimes \rho'_B$, corresponding to a unitary transformation $U_A \otimes U_B$). Not even a classical "controlled-NOT gate," for example, could ever be implemented.

4. **No Simultaneous Encoding**. If a universal bit is used to perfectly encode one classical bit of information, it cannot be used to simultaneously encode any further information.

This postulate says that the name "universal bit" is justified, in the sense that the information unit can carry exactly one binary alternative and not more. However, its detailed meaning is somewhat subtle, since interesting non-classical phenomena can happen in the GPT framework. This is most easily expressed by means of an example, namely the gbit from Section 12.3.1 (for more details of which, see Masanes et al. [2013]). Recall that a gbit state is given by $\omega = (p(\uparrow \; |0), p(\uparrow \; |1))$. Now we can encode one bit $b \in \{0, 1\}$ by setting $p(\uparrow \; |0) = b$. This bit can easily be read out: By performing the 0-measurement on the state ω, we get outcome $\uparrow$ with certainty if $b = 1$, and $\downarrow$ with certainty if $b = 0$. But we have some more remaining freedom: We can encode an *additional bit* $b' \in \{0, 1\}$ into the state by setting $p(\uparrow \; |1) = b'$. The bits b and b' determine ω uniquely; b' can be read out reliably by performing the 1-measurement.

However, the catch is that we can read out *either b or b'*, but *not both*: A simple GPT calculation shows that there exists no measurement that reads out both bits reliably at the same time. Moreover, reading out one of the bits introduces unavoidable disturbance that erases the other bit. We have "complementarity" in the sense of a "random access code": we can decide which *one* of the two bits we would like to read out, but we can never obtain both; hence the name "gbit." Yet we can definitely *encode* more than one bit into the state. It is this additional encoding capacity that Postulate 4 intends to forbid – the gbit violates "No Simultaneous Encoding." In addition, it also violates Continuous Reversibility.

Contrasting the quantum to the boxworld case (as is done further in Section 12.3.2) hence suggests that the interesting, *uniquely* quantum content of the postulates seems to be contained chiefly in Continuous Reversibility and the interaction of information units. As mentioned in Al-Safi and Short (2014), it has been conjectured more generally that tomographic locality and (not necessarily continuous!) reversibility are sufficient to pick out only theories embeddable within quantum theory. This would suggest that quantum theory is characterized as the maximal theory with reversible dynamics and tomographic locality. Even in the absence of a proof of this conjecture, it is clear that reversibility plays a key role in quantum theory, and that continuity is what distinguishes quantum theory from classical probability theory over a finite number of (qu)bits.[20]

Some further remarks regarding the significance of dynamics in singling out quantum theory are necessary. The key roles that continuous reversibility and interaction of information units play highlight the importance of the interplay between dynamics and kinematics in physical theory (Spekkens 2015). In order to arrive at the correct state space for a qubit (kinematics), we have to consider the possible interactions (dynamics) that are allowed. This emphasizes Spekkens's (2015) point that kinematics and dynamics only play a meaningful physical role when appropriately combined. Individually, each alone is inadequate for a real physical characterization. For example, analysis of static correlation tables from possible quantum experiments (Bub 2016) leads to postulates that allow for *almost-quantum correlations* (Navascués et al. 2015) that are slightly more general than those allowed by quantum theory. Perhaps this is a hint at the presence of new physics in regimes where the concept of dynamics becomes inapplicable, such as in quantum gravity, where time may not always be well-defined. In any case, the structure of quantum theory – as we currently know it – depends crucially on the form of permissible dynamics.

We contrast our discussion with that of Fuchs (2003), who anticipated that information-theoretic axiomatizations of quantum theory would show that much – *but not all* – of quantum theory was simply a consequence of information theory. He hoped that what was left would be the true physical kernel of quantum theory, an ambition also expressed by Brukner (2017, p. 114). Much of the discussion in Fuchs (2003) anticipates that contextuality is a genuine physical feature of quantum theory, and will turn out to be the non-information-theoretic kernel. After much progress in axiomatizing quantum theory, Fuchs showed his disappointment with the lack of surprising physical content revealed (Fuchs and Stacey 2016). With a correct axiomatization, he hoped that *"[t]he distillate that remains – the piece of quantum theory with no information theoretic significance – will be our first unadorned glimpse of 'quantum reality.' Far from being the end of the journey, placing this conception of nature in open view will be the start of a great adventure"* (Fuchs 2003, p. 990, emph. added).

We suggest that (continuous) reversibility may be the postulate which comes closest to being a candidate for a glimpse of the genuinely physical kernel of "quantum reality." Even though Fuchs may want to set a higher threshold for a "glimpse of quantum reality," this postulate is quite surprising from the point of view of classical physics: When we have a *discrete* system that can be in a *finite*

---

[20] This is a remarkable inversion of the standard view that infinite-dimensional *continuous* classical systems can exhibit *discreteness* after quantization.

number of perfectly distinguishable alternatives, then one would classically expect that reversible evolution must be discrete too. For example, a single bit can only ever be flipped, which is a discrete indivisible operation. Not so in quantum theory: the state $|0\rangle$ of a qubit can be continuously reversibly "moved over" to the state $|1\rangle$. For people without knowledge of quantum theory (but of classical information theory), this may appear as surprising or "paradoxical" as Einstein's light postulate sounds to people without knowledge of relativity.

This viewpoint notwithstanding, our arguments from Section 12.2, in combination with the result above, support the hypothesis that *quantum theory is a principle theory of information*, with continuously-reversible evolution in time as a characteristic property. Any further insights into an underlying "quantum reality" (if it exists), or into the question "information about what?" (if it has an answer), should not be expected to arise directly from these principles, or from quantum theory itself, but from a novel, yet-to-be-found constructive theory with additional beyond-quantum predictive power (if it exists). This is comparable to thermodynamics, whose formulation in terms of principles does not in itself imply the existence of atoms, in contrast to statistical physics, which in addition yields novel predictions.

12.5   Conclusions

We take it that information-theoretic reconstructions of quantum theory provide a fruitful – albeit only a *partial* – interpretation of quantum theory. By reconstruing the formalism of quantum theory in terms of operational constraints, one can cast quantum theory as a *principle theory*, and thereby gain explanatory power regarding structural features of a quantum world. In particular, the postulates of most of the current reconstructions (in particular of the one that we have outlined in Section 12.4) are broadly computational and information-theoretic in nature, emphasizing the role that this terminology plays in the formalism of quantum theory. Continuous reversible transformations, which can either be interpreted as computational processes or physical time evolutions, play a major role in singling out quantum theory from the space of GPTs. This indicates that interaction and reversibility of dynamics may be physically characteristic of quantum theory, providing a partial response to the hope that there would be bare physical content separable from the information-theoretic content of quantum theory. However, it is clear that these postulates do not provide a full interpretation; we learn what combination of operational constraints make necessary the structure of quantum theory, but we do not gain a constructive account of ontological structure. Information-theoretic reconstructions can be augmented with further metaphysical claims to arrive at a full interpretation; we outlined three possibilities in Section 12.2. We also argued that the information-theoretic reconstructions pose a challenge to existing ψ-ontic interpretations by highlighting a lack of explanatory power: these interpretations neither provide a similarly illuminating derivation of the formalism from principles, nor do they yield additional predictions or unification in the empirically accessible realm of our world as one might hope to obtain from a constructive theory.

Acknowledgments

We are grateful to Rob Spekkens for helpful discussions. We acknowledge funding from the Canada Research Chairs Program, from the Natural Sciences and Engineering Research Council of Canada (NSERC) Discovery Grants program, and from the Social Sciences and Humanities Research Council of Canada (SSHRC) Joseph-Armand Bombardier Canada Graduate Scholarships. This research was supported in part by Perimeter Institute for Theoretical Physics. Research at Perimeter Institute is supported by the Government of Canada through the Department of Innovation, Science and Economic Development Canada and by the Province of Ontario through the Ministry of Research, Innovation and Science.

Bibliography

Aaronson, S. (2005). Guest column: NP-complete problems and physical reality. *ACM SIGACT News*, *36*, 30–52.

(2013). *Quantum Computing Since Democritus*. Cambridge University Press.

Aaronson, S., and Watrous, J. (2009). Closed timelike curves make quantum and classical computing equivalent. *Proceedings of the Royal Society of London A: Mathematical, Physical and Engineering Sciences*, *465*(2102), 631–647.

Abbott, B. P., et al. (2016). GW151226: Observation of gravitational waves from a 22-solar-mass binary black hole coalescence. *Physical Review Letters*, *116*, 241103.

Adamatzky, A. (2010). *Physarum Machines: Computers from Slime Mould*. Hackensack, NJ: World Scientific.

Aharonov, D., and Ben-Or, M. (2008). Fault-tolerant quantum computation with constant error rate. *SIAM Journal on Computing*, *38*(4), 1207–1282.

Al-Safi, S. W., and Short, A. J. (2014). Reversible dynamics in strongly non-local boxworld systems. *Journal of Physics A: Mathematical and Theoretical*, *47*, 325303.

Amit, D. J., Gutfreund, H., and Sompolinsky, H. (1984). Spin-glass models of neural networks. *Physical Review A*, *32*, 1007–1018.

Anderson, N. G. (2010). On the physical implementation of logical transformations: Generalized L-machines. *Theoretical Computer Science*, *411*, 4179–4199.

(2017). Information as a physical quantity. *Information Sciences*, *415–416*, 397–413.

Anderson, N. G., and Piccinini, G. (2017). Pancomputationalism and the computational description of physical systems. Pitt PhilSci Archive preprint 12812.

Andraus, A. (2016). Zeno-machines and the metaphysics of time. *Unisinos Journal of Philosophy*, *17*, 161–167.

Andréka, H., Németi, I., and Németi, P. (2009). General relativistic hypercomputing and foundation of mathematics. *Natural Computing*, *8*, 499–516.

Andréka, H., Németi, I., and Székely, G. (2012). Closed timelike curves in relativistic computation. *Parallel Processing Letters*, *22*, 1240010.

Arora, S., and Barak, B. (2009). *Computational Complexity: A Modern Approach*. Cambridge University Press.

Ascher, U., Matheij, R., and Russell, R. (1995). *Numerical Solution of Boundary Value Problems for Ordinary Differential Equations*. Philadelphia: SIAM.

Avrachenkov, K. E., Filar, J. A., and Howlett, P. G. (2013). *Analytic Perturbation Theory and Its Applications*. Philadelphia: SIAM.

Bacon, D. (2004). Quantum computational complexity in the presence of closed timelike curves. *Physical Review A, 70*, 032309.

Bailey, J. (1977). Measurements of relativistic time dilatation for positive and negative muons in a circular orbit. *Nature, 268*, 301–305.

Baker, T., Gill, J., and Solovay, R. (1975). Relativizations of the $\mathcal{P} =? \mathcal{NP}$ question. *SIAM Journal on Computing, 4*, 431–442.

Ball, P. (2002). Universe is a computer. *Nature News*. Doi:10.1038/news020527-16.

(2013). A demon-haunted theory. *Physics World, 26*, 36–39.

Bar-Hillel, Y., and Carnap, R. (1953). Semantic information. *The British Journal for the Philosophy of Science, 4*(14), 147–157.

Barker-Plummer, D. (2016). Turing machines. In E. N. Zalta (Ed.) *The Stanford Encyclopedia of Philosophy*. Metaphysics Research Lab, Stanford University, Winter 2016 ed.

Barnum, H., Barrett, J., Orloff Clark, L., Leifer, M., Spekkens, R., Stepanik, N., Wilce, A., and Wilke, R. (2010). Entropy and information causality in general probabilistic theories. *New Journal of Physics, 12*, 033024.

Barrett, J. S. (2007). Information processing in generalized probabilistic theories. *Physical Review A, 75*, 032304.

Beggs, E., Costa, J. F., and Tucker, J. V. (2014). Three forms of physical measurement and their computability. *Review of Symbolic Logic, 17*, 618–646.

Beggs, E. J., and Tucker, J. V. (2007). Can Newtonian systems, bounded in space, time, mass and energy compute all functions? *Theoretical Computer Science, 371*, 4–19.

Bekenstein, J. D. (2007). Information in the holographic universe. *Scientific American, 17*, 66–73.

Bengtsson, I., and Życzkowski, K. (2006). *Geometry of Quantum States: An Introduction to Quantum Entanglement*. Cambridge University Press.

Benjamin, S. C., Lovett, B. W., and Smith, J. M. (2009). Prospects for measurement-based quantum computing with solid state spins. *Laser and Photonics Reviews, 3*, 556.

Bennett, C. H. (1982). The thermodynamics of computation—A review. *International Journal of Theoretical Physics, 21*, 905–940.

(1987). Demons, engines and the second law. *Scientific American, 257*, 108–116.

(2003). Notes on Landauer's principle, reversible computation, and Maxwell's demon. *Studies in History and Philosophy of Modern Physics, 34*(3), 501–510.

Berut, A., Arakelyan, A., Petrosyan, A., Ciliberto, S., Dillenschneider, R., and Lutz, E. (2012). Experimental verification of Landauer's principle linking information and thermodynamics. *Nature, 483*, 187–189.

Blanchette, P. (2014). The Frege-Hilbert controversy. In E. N. Zalta (Ed.) *The Stanford Encyclopedia of Philosophy*. Metaphysics Research Lab, Stanford University, Spring 2014 ed.

Block, N. (1978). Troubles with functionalism. In C. W. Savage (Ed.) *Perception and Cognition: Issues in the Foundations of Psychology*, vol. 9 of *Minnesota Studies in the Philosophy of Science* (pp. 261–325). Minneapolis: University of Minnesota Press.

Blum, L., Cucker, F., Shub, M., and Smale, S. (1998). *Complexity and Real Computation*. Berlin: Springer.

Blum, M. (1967). A machine-independent theory of the complexity of recursive functions. *Journal of the ACM, 14*, 322–336.

Blundell, S., and Blundell, K. (2010). *Concepts in Thermal Physics*. Oxford University Press.

Bostrom, N. (2003). Are we living in a computer simulation? *Philosophical Quarterly, 53*, 243–255.

Bourbaki, N. (1949). Foundations of mathematics for the working mathematician. *Journal of Symbolic Logic, 14*, 1–8.

Brillouin, L. (1951). Maxwell's demon cannot operate: Information and entropy I. *Journal of Applied Physics, 22*, 334–337.

Broersma, H., Stepney, S., and Wendin, G. (in press). Computability and complexity of unconventional computing devices. In S. Stepney, S. Rasmussen, and M. Amos (Eds.) *Computational Matter*. Cham: Springer.

Brown, H., and Timpson, C. (2006). Why special relativity should not be a template for a fundamental reformulation of quantum mechanics. In W. Demopoulos and I. Pitowsky (Eds.) *Physical Theory and Its Interpretation: Essays in Honor of Jeffrey Bub* (pp. 29–41). Dordrecht: Springer.

Brukner, Č. (2017). On the quantum measurement problem. In R. Bertlmann and A. Zeilinger (Eds.) *Quantum [Un]Speakables II: Half a Century of Bell's Theorem* (pp. 95–117). Cham: Springer.

Brukner, Č., and Zeilinger, A. (2003). Information and fundamental elements of the structure of quantum theory. In L. Castell and O. Ischebeck (Eds.) *Time, Quantum and Information* (pp. 323–354). Berlin: Springer.

(2009). Information invariance and quantum probabilities. *Foundations of Physics, 39*, 677–689.

Brun, T. A. (2003). Computers with closed timelike curves can solve hard problems efficiently. *Foundations of Physics Letters, 16*(3), 245–253.

Bub, J. (2005). Quantum mechanics is about quantum information. *Foundations of Physics, 35*, 541–560.

(2010). Quantum computation: Where does the speed-up come from? In A. Bokulich and G. Jaegger (Eds.) *Philosophy of Quantum Information and Entanglement* (pp. 231–246). Cambridge University Press.

(2016). *Bananaworld: Quantum Mechanics for Primates*. Oxford University Press.

Cabello, A. (2015). Interpretations of quantum theory: A map of madness. ArXiv preprint arXiv:1509.04711.

Carnap, R. (1947). *Meaning and Necessity: A Study in Semantics and Modal Logic*. University of Chicago Press.

(1950a). Empiricism, semantics, and ontology. *Revue Internationale de Philosophie, 4*, 20–40.

(1950b). *Logical Foundations of Probability*. University of Chicago Press.

Castelvecchi, D. (2015). Paradox at the heart of mathematics makes physics problem unanswerable. *Nature, 528*, 207.

Caves, C. M., Fuchs, C. A., and Schack, R. (2007). Subjective probability and quantum certainty. *Studies in History and Philosophy of Modern Physics, 38*, 255–274.

Chalmers, D. J. (1996). Does a rock implement every finite state automaton? *Synthese, 108*, 309–333.

(2011). A computational foundation for the study of cognition. *Journal of Cognitive Science*, *12*, 323–357.

Chen, C., Covanov, S., Mansouri, F., Moreno Maza, M., Xie, N., and Xie, Y. (2015). Basic polynomial algebra subprograms. *ACM Communications in Computer Algebra*, *48*(3/4), 197–201.

Chiribella, G., D'Ariano, G. M., and Perinotti, P. (2011). Informational derivation of quantum theory. *Physical Review A*, *84*, 012311.

Church, A. (1936). An unsolvable problem of elementary number theory. In Davis (1965, pp. 89–107).

Clark, J. A., Stepney, S., and Chivers, H. (2005). Breaking the model: Finalisation and a taxonomy of security attacks. *Electronic Notes in Theoretical Computer Science*, *137*, 225–242. Proceedings of the REFINE 2005 Workshop (REFINE 2005).

Clifton, R., Bub, J., and Halvorson, H. (2003). Characterizing quantum theory in terms of information-theoretic constraints. *Foundations of Physics*, *33*, 1561.

Cook, M. (2004). Universality in elementary cellular automata. *Complex Systems*, *15*, 1–40.

Cooper, S. B. (2004). *Computability Theory*. Boca Raton: Chapman and Hall.

(2006). How can nature help us compute? In J. Wiedermann, G. Tel, J. Pokorný, M. Bieliková, and J. Štuller (Eds.) *SOFSEM 2006: Theory and Practice of Computer Science*, vol. 3831 of *Lecture Notes in Computer Science* (pp. 1–13). Berlin: Springer. 32nd Conference on Current Trends in Theory and Practice of Computer Science, Merin, Czech Republic, January 21–27, 2006.

Coopersmith, J. (2010). *Energy: The Subtle Concept*. Oxford University Press.

Copeland, B. J. (1996). What is computation? *Synthese*, *108*, 335–359.

(1997). The broad conception of computation. *American Behavioral Scientist*, *40*, 690–716.

(1998a). Even Turing machines can compute uncomputable functions. In C. Calude, J. Casti, and M. Dinneen (Eds.) *Unconventional Models of Computation* (pp. 150–164). Berlin: Springer.

(1998b). Turing's O-machines, Penrose, Searle, and the brain. *Analysis*, *58*, 128–138.

(1999). A lecture and two radio broadcasts on machine intelligence by Alan Turing. In K. Furukawa, D. Michie, and S. Muggleton (Eds.) *Machine Intelligence 15* (pp. 445–476). Oxford University Press.

(2000). Narrow versus wide mechanism. *Journal of Philosophy*, *96*, 5–32. Repr. in Scheutz (2002, pp. 59–86).

(2002a). Accelerating Turing machines. *Minds and Machines*, *12*, 281–301.

(2002b). Hypercomputation. *Minds and Machines*, *12*, 461–502.

(2004a). Computable numbers: A guide. In Copeland (2004b, pp. 5–57).

(Ed.) (2004b). *The Essential Turing*. Oxford: Clarendon Press.

(2015). The Church-Turing thesis. In E. N. Zalta (Ed.) *The Stanford Encyclopedia of Philosophy*. Metaphysics Research Lab, Stanford University, Summer 2015 ed.

Copeland, B. J., and Proudfoot, D. (1999). Alan Turing's forgotten ideas in computer science. *Scientific American*, *280*, 99–103.

(2007). Artificial intelligence: History, foundations, and philosophical issues. In P. Thagard (Ed.) *Handbook of the Philosophy of Psychology and Cognitive Science* (pp. 429–482). Amsterdam: Elsevier.

Copeland, B. J., and Shagrir, O. (2007). Physical computation: How general are Gandy's principles for mechanisms? *Minds and Machines, 17*, 217–231.

(2011). Do accelerating Turing machines compute the uncomputable? *Minds and Machines, 21*, 221–239.

(2013). Turing versus Gödel on computability and the mind. In B. J. Copeland, C. J. Posy, and O. Shagrir (Eds.) *Computability: Turing, Gödel, Church, and Beyond* (pp. 1–33). Cambridge, MA: MIT Press.

Copeland, B. J., Sprevak, M., and Shagrir, O. (2017). Is the whole universe a computer? In B. J. Copeland, J. Bowen, M. Sprevak, and R. Wilson, *The Turing Guide* (pp. 445–462). Oxford University Press.

Corless, R. M., and Fillion, N. (2013). *A Graduate Introduction to Numerical Methods: From the Viewpoint of Backward Error*. New York: Springer.

Cousot, P., and Cousot, R. (1977). Abstract interpretation: A unified lattice model for static analysis of programs by construction or approximation of fixpoints. In *Conference Record of the Fourth Annual ACM SIGPLAN-SIGACT Symposium on Principles of Programming Languages* (pp. 238–252). New York: ACM.

Cubitt, T. S., Perez-Garcia, D., and Wolf, M. M. (2015). Undecidability of the spectral gap. *Nature, 528*, 207–211.

Cuffaro, M. E. (2012). Many worlds, the cluster-state quantum computer, and the problem of the preferred basis. *Studies in History and Philosophy of Modern Physics, 43*, 35–42.

(2013). *On the Physical Explanation for Quantum Speedup*. Ph.D. thesis, The University of Western Ontario, London, Ontario.

(2017). On the significance of the Gottesman-Knill theorem. *The British Journal for the Philosophy of Science, 68*, 91–121.

(in press). Universality, invariance, and the foundations of computational complexity in the light of the quantum computer. In S. O. Hansson (Ed.) *Technology and Mathematics: Philosophical and Historical Investigations*. Berlin: Springer.

Dahan, X., Moreno Maza, M., Schost, E., Wu, W., and Xie, Y. (2005). Lifting techniques for triangular decompositions. In *ISSAC '05: Proceedings of the 2005 International Symposium on Symbolic and Algebraic Computation* (pp. 108–115). New York: ACM.

Dakić, B., and Brukner, Č. (2011). Quantum theory and beyond: Is entanglement special? In H. Halvorson (Ed.) *Deep Beauty: Understanding the Quantum World through Mathematical Innovation* (pp. 365–392). Cambridge University Press.

Darrigol, O. (2005). *Worlds of Flow: A History of Hydrodynamics from the Bernoullis to Prandtl*. Oxford University Press.

Davies, E. B. (2001). Building infinite machines. *The British Journal for the Philosophy of Science, 52*, 671–682.

Davis, M. (1956). A note on universal Turing machines. In C. E. Shannon and J. McCarthy (Eds.) *Automata Studies*, vol. 34 of *Annals of Mathematics Studies* (pp. 167–175). Princeton University Press.

(1957). The definition of universal Turing machines. *Proceedings of the American Mathematical Society, 8*, 1125–1126.

(Ed.) (1965). *The Undecidable*. Hewlett, NY: Raven Press.

(1982). Why Gödel didn't have Church's thesis. *Information and Control, 54*, 3–24.

(2000). *The Universal Computer: The Road from Leibniz to Turing*. New York: W. W. Norton and Company.

(2006). The myth of hypercomputation. In C. Teuscher (Ed.) *Alan Turing: The Life and Legacy of a Great Thinker* (pp. 195–212). Berlin: Springer.

Davis, M., and Sieg, W. (2015). Conceptual confluence in 1936: Post and Turing. In G. Sommaruga and T. Strahm (Eds.) *Turing's Revolution: The Impact of His Ideas about Computability* (pp. 3–27). Berlin: Springer.

de Araújo, A., and Baravalle, L. (2017). The ontology of digital physics. *Erkenntnis*, 82, 1211–1231.

Dean, W. (2016a). Algorithms and the mathematical foundations of computer science. In L. Horsten and P. Welch (Eds.) *Gödel's Disjunction: The Scope and Limits of Mathematical Knowledge* (pp. 19–66). Oxford University Press.

(2016b). Computational complexity theory. In E. N. Zalta (Ed.) *The Stanford Encyclopedia of Philosophy*. Metaphysics Research Lab, Stanford University, Winter 2016 ed.

(2016c). Squeezing feasibility. In A. Beckmann, L. Bienvenu, and N. Jonoska (Eds.) *Pursuit of the Universal: Proceedings of the 12th Conference on Computability in Europe* (pp. 78–88). Cham: Springer.

Dennett, D. (1991). Real patterns. *Journal of Philosophy*, 88, 27–51.

Deutsch, D. (1985). Quantum theory, the Church-Turing principle and the universal quantum computer. *Proceedings of the Royal Society of London A: Mathematical and Physical Sciences*, 400(1818), 97–117.

(1989). Quantum computational networks. *Proceedings of the Royal Society of London A: Mathematical, Physical and Engineering Sciences*, 425(1868), 73–90.

(1991). Quantum mechanics near closed timelike lines. *Physical Review D*, 44, 3197–3217.

(1997). *The Fabric of Reality*. London: Penguin.

(2003). It from qubit. In J. Barrow, P. Davies, and C. Harper (Eds.) *Science and Ultimate Reality* (pp. 90–102). Cambridge University Press.

(2013). Constructor theory. *Synthese*, 190, 4331–4359.

d'Inverno, R. (1992). *Introducing Einstein's Relativity*. Oxford: Clarendon Press.

DiVincenzo, D. P. (1997). Topics in quantum computers. In L. L. Sohn, L. P. Kouwenhoven, and G. Schön (Eds.) *Mesoscopic Electron Transport* (pp. 657–677). Dordrecht: Springer Netherlands.

Dürr, D., and Teufel, S. (2009). *Bohmian Mechanics: The Physics and Mathematics of Quantum Theory*. Berlin: Springer.

Duwell, A. (2004). *How to Teach an Old Dog New Tricks: Quantum Information, Quantum Computing, and the Philosophy of Physics*. Ph.D. thesis, University of Pittsburgh.

(2007a). The many-worlds interpretation and quantum computation. *Philosophy of Science*, 74(5), 1007–1018.

(2007b). Re-conceiving quantum theories in terms of information-theoretic constraints. *Studies in History and Philosophy of Modern Physics*, 38, 181–201.

(2017). Exploring the frontiers of computation: Measurement based quantum computers and the mechanistic view of computation. In A. Bokulich and J. Floyd (Eds.) *Turing 100: Philosophical Explorations of the Legacy of Alan Turing*, vol. 324 of *Boston Studies in the Philosophy and History of Science* (pp. 219–232). Cham: Springer.

Dyson, F. (1972). Missed opportunities. *Bulletin of the American Mathematical Society*, 78, 635–652.

Earman, J. (1986). *A Primer on Determinism*. Dordrecht: D. Reidel.

(1995). *Bangs, Crunches, Whimpers, and Shrieks: Singularities and Acausalities in Relativistic Spacetimes*. Oxford University Press.

Earman, J., and Norton, J. D. (1993). Forever is a day: Supertasks in Pitowsky and Malament-Hogarth spacetimes. *Philosophy of Science*, 60, 22–42.

(1998). Exorcist XIV: The wrath of Maxwell's demon. Part I. From Maxwell to Szilard. *Studies in History and Philosophy of Modern Physics*, 29, 435–471.

(1999). Exorcist XIV: The wrath of Maxwell's demon. Part II. From Szilard to Landauer and beyond. *Studies in History and Philosophy of Modern Physics*, 30, 1–40.

Ehrenfest-Afanassjewa, T. (1925). Zur Axiomatisierung des zweiten Hauptsatzes der Thermodynamik? *Zeitschrift für Physik*, 33, 933–945.

(1956). *Grundlagen der Thermodynamic*. Leiden: J. Brill.

Einstein, A. (1954). What is the theory of relativity? In *Ideas and Opinions*. New York: Bonanza Books.

(1956 [1905]). On the movement of small particles suspended in a stationary liquid demanded by the molecular-kinetic theory of heat. Trans. A. D. Cowper. In R. Fürth (Ed.) *Investigations on the Theory of the Brownian Movement* (pp. 1–18). New York: Dover.

Eisert, J., Müller, M. P., and Gogolin, C. (2012). Quantum measurement occurrence is undecidable. *Physical Review Letters*, 108, 260501.

Epstein, R. L., and Carnielli, W. A. (2008). *Computability: Computable Functions, Logic, and the Foundations of Mathematics*. Socorro, NM: Advanced Reasoning Forum, 3rd ed.

Esfeld, M. (2013). Ontic structural realism and the interpretation of quantum mechanics. *European Journal for Philosophy of Science*, 3(1), 19–32.

Etesi, G. (2013). A proof of the Geroch-Horowitz-Penrose formulation of the strong cosmic censor conjecture motivated by computability theory. *International Journal of Theoretical Physics*, 52, 946–960.

(2016). Exotica or the failure of the strong cosmic censorship in four dimensions. *International Journal of Geometric Methods in Modern Physics*, 12, 1–11.

Etesi, G., and Németi, I. (2002). Non-Turing computations via Malament-Hogarth space-times. *International Journal of Theoretical Physics*, 41, 341–370.

Ewald, W. B. (1996). *From Kant to Hilbert: A Source Book in the Foundations of Mathematics*. Oxford: Clarendon Press.

Faye, J. (2014). Copenhagen interpretation of quantum mechanics. In E. N. Zalta (Ed.) *The Stanford Encyclopedia of Philosophy*. Metaphysics Research Lab, Stanford University, Fall 2014 ed.

Feferman, S. (1957). Degrees of unsolvability associated with classes of formalized theories. *Journal of Symbolic Logic*, 22, 161–175.

(1977). Inductive schemata and recursively continuous functionals. In R. O. Gandy and J. M. E. Hyland (Eds.) *Logic Colloquium '76*. Amsterdam: North Holland.

Feintzeig, B. (2014). Can the ontological models framework accommodate Bohmian mechanics? *Studies in History and Philosophy of Modern Physics*, 48, 59–67.

Felline, L. (2016). It's a matter of principle: Scientific explanation in information-theoretic reconstructions of quantum theory. *Dialectica*, 70, 549–575.

Fenstad, J. E. (1980). *General Recursion Theory: An Axiomatic Approach.* Berlin: Springer.

Feynman, R. P. (1982). Simulating physics with computers. *International Journal of Theoretical Physics, 21,* 467–488.

Feynman, R. P., Hey, T., and Allen, R. W. (1996). *Feynman Lectures on Computation.* Reading, MA: Addison-Wesley.

Feynman, R. P., Leighton, R. B., and Sands, M. (1966). *Feynman Lectures on Physics.* Reading, MA: Addison-Wesley.

Fillion, N., and Corless, R. M. (2014). On the epistemological analysis of modeling and computational error in the mathematical sciences. *Synthese, 191*(7), 1451–1467.

Fodor, J. A. (1975). *The Language of Thought.* Cambridge, MA: Harvard University Press.

Fortnow, L. (2003). One complexity theorist's view of quantum computing. *Theoretical Computer Science, 292,* 597–610.

Fraser, D. (2009). Quantum field theory: Underdetermination, inconsistency, and idealization. *Philosophy of Science, 76,* 536–567.

(2011). How to take particle physics seriously: A further defence of axiomatic quantum field theory. *Studies in History and Philosophy of Modern Physics, 42,* 126–135.

Fredkin, E. (1990). Digital mechanics: An informational process based on reversible universal cellular automata. *Physica D, 45,* 254–270.

(1992). A new cosmogony. In *PhysComp'92: Workshop on Physics and Computation* (pp. 116–121). IEEE.

(2003). An introduction to digital philosophy. *International Journal of Theoretical Physics, 42,* 189–247.

Fredkin, E., and Toffoli, T. (1982). Conservative logic. *International Journal of Theoretical Physics, 21*(3/4), 219–253.

Fresco, N. (2014). *Physical Computation and Cognitive Science.* Berlin: Springer.

Friedberg, R. M. (1957). Two recursively enumerable sets of incomparable degrees of unsolvability (solution of Post's problem, 1944). *Proceedings of the National Academy of Sciences, 43,* 236–238.

Friedman, H. M. (2008). Limitations on our understanding of the behavior of simplified physical systems. u.osu.edu/friedman.8/foundational-adventures/downloadable-manuscripts/.

(2001). Long finite sequences. *Journal of Combinatorial Theory, Series A, 95*(1), 102–144.

Frigg, R. (2006). Scientific representation and the semantic view of theories. *Theoria, 55,* 37–53.

Frigg, R., and Hartmann, S. (2016). Models in science. In E. N. Zalta (Ed.) *The Stanford Encyclopedia of Philosophy.* Metaphysics Research Lab, Stanford University, Winter 2016 ed.

Fuchs, C. A. (2003). Quantum mechanics as quantum information, mostly. *Journal of Modern Optics, 50,* 987–1023.

Fuchs, C. A., Mermin, N. D., and Schack, R. (2014). An introduction to QBism with an application to the locality of quantum mechanics. *American Journal of Physics, 82,* 749–754.

Fuchs, C. A., and Schack, R. (2013). Quantum-Bayesian coherence. *Reviews of Modern Physics*, *85*, 1693–1714.

Fuchs, C. A., and Stacey, B. C. (2016). Some negative remarks on operational approaches to quantum theory. In G. Chiribella and R. W. Spekkens (Eds.) *Quantum Theory: Informational Foundations and Foils* (pp. 283–305). Dordrecht: Springer Netherlands.

Gabor, D. (1964). Light and information. *Progress in Optics*, *1*, 111–153.

Gandy, R. (1980). Church's thesis and principles for mechanisms. In J. Barwise, H. J. Keisler, and K. Kunen (Eds.) *The Kleene Symposium* (pp. 123–148). Amsterdam: North Holland.

Gardner, M. (1970). Mathematical games: The fantastic combinations of John Conway's new solitaire game "life." *Scientific American*, *223*, 120–123.

Gawrilow, E., and Joswig, M. (2000). Polymake: A framework for analyzing convex polytopes. In G. Kalai and G. M. Ziegler (Eds.) *Polytopes—Combinatorics and Computation* (pp. 43–73). Basel: Birkhäuser.

Geddes, K. O., Czapor, S. R., and Labahn, G. (1992). *Algorithms for Computer Algebra*. Dordrecht: Kluwer.

Gerhard, J. (2004). *Modular Algorithms in Symbolic Summation and Symbolic Integration*. Berlin: Springer.

Geroch, R., and Hartle, J. B. (1986). Computability and physical theories. *Foundations of Physics*, *16*, 533–550.

Geroch, R., and Horowitz, G. T. (1979). Global structure of spacetimes. In S. W. Hawking and W. Israel (Eds.) *General Relativity: An Einstein Centenary Survey* (pp. 212–293). Cambridge University Press.

Ghirardi, G. (2014). Collapse theories. In E. N. Zalta (Ed.) *The Stanford Encyclopedia of Philosophy*. Metaphysics Research Lab, Stanford University, Fall 2014 ed.

Gödel, K. (193?). Undecidable Diophantine propositions. In Gödel (1995, pp. 164–175).

——— (1933). The present situation in the foundations of mathematics. In Gödel (1995, pp. 45–53).

——— (1934). Postscriptum to "On undecidable propositions of formal mathematical systems." In Gödel (1986, pp. 369–371).

——— (1946). Remarks before the Princeton bicentennial conference on problems in mathematics. In Gödel (1990, pp. 150–153).

——— (1951). Some basic theorems on the foundations of mathematics and their implications. In Gödel (1995, pp. 304–323).

——— (1972). Some remarks on the undecidability results. In Gödel (1990, pp. 305–306).

——— (1986). *Kurt Gödel: Collected Works*, vol. I. Oxford University Press.

——— (1990). *Kurt Gödel: Collected Works*, vol. II. Oxford University Press.

——— (1995). *Kurt Gödel: Collected Works*, vol. III. Oxford University Press.

Goldstine, H. H. (1977). *A History of Numerical Analysis from the 16th through the 19th Century*. New York: Springer.

Grøn, Ø., and Hervik, S. (2007). *Einstein's General Theory of Relativity: With Modern Applications in Cosmology*. New York: Springer.

Gross, D., Müller, M. P., Colbeck, R., and Dahlsten, O. C. O. (2010). All reversible dynamics in maximally nonlocal theories are trivial. *Physical Review Letters*, *104*, 080402.

Grothendieck, A. (1960). The cohomology theory of abstract algebraic varieties. In *Proceedings of the International Congress of Mathematicians (14–21 August 1958, Edinburgh)* (pp. 103–118). Cambridge University Press.

Gurevich, Y. (2012). What is an algorithm? In M. Bieliková, G. Friedrich, G. Gottlob, S. Katzenbeisser, and G. Turán (Eds.) *SOFSEM 2012: Theory and Practice of Computer Science* (pp. 31–42). Berlin: Springer.

Hagar, A., and Cuffaro, M. (2017). Quantum computing. In E. N. Zalta (Ed.) *The Stanford Encyclopedia of Philosophy*. Metaphysics Research Lab, Stanford University, Spring 2017 ed.

Hairer, E., Lubich, C., and Wanner, G. (2006). *Geometric Numerical Integration: Structure-Preserving Algorithms for Ordinary Differential Equations*. New York: Springer.

Hallett, M. (1990). Physicalism, reductionism & Hilbert. In A. D. Irvine (Ed.) *Physicalism in Mathematics* (pp. 183–258). Dordrecht: Kluwer.

Hameroff, S., and Penrose, R. (2014). Consciousness in the universe: A review of the "Orch OR" theory. *Physics of Life Reviews, 11*, 39–78.

Hardy, L. (2001). Quantum theory from five reasonable axioms. ArXiv preprint arXiv:quant-ph/0101012.

(2011). Reformulating and reconstructing quantum theory. ArXiv preprint arXiv:1104.2066.

Harrigan, N., and Spekkens, R. W. (2010). Einstein, incompleteness, and the epistemic view of quantum states. *Foundations of Physics, 40*, 125–157.

Harrington, L. A., Morley, M. D., Ščedrov, A., and Simpson, S. G. (Eds.) (1985). *Harvey Friedman's Research on the Foundations of Mathematics*. Amsterdam: Elsevier.

Hawking, S. W., and Ellis, G. F. R. (1973). *The Large Scale Structure of Space-time*. Cambridge University Press.

Healey, R. (2012). Quantum theory: A pragmatist approach. *The British Journal for the Philosophy of Science, 63*(4), 729–771.

Hedlund, G. A. (1968). Transformations commuting with the shift. In J. Auslander and W. G. Gottschalk (Eds.) *Topological Dynamics* (pp. 258–289). New York: Benjamin.

Hewitt-Horsman, C. (2009). An introduction to many worlds in quantum computation. *Foundations of Physics, 39*(8), 869–902.

Higman, G. (1952). Ordering by divisibility in abstract algebras. *Proceedings of the London Mathematical Society, s3-2*, 326–336.

Hilbert, D. (1900). Mathematical problems. In Ewald (1996, pp. 1096–1105).

(1918). Axiomatic thought. In Ewald (1996, pp. 1105–1115).

(1923). Foundations of mathematics. In Ewald (1996, pp. 1136–1148).

(1930). Logic and the knowledge of nature. In Ewald (1996, pp. 1157–1165).

(1950). *The Foundations of Geometry*. Chicago: Open Court.

(1967 [1926]). On the infinite. In J. van Heijenoort (Ed.) *From Frege to Gödel* (pp. 367–392). Cambridge, MA: Harvard University Press.

Hodges, A. (2004). What would Alan Turing have done after 1954? In C. Teuscher (Ed.) *Alan Turing: Life and Legacy of a Great Thinker* (pp. 43–58). Berlin: Springer.

(2012). Beyond Turing's machines. *Science, 336*, 163–164.

Hogarth, M. (1992). Does general relativity allow an observer to view an eternity in a finite time? *Foundations of Physics Letters, 5*, 173–181.

(1994). Non-Turing computers and non-Turing computability. In R. M. Burian, D. Hull, and M. Forbes (Eds.) *Proceedings of the Biennial Meeting of the Philosophy of Science Association*, vol. 1 (pp. 126–138). Philosophy of Science Association.

(1996). *Predictability, Computability, and Spacetime*. Ph.D. thesis, University of Cambridge.

(2004). Deciding arithmetic using SAD computers. *The British Journal for the Philosophy of Science*, *55*, 681–691.

Höhn, P. A. (2014). Toolbox for reconstructing quantum theory from rules on information acquisition. ArXiv preprint arXiv:1412.8323.

Höhn, P. A., and Wever, C. (2017). Quantum theory from questions. *Physical Review A*, *95*, 012102.

Holevo, A. S. (1982). *Probabilistic and Statistical Aspects of Quantum Theory*. New York: North-Holland.

Hooper, P. K. (1966). The undecidability of the Turing machine immortality problem. *Journal of Symbolic Logic*, *31*, 219–234.

Horsman, C., Stepney, S., Wagner, R. C., and Kendon, V. (2014). When does a physical system compute? *Proceedings of the Royal Society A: Mathematical, Physical and Engineering Sciences*, *470*(2169), 20140182.

Horsman, D., Kendon, V., Stepney, S., and Young, J. P. W. (2017). Abstraction and representation in living organisms: When does a biological system compute? In G. Dodig-Crnkovic and R. Giovagnoli (Eds.) *Representation and Reality: Humans, Animals and Machines* (pp. 91–116). Cham: Springer.

Horsman, D. C. (2015). Abstraction/representation theory for heterotic physical computing. *Philosophical Transactions of the Royal Society A: Mathematical, Physical and Engineering Sciences*, *373*, 20140224.

Howard, M., Wallman, J., Veitch, V., and Emerson, J. (2014). Contextuality supplies the "magic" for quantum computation. *Nature*, *510*, 351–355.

Huffman, C. (2015). Pythagoreanism. In E. N. Zalta (Ed.) *The Stanford Encyclopedia of Philosophy*. Metaphysics Research Lab, Stanford University, Spring 2015 ed.

Hughes, R. I. G. (1997). Models and representation. *Philosophy of Science*, *64*, S325–S336.

Immerman, N. (2016). Computability and complexity. In E. N. Zalta (Ed.) *The Stanford Encyclopedia of Philosophy*. Metaphysics Research Lab, Stanford University, Spring 2016 ed.

Jozsa, R. (2000). Quantum algorithms. In D. Boumeester, A. Ekert, and A. Zeilinger (Eds.) *The Physics of Quantum Information: Quantum Cryptography, Quantum Teleportation, Quantum Computation* (pp. 104–125). Berlin: Springer.

Jozsa, R., and Linden, N. (2003). On the role of entanglement in quantum-computational speed-up. *Proceedings of the Royal Society of London A: Mathematical, Physical and Engineering Sciences*, *459*, 2011–2032.

Jozsa, R., and Schumacher, B. (1994). A new proof of the quantum noiseless coding theorem. *Journal of Modern Optics*, *41*, 2343–2349.

Kalff, F. E., Rebergen, M. P., Fahrenfort, E., Girovsky, J., Toskovic, R., Lado, J. L., Fernández-Rossier, J., and Otte, A. F. (2016). A kilobyte rewritable atomic memory. *Nature Nanotechnology*, *11*, 926–929.

Kantor, F. W. (1982). An informal partial overview of information mechanics. *International Journal of Theoretical Physics*, *21*, 525–535.

Kleene, S. C. (1943). Recursive predicates and quantifiers. In Davis (1965, pp. 255–287).

Kleene, S. C., and Post, E. L. (1954). The upper semi-lattice of degrees of recursive unsolvability. *Annals of Mathematics*, *59*, 379–407.

Kline, M. (1972). *Mathematical Thought from Ancient to Modern Times*, vol. I–III. Oxford University Press.

Kocher, P. (1996). Timing attacks on implementations of Diffie-Hellman, RSA, DSS, and other systems. In *Advances in Cryptology—CRYPTO '96: 16th Annual International Cryptology Conference*, vol. 1109 of *Lecture Notes in Computer Science* (pp. 104–113). Berlin: Springer.

Kolmogorov, A. N. (1933). *Grundbegriffe der Wahrscheinlichkeitsrechnung*. Berlin: Springer.

Komar, A. (1964). Undecidability of macroscopically distinguishable states in quantum field theory. *Physical Review*, *133*, B542–B544.

Kopeikin, S., Efroimsky, M., and Kaplan, G. (2011). *Relativistic Celestial Mechanics of the Solar System*. Weinheim: Wiley.

Kreisel, G. (1965). Mathematical logic. In T. L. Saaty (Ed.) *Lectures on Modern Mathematics*, vol. III (pp. 95–195). New York: Wiley.

(1967). Mathematical logic: What has it done for the philosophy of mathematics? In R. Schoenman (Ed.) *Bertrand Russell: Philosopher of the Century* (pp. 201–272). London: Allen and Unwin.

Ladyman, J. (1998). What is structural realism? *Studies in History and Philosophy of Science*, *29*(3), 409–424.

(2009). What does it mean to say that a physical system implements a computation? *Theoretical Computer Science*, *410*(4–5), 376.

Ladyman, J., Presnell, S., and Short, A. (2008). The use of the information-theoretic entropy in thermodynamics. *Studies in History and Philosophy of Modern Physics*, *39*, 315–324.

Ladyman, J., Presnell, S., Short, A. J., and Groisman, B. (2007). The connection between logical and thermodynamic irreversibility. *Studies in History and Philosophy of Modern Physics*, *38*(1), 58–79.

Ladyman, J., and Robertson, K. (2013). Landauer defended: Reply to Norton. *Studies in History and Philosophy of Modern Physics*, *44*, 263–271.

(2014). Going round in circles: Landauer vs. Norton on the thermodynamics of computation. *Entropy*, *16*, 2278–2290.

Ladyman, J., and Ross, D. (2007). *Every Thing Must Go: Metaphysics Naturalized*. Oxford University Press.

Landauer, R. (1961). Irreversibility and heat generation in the computing process. *IBM Journal of Research and Development*, *5*, 183–191.

(1991). Information is physical. *Physics Today*, *44*, 23–29.

(1996). The physical nature of information. *Physics Letters A*, *217*, 188–193.

Leff, H. S., and Rex, A. (Eds.) (2003). *Maxwell's Demon 2: Entropy, Classical and Quantum Information, Computing*. Philadelphia: Institute of Physics Publishing.

Leibniz, G. W. (1678–1679). Preface to a universal characteristic. In Leibniz (1989, pp. 5–10).

(1714). The principles of philosophy, or, the Monadology. In Leibniz (1989, pp. 213–225).

(1989). *Philosophical Essays*. Trans. R. Ariew and D. Garber. Indianapolis: Hackett.

Leifer, M. S. (2014). Is the quantum state real? An extended review of ψ-ontology theorems. *Quanta*, *3*, 67–155.

Lenstra, A. K., Lenstra Jr., H. W., Manasse, M. S., and Pollard, J. M. (1990). The number field sieve. In *STOC '90: Proceedings of the Twenty-second Annual ACM Symposium on Theory of Computing* (pp. 564–572). New York: ACM.

Lerman, M. (1973). Admissible ordinals and priority arguments. In A. R. D. Mathias and H. Rogers (Eds.) *Cambridge Summer School in Mathematical Logic*, vol. 337 of *Lecture Notes in Mathematics* (pp. 311–344). Berlin: Springer.

Li, M., and Vitányi, P. (1997). *An Introduction to Kolmogorov Complexity and Its Applications*. Berlin: Springer.

Liboff, R. L. (2003). *Introductory Quantum Mechanics*. Reading, MA: Addison-Wesley, 4th ed.

Linton, C. M. (2004). *From Eudoxus to Einstein: A History of Mathematical Astronomy*. Cambridge University Press.

Lloyd, S. (1996). Universal quantum simulators. *Science*, *273*, 1073.

(2000). Ultimate physical limits to computation. *Nature*, *406*, 1047–1054.

(2006a). A theory of quantum gravity based on quantum computation. ArXiv preprint arXiv:quant-ph/0501135.

(2006b). *Programming the Universe*. London: Vintage Books.

(2013). The universe as a quantum computer. In (Zenil 2013, pp. 567–582).

Lovelace, A. (1843). Translator's notes to M. Menabrea's memoir on Babbage's Analytical Engine. In R. Taylor (Ed.) *Scientific Memoirs*, vol. III (pp. 691–731). London: Richard and John E. Taylor.

Lucas, J. R. (1961). Minds, machines and Gödel. *Philosophy*, *36*, 112–127.

MacDonald, G., and Papineau, D. (2006). *Teleosemantics*. Oxford: Clarendon Press.

Mach, E. (1911). *The History and Root of the Principle of Conservation of Energy*. Chicago: Open Court.

Mackay, D. M. (1969). *Information, Mechanism, and Meaning*. Cambridge, MA: MIT Press.

Macrae, N. (1992). *John von Neumann*. New York: Random House.

Madarász, J. X., Németi, I., and Székely, G. (2007). First-order logic foundation of relativity theories. In D. M. Gabbay, M. Zakharyaschev, and S. S. Goncharov (Eds.) *Mathematical Problems from Applied Logic II: New Logics for the XXIst Century* (pp. 217–252). New York: Springer.

Manchak, J. B. (2010). On the possibility of supertasks in general relativity. *Foundations of Physics*, *40*, 276–288.

(2011). What is a physically reasonable space-time? *Philosophy of Science*, *78*, 410–420.

(2017). Malament-Hogarth machines. Pitt PhilSci Archive preprint 13061.

Manchak, J., and Roberts, B. W. (2016). Supertasks. In E. N. Zalta (Ed.) *The Stanford Encyclopedia of Philosophy*. Metaphysics Research Lab, Stanford University, Winter 2016 ed.

Margenau, H., and Park, J. L. (1967). Objectivity in quantum mechanics. In M. Bunge (Ed.) *Delaware Seminar in the Foundations of Physics* (pp. 161–187). New York: Springer.

Maroney, O. J. E. (2005). The (absence of a) relationship between thermodynamic and logical reversibility. *Studies in History and Philosophy of Modern Physics, 36,* 355–374.

(2009a). Information processing and thermodynamic entropy. In E. N. Zalta (Ed.) *The Stanford Encyclopedia of Philosophy.* Metaphysics Research Lab, Stanford University, Fall 2009 ed.

(2009b). Generalizing Landauer's principle. *Physical Review E, 79,* 031105.

Martin, C. B. (1997). On the need for properties: The road to Pythagoreanism and back. *Synthese, 112,* 193–231.

Masanes, L., and Müller, M. P. (2011). A derivation of quantum theory from physical requirements. *New Journal of Physics, 13,* 063001.

Masanes, L., Müller, M. P., Augusiak, R., and Pérez-García, D. (2013). Existence of an information unit as a postulate of quantum theory. *Proceedings of the National Academy of Sciences, 110,* 16373–16377.

Maudlin, T. (1989). Computation and consciousness. *Journal of Philosophy, 86*(8), 407–432.

(2011). *Quantum Non-Locality and Relativity.* Oxford: Wiley-Blackwell, 3rd ed.

Maxwell, J. C. (1871). *Theory of Heat.* London: Longmans, Green and Co.

Mermin, N. D. (2003). From Cbits to Qbits: Teaching computer scientists quantum mechanics. *American Journal of Physics, 71,* 23–30.

Miłkowski, M. (2013). *Explaining the Computational Mind.* Cambridge, MA: MIT Press.

Millikan, R. G. (1984). *Language, Thought and Other Biological Categories.* Cambridge, MA: MIT Press.

Minsky, M. (1961). *Computation: Finite and Infinite Machines.* Upper Saddle River, NJ: Prentice-Hall.

Moir, R. H. (2010). *Reconsidering Backward Error Analysis for Ordinary Differential Equations.* M.Sc. thesis, The University of Western Ontario.

(2013). *Structures in Real Theory Application: A Study in Feasible Epistemology.* Ph.D. thesis, The University of Western Ontario.

Muchnik, A. A. (1956). On the unsolvability of the problem of reducibility in the theory of algorithms. *Doklady Akademii Nauk, 108,* 194–197.

Mussardo, G. (1997). The quantum mechanical potential for the prime numbers. ArXiv preprint arXiv:cond-mat/9712010.

Myrvold, W. C. (2010). From physics to information theory and back. In A. Bokulich and G. Jaeger (Eds.) *Philosophy of Quantum Information and Entanglement* (pp. 181–207). Cambridge University Press.

(2011). Statistical mechanics and thermodynamics: A Maxwellian view. *Studies in History and Philosophy of Modern Physics, 42,* 237–243.

Nakagaki, T., Yamada, H., and Tóth, A. (2000). Maze-solving by an amoeboid organism. *Nature, 407*(6803), 470.

Navascués, M., Guryanova, Y., Hoban, M. J., and Acín, A. (2015). Almost quantum correlations. *Nature Communications, 6,* 6288.

Neary, R., and Woods, D. (2006a). On the time complexity of 2-tag systems and small universal Turing machines. In *47th Annual IEEE Symposium on Foundations of Computer Science (FOCS'06)* (pp. 439–448). IEEE.

(2006b). Small fast universal Turing machines. *Theoretical Computer Science, 362,* 171–195.

Németi, I., and Dávid, Gy. (2006). Relativistic computers and the Turing barrier. *Journal of Applied Mathematics and Computation, 178*, 118–142.

Niedzwiecki, A., and Miyakawa, T. (2010). General relativistic models of the X-ray spectral variability of MCG-6-30-15. *Astronomy and Astrophysics, 509*, A22.

Nielsen, M. A. (1997). Computable functions, quantum measurements, and quantum dynamics. *Physical Review Letters, 79*(15), 2915–2918.

Nielsen, M. A., and Chuang, I. L. (2000). *Quantum Computation and Quantum Information*. Cambridge University Press.

Nilsson, C. (2003). Heuristics for the traveling salesman problem. Tech. rep., Linköping University, Sweden.

Norton, J. D. (2005). Eaters of the lotus: Landauer's principle and the return of Maxwell's demon. *Studies in History and Philosophy of Modern Physics, 36*, 375–411.

(2011). Waiting for Landauer. *Studies in History and Philosophy of Modern Physics, 42*, 184–198.

(2013a). All shook up: Fluctuations, Maxwell's demon and the thermodynamics of computation. *Entropy, 15*, 4432–4483.

(2013b). Author's reply to Landauer defended. *Studies in History and Philosophy of Modern Physics, 44*, 272.

(2013c). The end of the thermodynamics of computation: A no-go result. *Philosophy of Science, 80*, 1182–1192.

(2014). The simplest exorcism of Maxwell's demon: The quantum version. Pitt PhilSci Archive preprint 10572.

Odifreddi, P. G. (1989). *Classical Recursion Theory*, vol. I. Amsterdam: Elsevier.

(1999). *Classical Recursion Theory*, vol. II. Amsterdam: Elsevier.

O'Neill, B. (1995). *The Geometry of Kerr Black Holes*. Wellesley, MA: A K Peters.

Pawlowski, M., Paterak, T., Kaszlikowski, D., Scarani, V., Winter, A., and Zukowski, M. (2009). Information causality as a physical principle. *Nature, 461*, 1101–1104.

Penrose, R. (1979). Singularities and time-asymmetry. In S. W. Hawking and W. Israel (Eds.) *General Relativity: An Einstein Centenary Survey* (pp. 581–638). Cambridge University Press.

(1989). *The Emperor's New Mind*. Oxford University Press.

(1990). Précis of The Emperor's New Mind. *Behavioral and Brain Sciences, 13*, 643–655, 692–705.

(1994). *Shadows of the Mind*. Oxford University Press.

(1996). Beyond the doubting of a shadow. *Psyche, 2*. psyche.cs.monash.edu.au.

(1997). *The Large, the Small and the Human Mind*. Cambridge University Press.

(2011). Gödel, the mind, and the laws of physics. In M. Baaz, C. H. Papadimitriou, D. S. Scott, H. Putnam, and C. L. Harper (Eds.) *Gödel and the Foundations of Mathematics: Horizons of Truth* (pp. 339–358). Cambridge University Press.

(2013). Foreword. In Zenil (2013, pp. xii–xxxvi).

(2016). On attempting to model the mathematical mind. In S. B. Cooper and A. Hodges (Eds.) *The Once and Future Turing: Computing the World* (pp. 361–378). Cambridge University Press.

Piccinini, G. (2008). Computation without representation. *Philosophical Studies, 137*, 205–241.

(2010). The mind as neural software? Understanding functionalism, computationalism, and computational functionalism. *Philosophy and Phenomenological Research*, *81*, 269–311.

(2011). The physical Church-Turing thesis: Modest or bold? *The British Journal for the Philosophy of Science*, *62*, 733–769.

(2015). *Physical Computation: A Mechanistic Account*. Oxford University Press.

(2017). Computation in physical systems. In E. N. Zalta (Ed.) *The Stanford Encyclopedia of Philosophy*. Metaphysics Research Lab, Stanford University, Summer 2017 ed.

Piechocinska, B. (2000). Information erasure. *Physical Review A*, *61*, 1–9.

Pippenger, N. (1997). *Theories of Computability*. Cambridge University Press.

Pitowsky, I. (1990). The physical Church thesis and physical computational complexity. *Iyyun*, *39*, 81–99.

(1996). Laplace's demon consults an oracle: The computational complexity of prediction. *Studies in History and Philosophy of Modern Physics*, *27*, 161–180.

(2002). Quantum speed-up of computations. *Philosophy of Science*, *69*, S168–S177.

(2006). Quantum mechanics as a theory of probability. In W. Demopoulos and I. Pitowsky (Eds.) *Physical Theory and Its Interpretation: Essays in Honor of Jeffrey Bub* (pp. 212–240). Dordrecht: Springer.

Pogge, R. W. (2017). Real-world relativity: The GPS navigation system. www.astronomy.ohio-state.edu/~pogge/Ast162/Unit5/gps.html.

Poincaré, H. (1952 [1902]). *Science and Hypothesis*. New York: Dover.

Popescu, S., and Rohrlich, D. (1994). Quantum nonlocality as an axiom. *Foundations of Physics*, *24*, 379–385.

Post, E. (1936). Finite combinatory processes: Formulation I. In Davis (1965, pp. 288–291).

(1943). Formal reductions of the general combinatorial decision problem. *American Journal of Mathematics*, *65*, 197–215.

(1965). Absolutely unsolvable problems and relatively undecidable propositions: Account of an anticipation. In Davis (1965, pp. 340–433).

Pour-El, M. B., and Richards, I. J. (1981). The wave equation with computable initial data such that its unique solution is not computable. *Advances in Mathematics*, *39*, 215–239.

(1989). *Computability in Analysis and Physics*. Berlin: Springer.

Putnam, H. (1960). Minds and machines. In S. Hook (Ed.) *Dimensions of Mind: A Symposium* (pp. 138–164). New York: Collier.

(1967). Psychological predicates. In W. H. Capitan, and D. D. Merrill (Eds.) *Art, Mind, and Religion* (pp. 37–48). University of Pittsburgh Press.

(1981). *Reason, Truth and History*. Cambridge University Press.

(1988). *Representation and Reality*. Cambridge, MA: MIT Press.

Quine, W. V. O. (1951). Two dogmas of empiricism. *The Philosophical Review*, *60*, 20–43.

(1971). Epistemology naturalized. *Akten des XIV. Internationalen Kongresses für Philosophie*, *6*, 87–103.

(1976). Whither philosophical objects? In R. S. Cohen and P. K. Feyerabend (Eds.) *Essays in Memory of Imre Lakatos* (pp. 497–504). Berlin: Springer.

Raussendorf, R., and Briegel, H. (2001). A one-way quantum computer. *Physical Review Letters*, *86*, 5188–5191.

Raussendorf, R., Browne, D. E., and Briegel, H. J. (2003). Measurement-based quantum computation on cluster states. *Physical Review A*, *68*, 022312.

Rescorla, M. (2014). A theory of computational implementation. *Synthese*, *191*, 1277–1307.

Rindler, W. (2001). *Relativity: Special, General, and Cosmological*. Oxford University Press.

Rogers, H. (1967). *Theory of Recursive Functions and Effective Computability*. New York: McGraw-Hill.

Rorty, R. (1979). *Philosophy and the Mirror of Nature*. Princeton University Press.

Rothstein, J. (1951). Information, measurement and quantum mechanics. *Science*, *114*, 171–175.

Russell, B. (1927). *The Analysis of Matter*. London: Kegan Paul, Trench, Trubner.

Sacks, G. E. (1964). The recursively enumerable degrees are dense. *Annals of Mathematics*, *80*, 300–312.

Sakarovitch, J. (2009). *Elements of Automata Theory*. Cambridge University Press.

Sarfatti, J. (2004). Wheeler's world: It from bit? In F. Columbus and V. Krasnoholovets (Eds.) *Developments in Quantum Physics* (pp. 41–84). New York: NOVA Science Publishers.

Saunders, S., Barrett, J., Kent, A., and Wallace, D. (2010). *Many Worlds? Everett, Quantum Theory, and Reality*. Oxford University Press.

Scarpellini, B. (1963). Zwei unentscheitbare Probleme der Analysis. *Zeitschrift für mathematische Logik und Grundlagen der Mathematik*, *9*, 265–289. English translation in B. Scarpellini (2003), Two undecidable problems of analysis. *Minds and Machines*, *13*, 49–77.

Scheutz, M. (1999). When physical systems realize functions.... *Mind and Machines*, *9*, 161–196.

(2002). *Computationalism: New Directions*. Cambridge University Press.

Schmidhuber, J. (1997). A computer scientist's view of life, the universe, and everything. In C. Freksa, M. Jantzen, and R. Valk (Eds.) *Foundations of Computer Science: Potential – Theory – Cognition*, vol. 1337 of *Lecture Notes in Computer Science* (pp. 201–208). Berlin: Springer.

(2000). Algorithmic theories of everything. ArXiv preprint arXiv:quant-ph/0011122.

(2013). The fastest way of computing all universes. In Zenil (2013, pp. 383–400).

Schumacher, B. (1995). Quantum coding. *Physical Review A*, *51*(4), 2738.

Searle, J. R. (1980). Minds, brains, and programs. *The Behavioral and Brain Sciences*, *3*, 417–457.

(1992). *The Rediscovery of the Mind*. Cambridge, MA: MIT Press.

Shagrir, O. (1999). What is computer science about? *The Monist*, *82*(1), 131–149.

(2006). Why we view the brain as a computer. *Synthese*, *153*, 393–416.

Shagrir, O., and Pitowsky, I. (2003). Physical hypercomputation and the Church-Turing thesis. *Minds and Machines*, *13*, 87–101.

Shannon, C. E. (1948). The mathematical theory of communication. *Bell Systems Technical Journal*, *27*, 379–423, 623–656.

Shepherdson, J. C., and Sturgis, H. E. (1963). Computability of recursive functions. *Journal of the ACM*, *10*, 217–255.

Shor, P. W. (1994). Algorithms for quantum computation: Discrete logarithms and factoring. In *Proceedings: 35th Annual IEEE Symposium on Foundations of Computer Science* (pp. 124–134).

(1995). Scheme for reducing decoherence in quantum computer memory. *Physical Review A*, *52*, R2493.

Sieg, W. (1999). Formal systems, properties of. In R. A. Wilson and F. Keil (Eds.) *The MIT Encyclopedia of the Cognitive Sciences* (pp. 322–324). Cambridge, MA: MIT Press.

(2002). Calculations by man and machine: Conceptual analysis. In W. Sieg, R. Sommer, and C. L. Talcott (Eds.) *Reflections on the Foundations of Mathematics* (pp. 396–415). Poughkeepsie, NY: Association for Symbolic Logic.

(2009). On computability. In A. Irvine (Ed.) *Philosophy of Mathematics* (pp. 535–630). Amsterdam: Elsevier.

Sieg, W., and Byrnes, J. (1999). An abstract model for parallel computations: Gandy's thesis. *The Monist*, *82*, 150–164.

Sinha, U., Couteau, C., Jennewein, T., Laflamme, R., and Weihs, G. (2010). Ruling out multi-order interference in quantum mechanics. *Science*, *329*, 418.

Skrzypczyk, P., Short, A. J., and Popescu, S. (2014). Work extraction and thermodynamics for individual quantum systems. *Nature Communications*, *5*, 1–8.

Smoluchowski, M. (1912). Experimentell nachweisbare, der üblichen Thermodynamik widersprechende Molekularphänomene. *Physikalische Zeitschrift*, *13*, 1069–1080.

Soare, R. I. (1972). The Friedberg-Muchnik theorem re-examined. *Canadian Journal of Mathematics*, *24*, 1070–1078.

(1987). *Recursively Enumerable Sets and Degrees*. Berlin: Springer.

(2004). Computability theory and differential geometry. *Bulletin of Symbolic Logic*, *10*, 457–486.

(2009). Turing oracle machines, online computing, and three displacements in computability theory. *Annals of Pure and Applied Logic*, *160*, 368–399.

Solovay, R. M. (2000). A version of Ω for which ZFC cannot predict a single bit. In C. S. Calude and G. Păun (Eds.) *Finite Versus Infinite: Contributions to an Eternal Dilemma* (pp. 323–334). London: Springer.

Sorkin, R. D. (1994). Quantum mechanics as quantum measure theory. *Modern Physics Letters A*, *9*, 3119–3128.

Spekkens, R. W. (2015). The paradigm of kinematics and dynamics must yield to causal structure. In A. Aguirre, B. Foster, and Z. Merali (Eds.) *Questioning the Foundations of Physics: Which of Our Fundamental Assumptions Are Wrong?* (pp. 5–16). Cham: Springer.

Sprevak, M. D. (2005). *Mind and World: A Realist Account of Computation in Cognitive Science*. Ph.D. thesis, Cambridge University.

(2010). Computation, individuation and the received view on representation. *Studies in History and Philosophy of Science*, *41*(3), 260–270.

Stannett, M. (2009). The computational status of physics: A computable formulation of quantum theory. *Natural Computing*, *8*, 517–538.

Stannett, M., and Németi, I. (2014). Using Isabelle/HOL to verify first-order relativity theory. *Journal of Automated Reasoning*, *52*, 361–378.

Steane, A. M. (1996). Multiple-particle interference and quantum error correction. *Proceedings of the Royal Society of London A: Mathematical, Physical and Engineering Sciences*, *452*(1954), 2551–2577.

(2003). A quantum computer only needs one universe. *Studies in History and Philosophy of Modern Physics*, *34*, 469–478.

Stillwell, J. (2010). *Roads to Infinity: The Mathematics of Truth and Proof*. Natick, MA: A K Peters.

Strong, H. R. (1970). Constructions of models for algebraically generalized recursive function theory. *Journal of Symbolic Logic, 35,* 401–409.

Struik, D. J. (1969). *A Source Book in Mathematics, 1200–1800*. Cambridge, MA: Harvard University Press.

Suppes, P. (1960). A comparison of the meaning and uses of models in mathematics and the empirical sciences. *Synthese, 12*(2), 287–301.

Sutner, K. (2011). Computational processes, observers and Turing incompleteness. *Theoretical Computer Science, 412,* 183–190.

(2013). Computational equivalence and classical recursion theory. In H. Zenil (Ed.) *Irreducibility and Computational Equivalence* (pp. 297–307). Berlin: Springer.

Svedberg, T. (1907). Über die Bedeutung der Eigenbewegung der Teilchen in kolloidalen Lösungen für die Beurteilung der Gültigkeitsgrenzen des zweiten Hauptsatzes der Thermodynamik. *Annalen der Physik, 59,* 451–458.

Syropoulos, A. (2008). *Hypercomputation: Computing Beyond the Church-Turing Barrier*. New York: Springer.

Szilard, L. (1972 [1929]). On the decrease of entropy in a thermodynamic system by the intervention of intelligent beings. In B. T. Feld and G. W. Szilard (Eds.) *The Collected Works of Leo Szilard: Scientific Papers* (pp. 120–129). Cambridge, MA: MIT Press.

Tarjan, R. E., and van Leeuwen, J. (1984). Worst-case analysis of set union algorithms. *Journal of the ACM, 31,* 245–281.

Taylor, E. F., and Wheeler, J. A. (2000). *Black Holes*. San Francisco: Addison, Wesley, Longman.

Tegmark, M. (2008). The mathematical universe. *Foundations of Physics, 38,* 101–150.

(2014). *Our Mathematical Universe: My Quest for the Ultimate Nature of Reality.* New York: Random House.

't Hooft, G. (2002). Looking at life with Gerardus 't Hooft. *Plus Magazine.* Interview by R. Thomas. plus.maths.org/issue18/features/thooft/.

(2016). *The Cellular Automaton Interpretation of Quantum Mechanics*. Cham: Springer.

Thorne, K. S. (1994). *Black Holes and Time Warps: Einstein's Outrageous Legacy*. New York: W. W. Norton and Co.

Timpson, C. G. (2013). *Quantum Information Theory and the Foundations of Quantum Mechanics*. Oxford University Press.

Tipler, F. J. (1994). *The Physics of Immortality*. New York: Anchor Books.

Toffoli, T. (1982). Physics and computation. *International Journal of Theoretical Physics, 21,* 165–175.

Turgut, S. (2009). Relations between entropies produced in non-deterministic thermodynamic processes. *Physical Review E, 79,* 041102.

Turing, A. M. (1936). On computable numbers, with an application to the *Entscheidungsproblem*. In Copeland (2004b, pp. 58–90).

(1939). Systems of logic based on ordinals. In Copeland (2004b, pp. 146–204).

(1948). Intelligent machinery. In Copeland (2004b, pp. 410–432).

(1950). Computing machinery and intelligence. In Copeland (2004b, pp. 441–464).

(1951). Can digital computers think? In Copeland (2004b, pp. 476–486).

(1954). Solvable and unsolvable problems. In Copeland (2004b, pp. 576–595).

van Atten, M., and Kennedy, J. (2003). On the philosophical development of Kurt Gödel. *Bulletin of Symbolic Logic*, *9*, 425–476.

van Dam, W. (2013). Implausible consequences of superstrong nonlocality. *Natural Computing*, *12*(1), 9–12.

Van Dyke, M. (1975). Perturbation methods in fluid mechanics. *NASA STI/Recon Technical Report A*, *75*.

van Fraassen, B. C. (2008). *Scientific Representation: Paradoxes of Perspective*. Oxford: Clarendon Press.

van Leeuwen, J., and Wiedermann, J. (2001). The Turing machine paradigm in contemporary computing. In B. Enquist and W. Schmid (Eds.) *Mathematics Unlimited – 2001 and Beyond* (pp. 1139–1155). Berlin: Springer.

Vedral, V. (2010). *Decoding Reality: The Universe as Quantum Information*. Oxford University Press.

von Neumann, J. (1932). *Mathematische Grundlagen der Quantenmechanik*. Berlin: Springer.

(1956). Probabilistic logics and the synthesis of reliable organisms from unreliable components. In C. E. Shannon and J. McCarthy (Eds.) *Automata Studies*, vol. 34 of *Annals of Mathematics Studies* (pp. 43–98). Princeton University Press.

(1961 [1927]). Zur Hilbertschen Beweistheorie. In *Collected Works*, vol. I (pp. 256–300). Oxford: Pergamon.

(1966 [1949]). The role of high and of extremely high complication. In A. Burks (Ed.) *Theory of Self-Reproducing Automata* (pp. 64–73). Urbana, IL: University of Illinois Press.

Wagner, E. G. (1969). Uniformly reflexive structures: An axiomatic approach to computability. *Information Sciences*, *1*, 343–362.

Wagon, S. (1985). *The Banach–Tarski Paradox*. Cambridge University Press.

Wald, R. (1984). *General Relativity*. University of Chicago Press.

Wallace, D. (2003). Everett and structure. *Studies in History and Philosophy of Modern Physics*, *34*, 87–105.

(2011). Taking particle physics seriously: A critique of the algebraic approach to quantum field theory. *Studies in History and Philosophy of Modern Physics*, *42*, 116–125.

(2014). Thermodynamics as control theory. *Entropy*, *16*, 699–725.

Wang, H. (1996). *A Logical Journey: From Gödel to Philosophy*. Cambridge, MA: MIT Press.

Wangersky, P. J. (1978). Lotka-Volterra population models. *Annual Review of Ecology and Systematics*, *9*, 189–218.

Wapner, L. M. (2005). *The Pea and the Sun: A Mathematical Paradox*. Wellesley, MA: A K Peters.

Welch, P. D. (2008). The extent of computation in Malament-Hogarth spacetimes. *The British Journal for the Philosophy of Science*, *59*, 659–674.

Wenzel, M. (2012). The Isabelle/Isar reference manual. isabelle.in.tum.de/doc/isar-ref.pdf.

Wesseling, P. (2009). *Principles of Computational Fluid Dynamics*, vol. 29 of *Springer Series in Computational Mathematics*. Heidelberg: Springer.

Weyl, H. (1949). *Philosophy of Mathematics and Natural Sciences*. Princeton University Press.

Wheeler, J. A. (1982). The computer and the universe. *International Journal of Theoretical Physics*, *21*, 557–572.

(1983). Law without law. In J. A. Wheeler and W. H. Zurek (Eds.) *Quantum Theory and Measurement* (pp. 182–213). Princeton University Press.

(1990). Information, physics, quantum: The search for links. In W. Zurek (Ed.) *Complexity, Entropy, and the Physics of Information* (pp. 3–28). Redwood City, CA: Addison-Wesley.

(2006). The anthropic universe. www.abc.net.au/radionational/programs/science show/the-anthropic-universe/3302686. Interview by R. Williams. *The Science Show*, ABC Radio National, Australia. February 18, 2006.

Wiedermann, J., and van Leeuwen, J. (2002). Relativistic computers and non-uniform complexity theory. In C. S. Calude, M. J. Dinneen, and F. Peper (Eds.) *Unconventional Models of Computation: Third International Conference, Kobe, Japan, October 15–19, 2002*, vol. 2509 of *Lecture Notes in Computer Science* (pp. 287–299). Berlin: Springer.

Wittgenstein, L. (1922). *Tractatus Logico-Philosophicus*. Trans. C. K. Ogden. London: Routledge and Kegan Paul.

Wolfram, S. (1985). Undecidability and intractability in theoretical physics. *Physical Review Letters*, *54*, 735–738.

(2002). *A New Kind of Science*. Champaign, IL: Wolfram Media.

Wüthrich, C. (2015). A quantum-information-theoretic complement to a general-relativistic implementation of a beyond-Turing computer. *Synthese*, *192*, 1989–2008.

Zahedi, R. (2015). On discrete physics: A perfect deterministic structure for reality – and "a (direct) logical derivation of the laws governing the fundamental forces of nature". *Archive for Studies in Logic*, *16*, 1–97.

Zauner, K.-P., and Conrad, M. (1996). Parallel computing with DNA: Toward the anti-universal machine. In H.-M. Voigt, W. Ebeling, I. Rechenberg, and H.-P. Schwefel (Eds.) *Parallel Problem Solving from Nature – PPSN IV*, vol. 1141 of *Lecture Notes in Computer Science* (pp. 696–705). Berlin: Springer.

Zeilinger, A. (1999). A foundational principle for quantum mechanics. *Foundations of Physics*, *29*, 631–643.

Zenil, H. (Ed.) (2013). *A Computable Planck Universe*. Singapore: World Scientific.

Zizzi, P. A. (2006). Space-time at the Planck scale: The quantum computer view. In C. Garola, A. Rossi, and S. Sozzo (Eds.) *The Foundations of Quantum Mechanics: Historical Analysis and Open Questions – Cesena 2014* (pp. 345–358). Singapore: World Scientific.

Zuse, K. (1967). *Rechnender Raum*. Braunschweig: Friedr. Vieweg + Sohn.

(1982). The computing universe. *International Journal of Theoretical Physics*, *21*, 589–600.

Index

Printed in the United States
By Bookmasters